Decision Making Optimization Models for Business Partnerships

Efficiency and productivity improvement are imperative for businesses to remain competitive in an increasingly dynamic marketplace. While business organizations have the potential to thrive independently, collaborating with others fosters a collective strength that can lead to greater innovation, expanded reach, and shared success.

Decision making optimization models for business partnerships are essential, as businesses seldom have all the resources they need, and thus, they require alliances and partnerships with others to enable them to meet their goals. *Decision Making Optimization Models for Business Partnerships* extends non-parametric data envelopment analysis (DEA) and parametric econometrics approaches to better understand how economic efficiency and market competitiveness are achieved for different types of partnerships and strategic alliances.

Features

- Global contributions for a wide range of professionals and academics
- Invaluable resources for businesses, analysts, and academics interested in DEA, optimization, and operations research more widely
- Introduces readers to novel approaches, models, and decision making techniques on performance evaluation and business partnerships via the medium of parametric and nonparametric optimization.

Gholam R. Amin is Associate Professor of Management Science in the Faculty of Business at the University of New Brunswick, Saint John, Canada. He is an associate editor of the *IMA Journal of Management Mathematics* at Oxford University Press. Dr. Amin's research interests include performance measurement, productivity and efficiency analysis through data envelopment analysis, optimization, and inverse optimization approaches. Dr. Amin has published in several leading international journals including *Operations Research* (FT-50), *European Journal of Operational Research, Journal of Productivity Analysis, Annals of Operations Research, International Journal of Production Research, Journal of the Operational Research Society, Computers and Operations Research, Computers & Industrial Engineering, Applied Mathematical Modeling, International Journal of Approximate Reasoning, International Journal of Intelligent Systems, ABACUS, Soft Computing, International Journal of Performance Analysis in Sport, Health Economics, Policy and Law, European Journal of Sport Science, Journal of Intelligent Manufacturing, International Journal of Computer Mathematics*, and *IMA Journal of Management Mathematics* among others.

Mustapha Ibn Boamah is Professor of Economics in the Faculty of Business at the University of New Brunswick, Saint John, Canada. Dr. Ibn Boamah's research interests include open-economy macroeconomics, monetary economics, international finance, and the economics of financial institutions. He has published in various peer-reviewed journals including publications in the *Review of Financial Economics, Atlantic Economic Journal, Strategic Change, Social Responsibility Journal, International Journal of Organizational Analysis, International Journal of Social Economics, Managerial and Decision Economics, Annals of Operations Research*, and *the European Journal of Operational Research.*

Decision Making Optimization Models for Business Partnerships

Edited by
Gholam R. Amin and Mustapha Ibn Boamah

CRC Press
Taylor & Francis Group
Boca Raton London New York

CRC Press is an imprint of the
Taylor & Francis Group, an **informa** business
A CHAPMAN & HALL BOOK

Designed cover image: Gholam R. Amin and Mustapha Ibn Boamah

First edition published 2025
by CRC Press
2385 NW Executive Center Drive, Suite 320, Boca Raton FL 33431

and by CRC Press
4 Park Square, Milton Park, Abingdon, Oxon, OX14 4RN

CRC Press is an imprint of Taylor & Francis Group, LLC

ISBN: 978-1-032-38248-7 (hbk)
ISBN: 978-1-032-38468-9 (pbk)
ISBN: 978-1-003-34520-6 (ebk)

DOI: 10.1201/9781003345206

Typeset in Minion
by SPi Technologies India Pvt Ltd (Straive)

Contents

Preface

PARTNERSHIPS AND STRATEGIC ALLIANCES are essential business relationships that are established for a variety of reasons including to gain a competitive edge. Resources are always limited, so firms search for structures and processes to reduce production cost, increase revenue, access new technology, or strengthen supply chains to obtain inputs and access new markets. Improving economic efficiency and productivity is possible under business partnerships that ensure that partners achieve synergies to make them competitive and collectively thrive in the marketplace.

Data envelopment analysis (DEA) has been used to explore the economic efficiency of business partnerships and alliances, including mergers and acquisitions, by determining the gains from partnerships in terms of increased collective output and reduced cost from the use of less inputs by decision making units (DMUs). Amin and Ibn Boamah (2023), modeling business partnerships: a data envelopment analysis approach, *European Journal of Operational Research*, 305(1), 329-337, are the first to develop mathematical models under the umbrella of DEA for different partnership types to show the potential benefits of business partnerships that seek to enhance partners performance.

Decision Making Optimization Models for Business Partnerships is the first book of its kind that focuses on non-parametric DEA and parametric econometrics approaches to extend the understanding of partnerships and their role in improving economic efficiency and competitiveness. The book contains 13 chapters.

Chapter 1 by Amin and Ibn Boamah briefly explains basic concepts and models required to understand the advanced DEA models in the book. Ibn Boamah et al. review the literature on parametric econometric methods to understand the often complex and different business partnerships in Chapter 2.

Chapter 3 by Garner et al. reviews the concepts of uncertainty and robust optimization that promotes uncertain data envelopment analysis in identifying peer groups that might form beneficial partnerships. Chapter 4 by Oukil proposes a hybrid DEA methodology that unfolds in two stages to evaluate the potential gains achievable from pairwise hotel mergers and determine the optimal partnerships for prospective mergers. Yu et al. examine various models within a centralized DEA approach for evaluating the interactions among different partners and changes in resource reallocation over time within an organization in Chapter 5.

In Chapter 6, Soltanifar et al. utilize ratio inverse DEA models to assess unit performance, to estimate inputs and outputs, and to facilitate processes such as mergers, acquisitions, and resource allocation. Chapter 7 by Amin and Ibn Boamah develops inverse DEA models to identify the optimal level of inputs and outputs from merging units that would minimize undesirable outputs such as greenhouse gas emissions. Arabi and Choudhary emphasize the economic and environmental performance DEA models that help achieve sustainability objectives in Chapter 8.

Chapter 9 by Huang et al. applies DEA to evaluate relative investment efficiencies and a generalized structural equation model to investigate investment performance factors in a public-private partnership. In Chapter 10, Ray formulates nonparametric optimization models using DEA for the measurement and decomposition of potential gains from a merger. Gattoufi et al. present a comprehensive map of the research landscape where DEA and its variants are applied to assess business mergers in Chapter 11.

Chapter 12 by Chilingerian and Glavin employs DEA in a two-stage approach where the second stage involves application of an assurance region to the best practice frontier identified in the first stage to identify DMUs that can be targets for partnerships. Finally, Amin and Ibn Boamah discuss the origin and recent development of inverse DEA in business mergers in Chapter 13.

We extend our heartfelt gratitude to all contributors and authors whose expertise and dedication have shaped this book. Their invaluable insights and collaborative spirit have enriched our work and made this project a success.

Gholam R. Amin and Mustapha Ibn Boamah
Saint John, NB, Canada

CHAPTER 1

Introduction to Data Envelopment Analysis Models

Gholam R. Amin and Mustapha Ibn Boamah
Faculty of Business, University of New Brunswick, Saint John, Canada

1.1 INTRODUCTION

The concept of productive efficiency was introduced empirically in Farrell (1957). Charnes, Cooper and Rhodes (1978) developed data envelopment analysis (DEA) using linear programming for measuring the relative efficiency of decision making units (DMUs). As a non-parametric approach, DEA does not require specification of a functional form for the production processes and can handle multiple different inputs and multiple different outputs simultaneously. This important property of DEA makes it an important analytical tool that lends itself to countless applications.

DEA models are used to measure productivity change in many industries, including the food industry in the European Union (Kapelko, 2019). DEA is also a powerful performance measure tool in supply chain performance evaluations (Li et al. 2019) and in environmental efficiency assessments (Wegener and Amin 2019), just to mention a few. As a pre-merger planning tool, DEA is useful in estimating gains from mergers and acquisitions for a given set of DMUs (Gattoufi et al. 2014; Amin and Ibn Boamah, 2020). DEA models can guide business partners, under various types of

DOI: 10.1201/9781003345206-1

partnerships, in the redistribution of inputs and outputs to enhance performance and competitiveness (Amin and Ibn Boamah, 2023). In this chapter, we explain the standard DEA models and their mathematical basis. The chapter is intended to be especially helpful for readers who are not familiar with DEA. It helps them understand advanced DEA models in the subsequent chapters of this book.

1.2 STANDARD DEA MODELS

The mathematical models of DEA can be derived directly from the definition of efficiency where the ratio of sum of weighted outputs to sum of weighted inputs should be maximized for each DMU under assessment. These models are known as multiplier DEA models. Alternatively, DEA models are envelopment models if they are formulated from the concept of production possibility set (PPS). Interestingly, the conventional multiplier and envelopment DEA models are primal and dual of each other though they were formulated separately from two different methods. Let us start with the multiplier DEA models. The original multiplier DEA model was developed in Charnes, Cooper and Rhodes (1978). They introduced the following optimization model for measuring the comparative efficiency score of $\text{DMU}_{\text{k}}\left(k=1,2,..,n\right)$.

$$
\begin{aligned}
&\max \frac{\sum_{r=1}^{s} w_r y_{rk}}{\sum_{i=1}^{m} \mu_i x_{ik}} \\
&s.t. \\
&\quad \frac{\sum_{r=1}^{s} w_r y_{rj}}{\sum_{i=1}^{m} \mu_i x_{ij}} \leq 1 \; j=1,\ldots,n \\
&\quad w_r \geq 0, \mu_i \geq 0 \; r=1,\ldots,s, i=1,\ldots,m
\end{aligned} \tag{1.1}
$$

where x_{ij} and y_{rj} are the ith input and r^{th} output of the jth DMU, and w_i and μ_r are the decision variables representing the input and output weights to be determined by the model (for each $i=1,\ldots,m$, $r=1,\ldots,s, \& j=1,\ldots,n$). The above nonlinear programming model (1.1) can be linearized using the Charnes-Cooper transformation (Charnes and Cooper, 1962) as follows:

$$
\frac{1}{\sum_{i=1}^{m} \mu_i x_{ik}} = t
$$

$$
u_r = t w_r \;\&\; v_i = t\mu_i
$$

Substituting the above variable changes in model (1.1), we obtain the following equivalent linear programming (LP) model.

$$\max \sum_{r=1}^{s} u_r y_{rk}$$

$$\text{s.t.}$$

$$\sum_{i=1}^{m} v_i x_{ik} = 1 \tag{1.2}$$

$$\sum_{r=1}^{s} u_r y_{rj} - \sum_{i=1}^{m} v_i x_{ij} \le 0 \; j = 1,\ldots,n$$

$$u_r \ge 0, v_i \ge 0 \; r = 1,\ldots,s, i = 1,\ldots,m$$

where v_i and u_r are the decision variables indicating the input and output weights $(i = 1,\ldots,m \,\&\, r = 1,\ldots,s)$. The LP model (1.2) is known as the Charnes, Cooper and Rhodes or CCR model (Charnes et al. 1978), and it is appropriate when the return to scale is constant. According to the theory of production in economics, return to scale refers to a long-run rate at which a change in inputs leads to a change in outputs. A production function exhibits constant return to scale (CRS) if the rate is proportional.

As stated, an equivalent DEA model can be formulated through the concept of PPS. In general, the PPS can be defined as follows:

$$P = \{(\boldsymbol{x}, \boldsymbol{y}) : \boldsymbol{x} \text{ can produce } \boldsymbol{y}\}$$

where $\boldsymbol{x} = (x_1,\ldots,x_m)$ denotes the vector of inputs and $\boldsymbol{y} = (y_1,\ldots,y_s)$ represents the vector of outputs. Set P contains all feasible production units $(\boldsymbol{x}, \boldsymbol{y})$ where the inputs $\boldsymbol{x}$ are consumed to produce the outputs $\boldsymbol{y}$. This set can be obtained empirically through the following assumptions:

(1) All observed units $(\boldsymbol{x}_j, \boldsymbol{y}_j)$ belong to P (for each $j = 1,\ldots,n$).

(2) For any activity $(\boldsymbol{x}, \boldsymbol{y}) \in P$, the activity $t(\boldsymbol{x}, \boldsymbol{y}) = (t\boldsymbol{x}, t\boldsymbol{y})$ belongs to P, for each $t > 0$. This assumption guarantees the CRS property.

(3) For any activity $(\boldsymbol{x}, \boldsymbol{y}) \in P$, if $\bar{\boldsymbol{x}} \ge \boldsymbol{x}$ and $\bar{\boldsymbol{y}} \le \boldsymbol{y}$, then $(\bar{\boldsymbol{x}}, \bar{\boldsymbol{y}})$ is included in P. This assumption simply indicates that for a feasible production unit, all dominated points by the unit are included in P.

(4) Any semipositive linear combination of activities in P belongs to P.

Mathematically, the PPS created by n observed DMUs $\left(\left(\boldsymbol{x}_j, \boldsymbol{y}_j\right), j=1,\ldots,n\right)$ satisfying the above four assumptions can be written as follows:

$$P_{\text{CRS}} = \left\{ (\boldsymbol{x}, \boldsymbol{y}) : \sum_{j=1}^{n} \lambda_j \boldsymbol{x}_j \leq \boldsymbol{x}, \sum_{j=1}^{n} \lambda_j \boldsymbol{y}_j \geq \boldsymbol{y}, \lambda_j \geq 0, j=1,\ldots,n \right\}$$

where $\lambda = (\lambda_1,\ldots,\lambda_n)$ is an n-dimensional semipositive vector. We use a two-dimensional numerical example to illustrate the PPS graphically. Table 1.1 presents an example containing ten DMUs each with a single input and a single output.

Figure 1.1 presents the PPS associated with the DMUs in Table 1.1.

As the graph shows, only one DMU (B) is on the efficiency frontier, and the other nine DMUs are not on the efficiency frontier and hence are inefficient. The best practice efficiency frontier is the basis for efficiency evaluation in DEA. There are different ways to evaluate the efficiency of a DMU depending on how to project the unit on the efficiency frontier. We explain the input and output orientation models where the objective is to minimize

TABLE 1.1 Data for Ten DMUs

DUMs	A	B	C	D	E	F	G	H	I	J
Input	5	2	2.5	3	4.5	4	6.5	6	5.5	3.5
Output	2	4	3	5	2	1	5	1	4	3

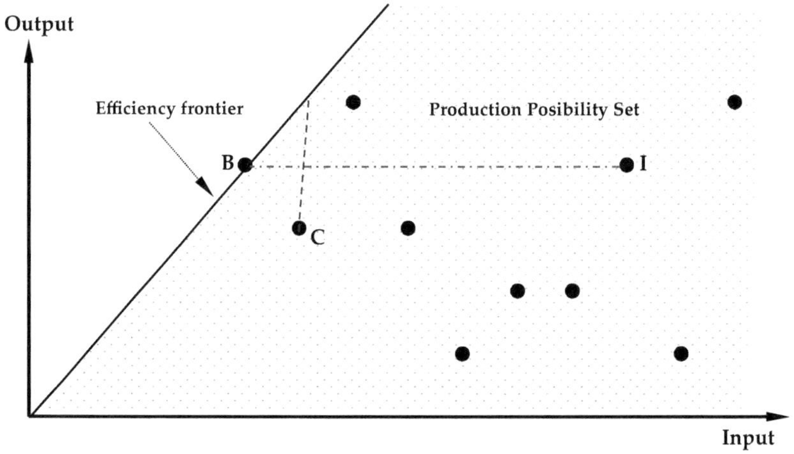

FIGURE 1.1 PPS under the CRS assumption.

the inputs in the input-orientation and to maximize the outputs in the output orientation. The simplest way is to decrease all inputs proportionally in the input-orientation and to increase all outputs proportionally in the output-orientation. In these cases, the vector of inputs (outputs) needs to be multiplied by a single scalar. Let us consider the input-orientation scenario. In this case, the DEA model to calculate the efficiency score of DMU_k $(k = 1,\ldots,n)$ is as follows:

$$\begin{aligned} &\min \theta \\ &s.t. \\ &(\theta \boldsymbol{x}_k, \boldsymbol{y}_k) \in P_{\text{CRS}} \end{aligned}$$

which can be rewritten as follows:

$$\begin{aligned} &\min \theta \\ &s.t. \\ &\sum_{j=1}^{n} \lambda_j \boldsymbol{x}_j \le \theta \boldsymbol{x}_k \\ &\sum_{j=1}^{n} \lambda_j \boldsymbol{y}_j \ge \boldsymbol{y}_k \\ &\lambda_j \ge 0, j = 1,\ldots,n \end{aligned} \tag{1.3}$$

It is insightful to note that the LP models (1.2) and (1.3) are indeed primal and dual of each other. However, these models were formulated separately from two different methods. The duality theorem (Bazaraa et al. 2011) concludes that the optimal value of model (1.2) is the same as the optimal value of model (1.3). In fact, the optimal value of these models determines the efficiency score of the DMU under assessment. In the input-oriented model, the efficiency score is defined by

$$\frac{\text{target input } i}{\text{actual input } i} = \frac{\theta^* x_{ik}}{x_{ik}} = \theta^* \left(\text{for each input } i = 1,\ldots,m\right)$$

where θ^* is the optimal value of model (1.3). The above definition guarantees that a unit is efficient in the input orientation if and only if the DMU consumes the minimum quantity of inputs in making its outputs. To obtain the efficiency score of all DMUs, model (1.3) needs to be solved at most n times.

Let us apply the above model to the numerical example shown in Figure 1.1. The optimal value of model (1.3) for DMU-B is $\theta^* = 1$, indicating the input reduction is not possible for this unit because it is on the efficiency frontier and fully efficient. Now, we employ the input-oriented DEA model (1.3) for DMU-I. The model becomes as follows:

$$\begin{gathered}\min \theta \\ s.t. \\ 5\lambda_A + 2\lambda_B + 2.5\lambda_C + \ldots + 5.5\lambda_I + 3.5\lambda_J \leq 5.5\theta \\ 2\lambda_A + 4\lambda_B + 3\lambda_C + \ldots + 4\lambda_I + 3\lambda_J \geq 4 \\ \lambda_j \geq 0, j = A, B, \ldots, I, J\end{gathered}$$

The above model determines the minimum input that could make the output of DMU-I. The appropriate input-oriented projection is shown in Figure 1.1 by the broken line. The above model can be solved by any LP or DEA software. The optimal value of the model is $\theta^* = 0.364$, indicating DMU-I is 36.4% efficient. In the optimal solution of the above model, there is only one positive lambda $\left(\lambda_B^* = 1\right)$, implying DMU-B is the sole benchmark for DMU-J. This could be revealed from the PPS in Figure 1.1 as DMU-B is the only efficient unit located on the efficiency frontier, and it is the benchmark for all other nine inefficient DMUs.

Akin to the input orientation, the output-oriented DEA model for $\text{DMU}_k \left(k = 1, \ldots, n\right)$ can be generally written as

$$\begin{gathered}\max h \\ s.t. \\ \left(\boldsymbol{x}_k, h\boldsymbol{y}_k\right) \in P_{\text{CRS}}\end{gathered}$$

which is equivalent to

$$\begin{gathered}\max h \\ s.t. \\ \sum_{j=1}^{n} \lambda_j \boldsymbol{x}_j \leq \boldsymbol{x}_k \\ \sum_{j=1}^{n} \lambda_j \boldsymbol{y}_j \geq h\boldsymbol{y}_k \\ \lambda_j \geq 0, j = 1, \ldots, n\end{gathered} \tag{1.4}$$

Model (1.4) determines the maximum producible output by a feasible unit in the PPS consuming the input of DMU_k. It projects the DMU under evaluation on the frontier by multiplying all outputs to the same scalar h. It is worth noting that the proportional output augmentation in model (1.4) can be replaced by a more general non-proportional projection where the decision maker gives different preferences to different output augmentations.

As an example, to the output orientation, Figure 1.1 shows the projection of DMU-C graphically. The efficiency score of the DMU under evaluation in the output-oriented model (1.4) can be defined as follows:

$$\frac{\text{actual output } r}{\text{target output } r} = \frac{y_{rk}}{h^* y_{rk}} = \frac{1}{h^*} \left(\text{for each output } r = 1,\ldots,s\right)$$

where h^* is the optimal value of model (1.4). The above definition guarantees that a unit is efficient in the output orientation if and only if the DMU makes the maximum comparative outputs by using its inputs. In the case of a non-proportional projection, the efficiency score cannot be uniquely obtained as in the above proportional projection scenario.

As an example, the output-oriented DEA model for DMU-C is shown as follows:

$$\begin{aligned}
&\max h \\
&s.t. \\
&5\lambda_A + 2\lambda_B + 2.5\lambda_C + \ldots + 5.5\lambda_I + 3.5\lambda_J \le 2.5 \\
&2\lambda_A + 4\lambda_B + 3\lambda_C + \ldots + 4\lambda_I + 3\lambda_J \ge 3h \\
&\lambda_j \ge 0, j = A, B, \ldots, I, J
\end{aligned}$$

The optimal value of the above model is $h^* = 1.667$, and hence, DMU-C is 60% efficient in the output orientation model.

Now, we formulate the variable return to scale (VRS) DEA model which is known as the BBC model (Banker et al. 1984). The PPS for the VRS technology is defined as follows:

$$P_{\text{VRS}} = \left\{ (\boldsymbol{x}, \boldsymbol{y}) : \sum_{j=1}^{n} \lambda_j \boldsymbol{x}_j \le \boldsymbol{x}, \sum_{j=1}^{n} \lambda_j \boldsymbol{y}_j \ge \boldsymbol{y}, \sum_{j=1}^{n} \lambda_j = 1, \lambda_j \ge 0, j = 1,\ldots,n \right\}$$

The difference between P_{VRS} and P_{CRS} is that in P_{VRS}, any convex combination of any set of feasible units in the PPS should be feasible and producible. The input and output orientation DEA models under the VRS assumption are, respectively, as follows:

$$\begin{aligned}
&\min \theta \\
&s.t. \\
&\sum_{j=1}^{n} \lambda_j \boldsymbol{x}_j \leq \theta \boldsymbol{x}_k \\
&\sum_{j=1}^{n} \lambda_j \boldsymbol{y}_j \geq \boldsymbol{y}_k \\
&\sum_{j=1}^{n} \lambda_j = 1 \\
&\lambda_j \geq 0, j = 1,\ldots,n
\end{aligned} \tag{1.5}$$

and

$$\begin{aligned}
&\max h \\
&s.t. \\
&\sum_{j=1}^{n} \lambda_j \boldsymbol{x}_j \leq \boldsymbol{x}_k \\
&\sum_{j=1}^{n} \lambda_j \boldsymbol{y}_j \geq h \boldsymbol{y}_k \\
&\sum_{j=1}^{n} \lambda_j = 1 \\
&\lambda_j \geq 0, j = 1,\ldots,n
\end{aligned} \tag{1.6}$$

To illustrate the above VRS DEA models (1.5) and (1.6) graphically, we use a numerical example. Table 1.2 presents six DMUs each using a single input to make a single output.

Figure 1.2 shows P_{VRS} for the above example.

In the above PPS, there are three efficient DMUs located on the efficiency frontier (A, B and D), and three units are inefficient. To illustrate

TABLE 1.2 Data for Six DMUs

DUMs	A	B	C	D	E	F
Input	1	2	2.5	4	4	6.5
Output	2	5	3.0	6	4	5.5

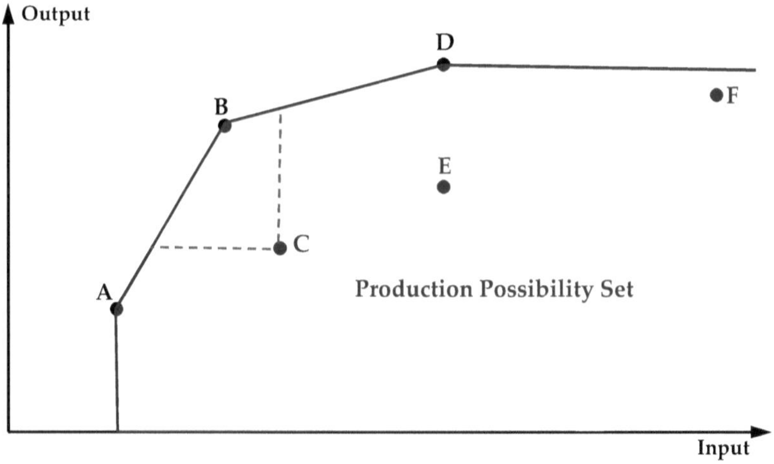

FIGURE 1.2 PPS under the VRS assumption.

the VRS models, let us assume DMU-C is the unit under evaluation. The associated input-oriented DEA model (1.5) can be written as follows:

$$\min \theta$$
$$s.t.$$
$$\lambda_A + 2\lambda_B + 2.5\lambda_C + 4\lambda_D + 4\lambda_E + 6.5\lambda_F \leq 2.5\theta$$
$$2\lambda_A + 5\lambda_B + 3\lambda_C + 6\lambda_D + 4\lambda_E + 5.5\lambda_E \geq 3$$
$$\lambda_A + \lambda_B + \lambda_C + \lambda_D + \lambda_E + \lambda_F = 1$$
$$\lambda_j \geq 0, j = A, B, \ldots, F$$

The optimal value of the above model is $\theta^* = 0.53$, indicating DMU-C is about 53% efficient under the VRS input-oriented DEA model. The reference set or benchmarks of DMU-C are DMUs A and B according to the optimal solution given by $\lambda_A^* = 0.6667$ and $\lambda_B^* = 0.3333$. The above model can also be solved graphically. As shown in Figure 1.2, the target of DMU-C falls on the piece of the efficiency frontier that connects DMUs A and B. Hence, $(2.5\theta, 3)$ is on the line segment connecting A and B, and so θ can be obtained.

Now, let us evaluate DMU-C under the output orientation. The associate DEA model (1.6) is as follows:

$$\begin{gathered}
\max h \\
s.t. \\
\lambda_A + 2\lambda_B + 2.5\lambda_C + 4\lambda_D + 4\lambda_E + 6.5\lambda_F \leq 2.5 \\
2\lambda_A + 5\lambda_B + 3\lambda_C + 6\lambda_D + 4\lambda_E + 5.5\lambda_E \geq 3h \\
\lambda_A + \lambda_B + \lambda_C + \lambda_D + \lambda_E + \lambda_F = 1 \\
\lambda_j \geq 0, j = A, B, \ldots, F
\end{gathered}$$

The optimal solution of the above model is as follows:

$$\lambda_B^* = 0.75, \lambda_D^* = 0.25, \lambda_j^* = 0, j \neq B \,\&\, D \text{ and } h^* = 1.75$$

which indicates that DMU-C is 57% under the output orientation model. The efficiency score of DMU-C could also be obtained graphically.

There are two special cases where the PPS can be plotted in a two-dimensional space, (1) two inputs and one output where the quantity of output is the same for all DMUs and (2) two outputs and one input where the value of input is the same for all units. These cases provide valuable insights when discussing DEA concepts.

Table 1.3 shows a numerical example consisting of eight suppliers in a supply chain management context. Each supplier is defined by two inputs and one output. The two inputs are the staff cost and R&D cost, and the single output is the service provided to customers. Table 1.3 shows the data.

All suppliers provide the same number of service (1000); hence, a two-dimensional PPS comparing only the two inputs can be fairly plotted for performance evaluation of the DMUs as illustrated in Figure 1.3.

TABLE 1.3 Data for Eight Suppliers

DMUs	Staff Cost	R&D Cost	Service
S1	225	85	1000
S2	135	145	1000
S3	45	160	1000
S4	115	104	1000
S5	155	15	1000
S6	90	30	1000
S7	120	190	1000
S8	55	75	1000

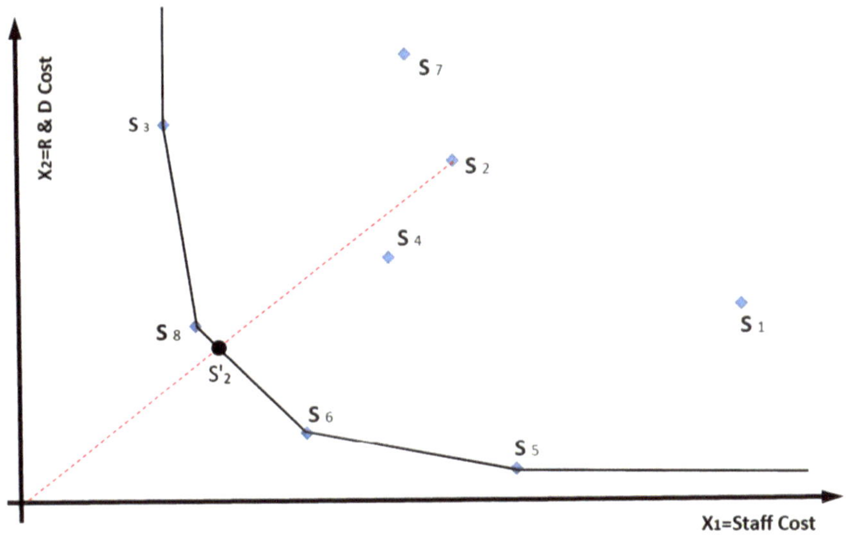

FIGURE 1.3 PPS for eight suppliers.

There are four DMUs on the efficiency frontier (S_3, S_5, S_6, and S_8); therefore, they are efficient. The efficiency score of an inefficient supplier, say S_2, can be obtained using the input-oriented DEA model (1.5). As shown in Figure 1.3, the target of supplier S_2 falls on the line segment of the piecewise linear efficiency frontier connecting DMUs S_6 and S_8, or $\left(135\times\theta,145\times\theta\right)$ is on the line

$$X_2 = -\frac{45}{35}X_1 + \frac{1020}{7}$$

Substituting the inputs of the target unit into the above equation implies that $\theta = 0.4574$, indicating supplier S_2 could save about 54% of its inputs. Table 1.4 presents an example consisting of ten hospitals, denoted by H_1 to H_{10}, each defined by two outputs and one input. The two outputs are inpatient admissions and outpatient visits, and the single input is nursing hours.

As shown, the nursing hours is 100 for all hospitals. Hence, a two-dimensional PPS comparing only the two outputs can be plotted as shown in Figure 1.4.

As shown, four DMUs are efficient (H_9, H_3, H_1, and H_5) in the above PPS, and six are inefficient. For instance, H_2 is an inefficient unit. This

TABLE 1.4 Data for Ten Hospitals

DMUs	Nursing Hours	Inpatient Admissions	Outpatient Visits
H_1	100	74	35
H_2	100	26	24
H_3	100	62	48
H_4	100	25	38
H_5	100	78	18
H_6	100	52	25
H_7	100	65	34
H_8	100	38	11
H_9	100	34	55
H_{10}	100	32	23

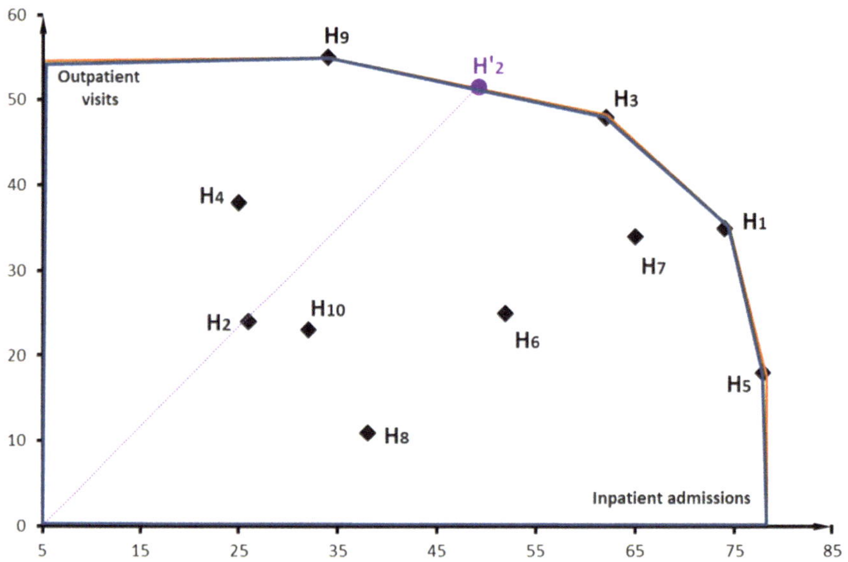

FIGURE 1.4 PPS for ten hospitals.

DMU is projected on the frontier by multiplying its outputs by h as per the output orientation DEA model (1.6). As shown in Figure 1.4, the target of H_2 is denoted by H'_2, or equivalently $H'_2 = h \times H_2 = (26h, 24h)$. The maximum value of h and the efficiency score of H_2 can be obtained graphically as H'_2 is on the line segment of the piecewise linear efficiency frontier connecting two hospitals H_3 and H_9. The equation of the line is

$$\text{Outpatient visits} = 63.5 - 0.25 \times (\text{Inpatient admissions})$$

This would result in

$$24 \times h = 63.5 - 0.25 \times (26h), h = \frac{63.5}{30.5} = 2.08$$

This means hospital H_2 is about 48% efficient. Equivalently, they could provide about 52% additional outputs according to their target DMU obtained from the benchmarks H_3 and H_9.

The efficiency discussed in all the above DEA models is referred to as technical efficiency. It refers to the ability of a DMU to convert inputs into outputs aiming to achieve the maximum outputs for a given set of inputs, or the minimum level of inputs necessary to produce a given level of outputs. There are other types of efficiency measurements in DEA, including cost efficiency, revenue efficiency, profit efficiency, allocative efficiency, and scale efficiency. It is evident that cost reduction is crucial to a DMU's long-term profitability and sustainability.

As explained above, technical efficiency measurement needs only the quantity of inputs and outputs. In addition to this information, cost efficiency requires the cost/price of inputs. Table 1.5 presents seven DMUs, denoted by A to G, each defined by two inputs. The quantity of output is the same for all DMUs. Table 1.5 also presents the price of inputs for each unit.

Figure 1.5 presents the PPS for the above data. The equation for each line segment of the efficiency frontier is presented next to the corresponding line.

As shown, DMU B is technically inefficient. The cost function of this unit can be written as

$$z = 11X_1 + X_2$$

TABLE 1.5 Data for Seven DMUs

DMUs	Input 1	Input 2	Price of Input 1	Price of Input 2
A	2.00	4.25	\$9	\$4
B	3.50	6.00	\$11	\$1
C	3.25	2.00	\$8	\$7
D	1.75	7.00	\$5	\$8
E	3.00	5.50	\$12	\$2
F	7.00	1.00	\$2	\$10
G	8.00	2.75	\$7	\$8

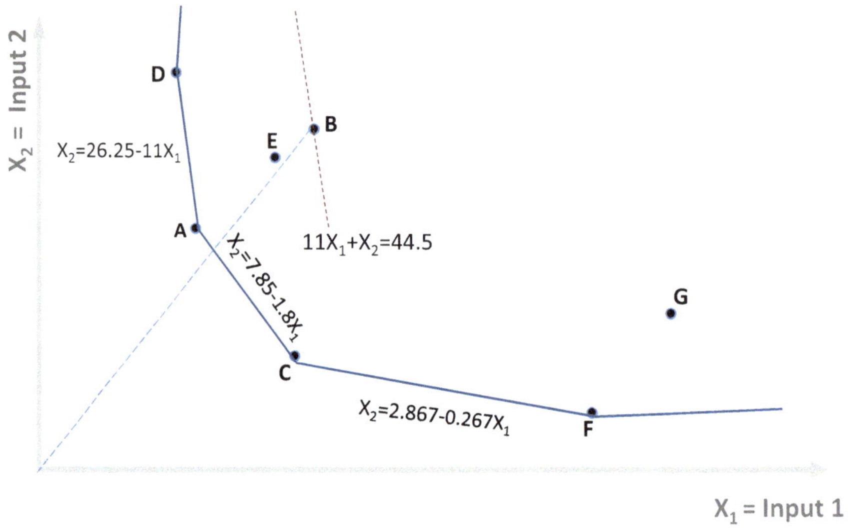

FIGURE 1.5 PPS for seven DMUs.

where X_1 and X_2 present inputs 1 and 2, respectively. It is easy to verify that the actual and target costs of B are, respectively, \$44.5 and \$26.25. The target cost of a DMU can be obtained by projecting its cost function on the efficiency frontier. For unit B, its cost function is parallel to the line segment connecting DMUs A and D. The cost efficiency of B can be defined as follows:

$$\text{Cost Efficiency} = \frac{\text{Target Cost}}{\text{Actual Cost}} = \frac{26.25}{44.5} = 0.59$$

Hence, DMU B is about 59% cost efficient. It's worth noting that under the input price of DMU B, there are only two cost-efficient DMUs, A and D; however, each unit has its own input price, as shown in Table 1.5, and hence, the corresponding cost efficient units will be different. The technical efficiency of DMU B can be obtained by projecting the unit on the efficiency frontier. As shown in Figure 1.5, its target falls on the line segment $X_2 = 7.85 - 1.8X_1$. Therefore, its technical efficiency can be obtained by substituting $(3.5\theta, 6\theta)$ in the equation below:

$$6\theta = 7.85 - 1.8 * (3.5\theta)$$

Equivalently,

$$\theta = \frac{7.85}{12.3} = 0.64$$

Hence, B is 64% technically efficient. The cost efficiency DEA model can be generally written as follows:

$$\begin{aligned} & \min \boldsymbol{c}_k \bar{\boldsymbol{x}}_k \\ & s.t. \\ & \sum_{j=1}^{n} \lambda_j \boldsymbol{x}_j \le \bar{\boldsymbol{x}}_k \\ & \sum_{j=1}^{n} \lambda_j \boldsymbol{y}_j \ge \boldsymbol{y}_k \\ & \sum_{j=1}^{n} \lambda_j = 1 \\ & \lambda_j \ge 0, j = 1, \ldots, n \end{aligned} \tag{1.7}$$

where $\bar{\boldsymbol{x}}_k$ is the vector of target inputs and $\boldsymbol{c}_k$ is the vector presenting the price of inputs for DMU_k. For instance, model (1.7) associated with DMU B is as follows:

$$\begin{aligned} & \min 11X_1 + X_2 \\ & s.t. \\ & 2\lambda_A + 3.5\lambda_B + 3.25\lambda_C + 1.75\lambda_D + 3\lambda_E + 7\lambda_F + 8\lambda_G \le X_1 \\ & 4.25\lambda_A + 6\lambda_B + 2\lambda_C + 7\lambda_D + 5.5\lambda_E + \lambda_F + 2.75\lambda_G \le X_2 \\ & \lambda_A + \lambda_B + \lambda_C + \lambda_D + \lambda_E + \lambda_F + \lambda_G = 1 \\ & \lambda_j \ge 0, j = A, B, \ldots, G \end{aligned}$$

An optimal solution of the above model is

$$\lambda_A^* = 1, \lambda_j^* = 0 \text{ for all } j = B, C, D, E, F, G$$

$$X_1^* = 2, X_2^* = 4.25$$

The minimum cost or the objective value of the above model is also \$26.25. Hence, the cost efficiency is

$$\frac{11\times 2+1\times 4.25}{11\times 3.5+1\times 6}=\frac{26.25}{44.5}=0.59$$

which is the same cost efficiency score obtained above graphically.

Allocative efficiency is another measure of efficiency in DEA. It is a measure that decomposes technical and cost efficiencies and highlights the optimal input mix. Allocative efficiency is defined as the ratio of cost efficiency to technical efficiency as shown below:

$$\text{Allocative Efficiency}=\frac{\text{Cost Efficiency}}{\text{Technical Efficiency}}$$

To illustrate this efficiency measurement, we use DMU B presented in Figure 1.5. As shown above, this unit is 59% cost-efficient and 64% technical efficient. Hence,

$$\text{Allocative Efficiency}=\frac{0.59}{0.64}=0.92$$

DMU B is about 92% allocatively efficient. These three efficiency measures indicate that a 36% cost reduction is possible by eliminating the technical inefficiency of DMU B without changing the input ratio. An additional 8% cost reduction could be attained by modifying the input mix.

Akin to the cost efficiency, revenue efficiency is another helpful efficiency measure. It requires the price of outputs to be given along with the data. Let us assume that the price of outputs 1 and 2 for hospital H_2 in Table 1.4 are \$15 and \$17, respectively. Then, the following DEA model can be applied to obtain the maximum level of outputs (Y_1, Y_2) for hospital H_2.

$$\begin{gathered}
\max 15Y_1+17Y_2 \\
s.t. \\
74\lambda_1+26\lambda_2+62\lambda_3+\ldots+34\lambda_9+32\lambda_{10}\geq Y_1 \\
35\lambda_1+24\lambda_2+48\lambda_3+\ldots+55\lambda_9+23\lambda_{10}\geq Y_2 \\
\lambda_1+\lambda_2+\lambda_3+\ldots+\lambda_9+\lambda_{10}=1 \\
\lambda_j\geq 0, j=1,2,\ldots,10
\end{gathered}$$

All the DEA models explained in this chapter can be solved by any optimization software with a linear programming solver. They include specialized software like the Advanced Interactive Multidimensional Modeling System (AIMMS) to a more general one such as Excel-Solve. An optimal solution of the above model is

$$\lambda_3^* = 1, \lambda_j^* = 0 \text{ for all } j \neq 3, Y_1^* = 62, Y_2^* = 48$$

The target revenue for hospital H_2 is $\$15 \times 62 + \$17 \times 48 = \$1746$. Therefore, the revenue efficiency can be obtained as follows:

$$\text{Revenue Efficiency} = \frac{\text{Actual Revenue}}{\text{Target Revenue}} = \frac{798}{1746} \approx 0.46$$

where the actual revenue is obtained from $\$15 \times 26 + \$17 \times 24 = \$798$. This efficiency score could also be obtained graphically by projecting H_2 on the efficiency frontier along with the revenue function of $z = 15Y_1 + 17Y_2$ in the PPS depicted in Figure 1.4.

Profit efficiency is also an efficiency measurement in DEA that can be defined when the cost of inputs and price of outputs are given along with the inputs and outputs. The objective function in the profit efficiency DEA model is to maximize the profit of the unit. Scale efficiency is another valuable metric in DEA. A DMU is scale-efficient if its size of operations is optimal so that any modifications to its size will result in the unit being less efficient. The value for scale efficiency is obtained by dividing the technical efficiency under the CRS to the technical efficiency under the VRS assumption.

The following section contains a real-world application of DEA in banking. We use this application to review the basic DEA models explained in this chapter.

1.3 AN APPLICATION IN BANKING

We apply some of the basic DEA models explained in the previous section to a real-world banking application. Data on 28 US banks was obtained from the Federal Deposit Insurance Corporation to illustrate how DEA models can be used to determine the efficiency of these banks. Banks earn profit by transforming inputs such as borrowed funds into earning assets such as loans and securities in a production process. Table 1.6 presents the dataset consisting of two inputs: (1) total employees at each bank (X_1) and (2) total liabilities including deposits (in thousands of dollars) at each bank (X_2). The three financial products or outputs are (1) balances due

TABLE 1.6 Data for 28 US Banks

DMUs	X_1	X_2	Y_1	Y_2	Y_3	Technical Efficiency
DMU_1	10,007	85,737,421	15,022	359,952	1,447,538	**0.767**
DMU_2	133,812	2,283,098,000	1,167,000	11,596,000	26,327,000	**1.000**
DMU_3	218,271	1,608,362,000	1,093,000	6,454,000	24,574,000	**1.000**
DMU_4	4471	45,139,990	5478	190,939	645,474	**0.749**
DMU_5	59,637	499,313,543	291,990	1,888,868	7,807,346	**0.936**
DMU_6	67,684	512,051,336	218,411	2,379,791	8,899,877	**0.949**
DMU_7	19,232	154,117,699	31,650	681,096	3,219,835	**0.845**
DMU_8	19,086	186,732,497	56,516	1,056,173	3,341,061	**1.000**
DMU_9	6850	86,027,456	6622	243,490	1,398,280	**0.581**
DMU_{10}	17,297	166,958,118	42,275	706,134	2,861,580	**0.809**
DMU_{11}	24,192	376,579,758	152,621	2,112,445	4,185,959	**1.000**
DMU_{12}	12,321	111,111,505	68,253	411,876	2,072,659	**1.000**
DMU_{13}	8454	89,793,993	61,291	213,648	1,484,863	**1.000**
DMU_{14}	17,096	1,65,327,770	51,974	553,545	4,128,652	**0.903**
DMU_{15}	3152	22,731,520	2197	79,819	628,332	**0.801**
DMU_{16}	3773	34,222,882	3732	115,458	867,792	**0.757**
DMU_{17}	3370	37,665,704	1725	67,742	1,228,911	**0.967**
DMU_{18}	2816	31,368,512	1875	176,491	760,681	**1.000**
DMU_{19}	956	19,703,731	1813	68,652	486,578	**1.000**
DMU_{20}	1439	24,354,369	872	108,679	795,882	**1.000**
DMU_{21}	562	8,688,552	1243	7339	221,855	**0.879**
DMU_{22}	1470	32,220,417	1054	147,628	784,897	**1.000**
DMU_{23}	891	5,588,867	656	13,260	84,013	**0.495**
DMU_{24}	1033	5,081,584	1041	7979	157,822	**0.849**
DMU_{25}	1674	8526,070	879	26,466	307,110	**0.985**
DMU_{26}	587	6,971,045	1060	12,682	153,997	**0.734**
DMU_{27}	1140	6,699,818	1993	27,984	245,109	**1.000**
DMU_{28}	466	3,336,556	946	14,231	83,888	**0.895**

representing short-term loans to other banks and financial institutions (Y_1), (2) securities held by the bank as assets (Y_2), and (3) all loans made by the bank to its customers (Y_3).

The last column in Table 1.6 shows the technical efficiency of the 28 banks. We apply the CRS input-oriented DEA model (1.3) to obtain the overall efficiency. As shown, there are 11 efficient banks and 17 inefficient ones. Table 1.7 presents the peers and benchmarks for each bank. The columns display the efficient banks and the DMUs for which each efficient bank is a benchmark; 'yes' means DMU-j is a benchmark or $\lambda_j^* > 0$ in the optimal solution of model (1.3).

TABLE 1.7 Peers and Benchmarks for 28 US Banks

Name	DMU_2	DMU_3	DMU_8	DMU_{11}	DMU_{12}	DMU_{13}	DMU_{18}	DMU_{19}	DMU_{20}	DMU_{22}	DMU_{27}
DMU_1			Yes				Yes				Yes
DMU_2	Yes										
DMU_3		Yes									
DMU_4			Yes	Yes			Yes				
DMU_5	Yes	Yes			Yes						Yes
DMU_6		Yes	Yes								Yes
DMU_7			Yes				Yes				Yes
DMU_8			Yes								
DMU_9			Yes				Yes		Yes		Yes
DMU_{10}	Yes		Yes						Yes		Yes
DMU^{11}				Yes							
DMU_{12}					Yes						
DMU_{13}						Yes					
DMU_{14}					Yes	Yes			Yes		Yes
DMU_{15}							Yes		Yes		Yes
DMU_{16}							Yes		Yes		Yes
DMU_{17}									Yes		Yes
DMU_{18}							Yes				
DMU_{19}								Yes			
DMU_{20}									Yes		
DMU_{21}	Yes					Yes			Yes		
DMU_{22}										Yes	

(Continued)

TABLE 1.7 (Continued)

Name	DMU_2	DMU_3	DMU_8	DMU_{11}	DMU_{12}	DMU_{13}	DMU_{18}	DMU_{19}	DMU_{20}	DMU_{22}	DMU_{27}
DMU_{23}			Yes				Yes				Yes
DMU_{24}											Yes
DMU_{25}											Yes
DMU_{26}						Yes			Yes		Yes
DMU_{27}											Yes
DMU_{28}	Yes	Yes	Yes								Yes

As shown, DMU_{27} appears as the benchmark for 15 inefficient banks, while DMU_{19} and DMU_{22} are not benchmarks for any other units. Now, we apply the cost efficiency DEA model for assessing the 28 US banks. The cost per unit of inputs used in transforming a bank's inputs into outputs is required to determine its cost efficiency. The two input prices in the cost efficiency DEA model are the average employee cost (C_1) and the average cost of liabilities (C_2).

As shown in Table 1.8, the number of efficient banks was significantly reduced to only four efficient banks when the cost efficiency DEA model (1.7)

TABLE 1.8 Input Prices, Cost Efficiency, and Allocative Efficiency

DMUs	C_1	C_2	Cost Efficiency	Allocative Efficiency
DMU_1	112657.3	0.000817	**0.379**	**0.495**
DMU_2	138059.4	0.001056	**1.000**	**1.000**
DMU_3	123626.1	0.001371	**0.574**	**0.574**
DMU_4	142582.6	0.001767	**0.441**	**0.589**
DMU_5	113847.8	0.00171	**0.596**	**0.637**
DMU_6	124476.8	0.00248	**0.469**	**0.494**
DMU_7	113557.2	0.001581	**0.414**	**0.489**
DMU_8	137717.8	0.0015	**0.595**	**0.595**
DMU_9	154823.9	0.001703	**0.424**	**0.730**
DMU_{10}	121217.8	0.001864	**0.471**	**0.582**
DMU_{11}	103724.5	0.002275	**0.966**	**0.966**
DMU_{12}	210382.6	0.002531	**0.708**	**0.708**
DMU_{13}	122239.4	0.002079	**0.852**	**0.852**
DMU_{14}	118700.1	0.002568	**0.639**	**0.708**
DMU_{15}	100854.4	0.00145	**0.389**	**0.486**
DMU_{16}	106394.1	0.002502	**0.462**	**0.611**
DMU_{17}	111307.7	0.002672	**0.666**	**0.689**
DMU_{18}	148541.5	0.000903	**0.628**	**0.628**
DMU_{19}	176011.5	0.002324	**1.000**	**1.000**
DMU_{20}	161577.5	0.001235	**1.000**	**1.000**
DMU_{21}	178669	0.00219	**0.833**	**0.947**
DMU_{22}	147053.7	0.003961	**1.000**	**1.000**
DMU_{23}	62289.56	0.000487	**0.215**	**0.434**
DMU_{24}	145040.7	0.001872	**0.334**	**0.393**
DMU_{25}	96681	0.000974	**0.351**	**0.357**
DMU_{26}	137671.2	0.002624	**0.579**	**0.789**
DMU_{27}	137705.3	0.000186	**0.495**	**0.495**
DMU_{28}	100442.1	0.00239	**0.458**	**0.512**

is applied. Also, allocative efficiency indicates that the same four banks were efficient. In fact, allocative efficiency is the ratio of cost efficiency to technical efficiency, so banks that were both technical and cost-efficient were also allocative-efficient.

1.4 CONCLUSION

DEA is a popular non-parametric optimization approach for measuring the comparative efficiency and productivity of decision making units (DMUs). Chapter 1 explains standard DEA models and their mathematical basis. Numerical examples were used to highlight the practicality of the various input and output orientation DEA models under both CRS and VRS. Real-world banking data was also used to reinforce the usefulness of the approach in determining efficiencies in decision making units.

REFERENCES

Amin, G. R., & Ibn Boamah, M. (2020). A new inverse DEA cost efficiency model for estimating potential merger gains: A case of Canadian banks, *Annals of Operations Research*, 295 (1), 21–36.

Amin, G.R., & Ibn Boamah, M. (2023). Modeling business partnerships: A data envelopment analysis approach, *European Journal of Operational Research*, 305(1), 329–337.

Banker, R.D., Charnes, A., & Cooper, W.W. (1984). Some models for estimating technical and scale inefficiencies in data envelopment analysis, *Management Science*, 30(9), 1078–1092.

Bazaraa, M., Jarvis, J., & Sherali, H. (2011). *Linear Programming and Network Flows*. 4th edition, John Wiley, USA.

Charnes, A., & Cooper, W.W. (1962). Programming with linear fractional functionals, *Naval Research Logistics Quarterly*, 9(3–4), 181–186.

Charnes, A., Cooper, W.W., & Rhodes, E. (1978). Measuring the efficiency of decision making units, *European Journal of Operational Research*, 2, 429–444.

Farrell, M.J. (1957). The measurement of productive efficiency, *Journal of the Royal Statistical Society Series A (General)*, 120(3), 253–290.

Gattoufi, S., Amin, G. R., & Emrouznejad, A. (2014). A new inverse DEA method for merging banks, *IMA Journal of Management Mathematics*, 25, 73–87.

Kapelko, M. (2019). Measuring productivity change accounting for adjustment costs: evidence from the food industry in the European Union, *Annals of Operations Research*, 278(1–2), 215–234.

Li, Y., Abtahi, A. R., & Seyedan, M. (2019). Supply chain performance evaluation using fuzzy network data envelopment analysis: A case study in automotive industry, *Annals of Operations Research*, 275(2), 461–484.

Wegener, M., & Amin, G. R. (2019). Minimizing GHG emissions using inverse DEA with an application in oil and gas. *Expert Systems with Applications*, 122, 369–375.

CHAPTER 2

Review of Econometrics Methods for Business Partnerships

Mustapha Ibn Boamah, Gholam R. Amin, and Thomas Patstone
Faculty of Business, University of New Brunswick, Saint John, Canada

2.1 INTRODUCTION

There are a variety of decision making models for business partnerships that seek to primarily enhance efficiency but also for other reasons such as market access and performance, product success, value creation, and cross ownership and knowledge transfer. Data envelopment analysis (DEA) and inverse DEA models are often used to determine the efficiency of decision making units (Amin and Ibn Boamah, 2020; Emrouznejad et al., 2019; Yan et al., 2002), potential gains from partnerships (Bogetoft and Wang, 2005; Gattoufi et al., 2014; Amin and Ibn Boamah, 2023), and the optimal number of partners in a business partnership (Lozano, 2013; Wu et al., 2016; Ibn Boamah and Amin, 2024).

Parametric econometric methods, on the other hand, are used to investigate the factors that contribute to business partnership objectives to ensure competitiveness. Various regression models and stochastic frontier analysis, among others, have presented an understanding of the compo-nents of partnership incentives such as enhancing productivity, market

DOI: 10.1201/9781003345206-2

performance, value creation, and knowledge transfer (Murray et al., 2005; Sjögren and Söderberg, 2011; Hoelz and Bataglia, 2022). This chapter briefly reviews literature on the econometric methods used to evaluate partnership objectives providing answers to core research questions in different business partnerships, including mergers and acquisitions (M&As), joint ventures that may or may not lead to a merger or an acquisition, and strategic alliances that impact on efficiency, market competition, and performance.

2.2 MERGERS AND ACQUISITIONS

Healy et al. (1990) analyzed the relationship between corporate and workers' performance in mergers by examining the post-merger cash flow performance of acquiring and target firms. Operating cash flows were defined as sales minus the cost of goods sold, and selling and administrative expenses, plus depreciation and goodwill expenses. Their study created two cross-sectional regression models for the extensive research on corporate performance in pre-/post-mergers, using data collected from year-end financial reports, proxy statements, and analysts' reports for 50 large mergers between US firms. The results indicated that merged firms had significant improvement in operating cash flows and worker performance for the majority of sample firms, resulting in a skyrocket of asset productivity relative to competitors in their industry. Thus, this study had significant findings concerning principal economic activities such as asset productivity improvements, long-term investments, and equity revaluations.

Cash flow as a performance measure was also emphasized in Ghosh (2001), as the study focused on merging firms' operating performance following corporate acquisitions. Using multiple linear regression and cross-sectional regression models, the paper examined the question of whether operating cash flow performance improves following corporate acquisitions as well as investigating whether operating performance was different across stock and cash acquisitions. Since performance comparisons with industry-median firms will probably lead to biased results when pre-event size and performance vary between sample and industry-median firms, the control variables in the regression models were an attempt to gauge the level of work needed to minimize bias. Operating cash flow increases significantly following cash acquisitions due to the liquidity of cash and how it can be reinvested into the company leading to higher sales growth. Results indicate that the operating performance of merging firms improved when compared to industry-median firms. This may be due to the fact that there were systematic differences in size across the two sets of firms and that merging firms often undertake acquisition following a period of superior performance.

Kim and Finkelstein (2009) investigated the effects of strategic and market complementarity on merger and acquisition performance noting that related businesses share similarities in management style, culture, and administrative processes, which would make it easy to leverage their resources in a merger or acquisition. To address complementarity differences between the acquirer and target and how exploitation of resources or strategies could lead to new business practices and thought processes throughout an organization, the study applied data from the US commercial banking industry. Multiple regression results supported the conclusion that merger and acquisition performance was positively associated with strategic complementarity and that the interaction of strategic complementarity and acquirer strategic focus was negatively related to market performance. Specifically, the lower the strategic focus, the more positive the relationship between strategic complementarity and acquisition performance.

Mergers and acquisitions are often motivated by the desire for industry stability and performance improvement. Du and Sim (2016) studied the relationship between M&As and efficiency in the banking sector in China, India, Indonesia, Malaysia, Russia, and Thailand. The study focused on the use of a cross-country panel regression that separates target banks from acquiring banks to better understand the relationship dynamics and to obtain disaggregate estimates. The regression model included interactive country-year fixed effects for time-invariant heterogeneity and country-specific factors such as policies and the effects of global shocks on the six emerging countries. Total assets and total equity were used to account for bank-specific differences, and DEA efficiency scores were regressed with the indicators of target and acquiring banks. The results indicated that indeed M&As affect target and acquiring banks differently and that M&As may lead to high efficiency in target banks. The asymmetry in bank efficiency between target and acquiring banks resulting from M&As is important to managers as they would like to know the distribution of efficiency gains for a merged entity.

A study on whether merger policies increase universities' efficiency was undertaken by Agasisti et al. (2021). They focused on merger policies in Russia that were intended to improve universities' performance using fuzzy regression discontinuity framework to analyze the impact on the efficiency level of the universities involved. The paper concluded that merged universities witnessed greater efficiency and that the merger policy was a good policy instrument for enhanced performance as it reduced average costs without impacting teaching and research. Zhang et al. (2018) explored the relationship between M&As and firm performance of listed Chinese pharmaceutical companies by running a partial least squares regression. The results indicated that technology-seeking and value-chain-extension M&As improve firm performance. Growth capacity, assets, and size were all important determinants of firm performance; thus, Chinese pharmaceutical firms at their growth stage could increase industry concentration and firm scale to enhance performance.

Furthermore, Sun et al. (2024) also investigated the impact of technological M&As on corporate innovation performance among Chinese manufacturing companies. In particular, the study examined the effect of acquisition behavior on enterprise innovation and whether technology M&As impact innovation through the scale of knowledge base. The results from various regression models confirmed that technology acquisition promotes innovation performance of enterprises and that the scale of the knowledge base is an important intermediate variable in the relationship process. In a similar study, Zhang et al. (2018) analyzed the effect of foreign M&As on firm innovation. Using fixed effect regressions, they found clear evidence of a positive relationship between foreign M&As and firm innovation. Foreign M&As also increased firm size, exports, human capital formation, and consequently firm competitiveness.

Carlton et al. (2019) evaluated the effects of mergers on competitiveness in the US airline industry. They conducted a comprehensive investigation of the impact of mergers on passenger fares and output to determine if they had a net pro-competitive or anti-competitive effect on consumers. Passenger fares are affected by many economic factors that may be changing over time, and changes in nominal fares are not a good measure of consumer welfare, especially if it is reasonable to believe that quality changes simultaneously. Therefore, the study employed difference-in-difference regression analysis that does not rely on the structural modeling of the effects of the merger on nominal fares but rather the impact of changes in fares on merger routes compared with non-merger routes. The results demonstrated that mergers were pro-competitive in output expansions, increasing passenger traffic but had no adverse impact on passenger fares.

2.3 JOINT VENTURES

Businesses may form joint ventures, a separate legal entity in which each decision making unit owns a share, prior to a merger or an acquisition. Watson et al. (1995) discussed the importance of a joint venture business and its impact on interpersonal process effectiveness and company performance as perceived by employees. A venture team is defined as two or more individuals who jointly participate in a business in which they have an equity/financial interest and work to grow the business and make it profitable. These teams need to be in sync with their communication within partnership integration which is critical if they are to succeed. Results from a logistic regression confirmed that venture partnership that indicated a more effective interpersonal process reported greater venture success and overall positively improved the joint venture's competitiveness. The choice of business partnership, especially for foreign market entry, depends on factors such as market structure and how they impact on risk-adjusted return on investment (Agarwal & Ramaswami, 1991). Firms must know, from an operation standpoint, the allocation of resources and the need for resources from partners for the partnership to achieve its objective. Firms with a higher number of equity alliances tend to have more products in development but fewer exploitation alliances. Using structural equation modeling, Rothaermel and Deeds (2004) indicated that in the biotech industry, the product development path runs from exploration alliances to products in development, creating a chain effect. Additionally, they found that a firm's exploration alliances are significant in predicting the firm's products in development, and a firm's products in progress in turn are a significant predictor of exploitation alliances.

Furthermore, Lane et al. (2001) concluded that the knowledge an international joint venture acquired from its foreign parents was positively associated with its performance. An emphasis that differentiation in an international joint venture's business strategy was positively associated with its performance; and its degree of competence in training and personnel development was also positively associated with its performance in the regression analyses. Luo (2002) delved into the intricate research of international joint venture by measuring the level cooperation, contract, and performance to shed light on critical factors that impact the success and economic performance of international joint ventures. Data on contract and cooperation were obtained from field interviews and a survey questionnaire, while those on performance were from archives. The study results from standardized multiple regressions found that there was a strong complementary relationship between cooperation and performance as well as contingency adaptability. Managers and senior executives need to understand the underlying effects of what poor cooperation, performance, and contingency adaptability can do to an organization and its partners.

As an international partner, it is best to have a company in a foreign market with a significant footprint in the area and can function as a parent to assist in the transition to currently unknown markets for the organization. Alignment with a partner who is of similar economic power and goals to perform at the highest rate in the business partnership is crucial; thus, business alignment or misalignment can lead to or dismantle a company and its business partnerships. The timing of entering a partnership is also vital to the success and direction of how to properly manage the business partnership. In foreign partnerships, the similarities between the benefits of foreign parent involvement in joint ventures and the changes occurring in both local and foreign environments as the economies evolve impact how foreign partners create value in joint ventures. Steensma et al. (2005) studied foreign partnerships in a transition economy to understand whether resource inputs provided by foreign parents were more critical to market performance and knowledge acquisition during the early phase of a country's transition to a market economy or later in the transition. Using data derived from a time series survey, their hierarchical moderation regression model concluded that foreign parent resource provision had a more positive influence on joint venture market performance in the early phase of an economic transition than in the late phase – emphasizing the timing and its importance in deciding to enter a transitioning economy. The study results also indicated that foreign parent resource provision had a more positive influence on joint venture knowledge acquisition in the late phase of economic transition than in the early phase. This means that foreign parent resources have a stronger influence on knowledge acquisition in the late phase of economic transition to a free market economy.

Anand and Khanna (2000) evaluated the drivers of value creation in alliances within the scope of whether firms can manage interim alliances as experience in the industry grows as well as identifying the driver(s) of value creation in alliances. The authors focused on joint ventures as the primary alliance because of their growing popularity at the time and concluded that learning effects were stronger in research and development joint ventures than in other categories of joint ventures such as production joint ventures and marketing joint ventures. The results of their multivariate regression analyses confirmed that differentiation of joint ventures was important in determining the learning effectiveness of creative value in partnerships.

2.4 STRATEGIC ALLIANCES

Christoffersen et al. (2014) analyzed different dimensions of strategic alliance performance and provided a comprehensive classification of performance measures. They applied logistic regression models to test hypotheses about how the choice of various performance measures may have on the strategic alliance performance. They found that conceptual differences between measures exist and were empirically associated with the likelihood of different results. Therefore, choosing the appropriate performance measures in strategic alliance research cannot be overstated.

Taylor (2005) examined the fundamental reasons why alliances succeed. They investigated whether structural, formation factors have more influence on alliance success than implementation or process factors. The study, based on informant interviews and surveys of alliance managers, highlighted the importance of alliance implementation process. It concluded that learning capacity and inter-partner leaning processes were important components of a successful alliance implementation strategy. Broadening the predictors of alliance success, Nielsen (2007) studied the relationship between multidimensional measures of international strategic alliance performance and predictors of success both before the alliance is formed and during the operation of the alliance. Three distinct multiple regression models were used to determine which combinations of factors that predicted pre and post alliance performance. The results showed that there was a relationship between alliance performance and host country risk in addition to partner reputation prior to alliance formation. Post-alliance formation factors that predicted performance included know-how, trust, complementarity, and cultural distance.

On the other hand, Hoelz and Bataglia (2022) investigated the influence of corporate reputation on alliance performance and examined the intermediate effects that elucidate the processes through which this phenomenon occurs. They tested how reputation influences alliance performance and how interorganizational trust functions as a mediating factor. The study used data from Brazilian pharmaceutical sector and applied structured equation modeling to analyze effect on alliance performance. The results confirmed that corporate reputation positively impacts the performance of strategic alliances and that interorganizational trust had a moderating effect on the relationship between corporate reputation and strategic alliance performance. Interorganizational trust, they concluded, implies minimum alliance transaction cost is needed to control a partner leading to an increase in alliances' performance.

Using stochastic frontier analysis for answers, Ludwig et al. (2009) questioned whether or not the Dutch government should make hospitals produce or outsource some standard incurring costs of running a hospital. The study first defined human asset specificity and physical asset specificity, and how this affects the costs of a hospital to determine the best option from an efficiency perspective. Human asset specificity is measured based on the experience and performance needed to reach an adequate performance level, and physical asset specificity relates the time equipment in the department is not routinely used in the provision of service. Since asset specificity is tremendously high in the health sector, it makes sense to only partner with suppliers or outsource non-medical parts of the business such as administration, cleaners, landscaping, laundry, and maintenance. The results confirmed that on average, hospitals were not cost-efficient and were wasting valuable taxpayers' money. Large hospitals were extremely inefficient as their costs were disproportionally high compared to small- and medium-sized hospitals. Notably, there were a couple of exceptions made in this frontier, as it would not be beneficial to outsource salary administration as it would cause more communication issues for the potential marginal savings. Additionally, it would be extremely cost-efficient to outsource physiotherapy as it would cut costs dramatically while providing the same care.

Mishra et al. (2020) examined sourcing structures of IT projects that cross boundaries of a firm, a country, or both to benefit from low labor cost and access to technical talent. This study considered four types of business projects based on the location of sourcing partners, domestic insourcing, domestic outsourcing, offshore insourcing, and offshore outsourcing. The dynamics of each partnership were significantly different from one another, and their coordination, communication, and project execution inevitably differ as well. The Cobb-Douglas model was used to express the stochastic production function showing utilization of inputs across all partnerships that lay below the efficiency frontier. Efficiency frontier mapping determines the most optimal solution to balance cost-effectiveness and performance execution. The results supported the overall hypothesis that project conversion, project input into project output, differed between each type of partnership, and reported the mean technical efficiency for each partnership.

Murray et al. (2005) relied on regression analysis to evaluate the impact of strategic alliance sourcing of components at the product level from foreign firms operating in China on market performance. The study focused on whether strategic alliance is necessary to provide a larger competitive advantage using key economic concepts such as resource complementarity and the resource dependence theory. The results concluded that market performance was influenced by product and uncertainty factors and that strategic alliance sourcing was positively related to market performance. This means that if organizations strategically collaborate, it can enhance their market outcomes. The paper shed light on strategic firms producing in China, an important economic powerhouse in production, and underscored the influence of contextual factors when making any strategic choices for a company.

Sjögren and Söderberg (2011) extended the strategic alliance research to the international airlines industry by investigating how deregulation, privatization, and the formation of strategic alliances have affected the productivity of international airlines. This study evaluated how productivity was affected by fundamental changes in the international airline industry using stochastic frontier analysis. A critical assumption made in this research which could cause skewness was that firms' behavior was assumed to be the result of economic optimization, that firms had access to the same technology, and that deviations from optimal performance followed a pre-specified distribution. Results found that carriers originating from North America are more efficient than the reference group, and that European-originated airlines have increased productivity over time after an international alliance. The future of airline productivity will likely continue to be modeled by partnerships, including strategic alliances, mergers and acquisitions, and technological advancements.

Companies find ways to build international operations capabilities through partnerships and networks to be efficient and competitive. Globalization is steadily becoming an integral part of trade and the basis for international business partnerships worldwide, and this makes it easier for new business start-ups who can differentiate based on price, product differentiation, or leading-edge technology to be effective in their market strategy and can tap into the local and global trade markets with the advising of a parental network. Elango and Pattnaik (2007) studied the value of international connections and also the relationships facilitated by parental network capabilities in the Indian economy; using a multiple linear regression model, they concluded that companies should have strong market power in the local economy before attempting to branch out to a wider global market. The results also indicated that operational efficiency was important and that companies should review their operations to ensure that they are efficient prior to seeking global partnerships and networks.

Su et al. (1996) also evaluated company's internal capabilities and external partnerships to understand how they contribute to its product and process innovativeness within the Taiwanese biotechnology industry, which is rapidly growing in the newly industrialized economy. They addressed two key questions: drawing on the theoretical framework of exploitation and exploration. The two sets of independent variables used to explain innovativeness in the regression model were internal capabilities and external partnerships. Internal capabilities were measured as R&D operations such as marketing and manufacturing. And external partnerships were defined as the extent to which firms interact with partners. The results found that R&D capability had positive effects on both product and process innovativeness. Marketing capability had a greater effect than manufacturing capability on product innovativeness, indicating that resources that were used in marketing a product were more efficiently allocated than into manufacturing. External partnerships with suppliers, customers, competitors, and universities and research institutes had positive effects on both product and process innovativeness, while partnerships with universities and research institutes had the greatest effect among the four types of partnerships on both product and process innovativeness. Finally, marketing capability had a significant positive interaction effect on product innovativeness as manufacturing capability had on process innovativeness. This study highlights important decisions management faces and how to better understand the internal capabilities and external factors in their organization.

Some partnerships benefit from the transfer of technology and knowledge sharing that enables companies working together to efficiently develop and introduce products and services that would not be possible if they were on their own. Using data European-based companies, Scuotto et al. (2020) chose this region as its solicitation and incentive to partner due to the absence of commercial barriers between countries in Europe. The authors defined technology transfer as the knowledge shared between parties, where knowledge sharing occurred as the activity of transferring or disseminating knowledge by managers from one decision making unit to another. This study turned to the Theory of Planned Behavior which is a framework to understand the social and psychological determinants of individual behavior that encompasses managers' intention to collaborate and quantify technology transfer. Multiple linear regression models were chosen as the parametric method of choice as the models allow the authors to assess the relationships between variables, and because the Theory of Planned Behavior relies on implicit assumptions. The dependent variables were technology transfer and perception of the value of collaboration. With governance and market, intellectual property and licensing, and types of cooperation as independent variables, the results showed that there was a linear and positive relationship between the implementation of technology behavior and the perception of the positive value of collaborative innovation. Meaning, that when knowledge is shared between the two partners, a level of trust is molded together creating a stronger business partnership that increases competitiveness.

Contractor and Kundu (1998) investigated how business partnerships set to determine the optimal form of a strategic business partnership in the hotel sector. The authors derived from a theory comprised of transaction costs theory, agency theory, corporate knowledge, and organizational capability theories to aid in understanding organizational forms and business partnerships. Ordinal logistic regression was used in this study because it enabled the identification of the independent variables that most strongly influence the selection of organizational mode. The dependent variable in this study was the modal choice for each hotel property. As a polytomous measure, it showed the rising levels of equity ownership, shareholder ownership, and overall control of the business. The independent variables on the other hand were country risk, cultural distance, level of development, penetration by foreign business, size, international experience, extent of foreign business, economies of scale, control quality, need for size, global reservation system, and investment in training. The results found that rising levels of equity and control were negatively associated with the country's political and economic risk and were also negatively associated with the level of economic development in the country where the hotel is located. Investors are attracted to or deterred from the hotel industry based on their personal risk/fixed equity ratio. The average hotel investor is eager to build hotels in emerging or solidified markets as market supremacy and domination are critical in the hotel industry.

Finally, Yang et al. (2015) examined the effect of strategic alliance on innovation capability and quality performance in Chinese manufacturing

TABLE 2.1 Summary of Business Partnerships Explored

Research Question/ Relationship	Partnership Objective	Econometric Method	Partnership Type
Corporate/workers' performance and cash flow	Market performance	Multiple linear and cross-sectional Regression analysis	M&As
Strategic complementarity and acquisition performance	Knowledge transfer	Random-effects generalized least squares regression analysis	M&As
Efficiency score and target firms	Productivity	Cross-country panel regression and DEA analysis	M&As
Cooperation and contract	Economic performance	Multiple moderated regressions regression analysis	Joint Ventures
Learning effectiveness and alliances	Value creation	Cross-sectional multivariate regression analysis	Joint Ventures
Size and cost efficiency	Productivity	Stochastic frontier analysis	Strategic Alliances

companies, and how these factors are related to the supply chain performance. They applied a series of econometrics approaches including multiple regression, logistic regression, and multivariate analysis of variance for assessing the hypothesized relationships. The study findings supported a causal relationship between quality performance and supply chain performance and that alliance formation benefited innovation capability and quality performance which were important determinants of the supply chain performance of Chinese manufacturing companies. It highlighted that companies with consistent goals are better motivated to innovate to satisfy their mutual relationships which results in better collective performance.

Table 2.1 shows typical research questions and relationships that were explored using different econometric methods for the three main business partnership types reviewed in this chapter. The methods used were not unique to the type of partnership reported, nor were the research questions/relationships and partnership objectives.

2.5 CONCLUSION

This chapter reviews literature on business partnerships through a parametric econometric methods perspective. It discusses different types of business partnerships, research questions, hypotheses, and their associated econometric models to understand the factors that contribute to performance and competitiveness. Some of the factors explored in the research questions and hypotheses include strategic complementarity, interpersonal process effectiveness, market structure, and corporate reputation and how they impact on measures of performance such as value creation, innovation and knowledge transfer, market performance, and productivity.

The research papers discussed in this chapter also revealed the measures of performance used in different business partnerships. These complex and different relationships include mergers and acquisitions where the pre-merger and post-merger partnership objectives are compared, joint ventures that could lead to mergers and acquisitions, and various strategic alliances involving companies that may or may not be in the same industry.

REFERENCES

Agarwal, S., & Ramaswami, S. N. (1991). Choice of foreign market entry mode: Impact of ownership, location and internalization factors. *Journal of International Business Studies, 23*(1), 1–27.

Agasisti T., Egorov A., & Maximova M. (2021). Do merger policies increase universities' efficiency? Evidence from a fuzzy regression discontinuity design. *Applied Economics, 53*(2),185–204.

Amin, G. R., & Ibn Boamah, M. (2020). A new inverse DEA cost efficiency model for estimating potential merger gains: A case of Canadian banks. *Annals of Operations Research, 295*(1), 21–36.

Amin, G.R., & Ibn Boamah, M. (2023). Modeling business partnerships: A data envelopment analysis approach. *European Journal of Operational Research, 305*(1), 329–337.

Anand, B. N., & Khanna, T. (2000). Do firms learn to create value? The case of alliances. *Strategic Management Journal, 21*(3), 295–315.

Bogetoft, P., & Wang, D. (2005). Estimating the potential gains from mergers. *Journal of Productivity Analysis, 23*, 145–171.

Carlton D., Israel M., MacSwain I., & Orlov E. (2019). Are legacy airline mergers pro- or anti-competitive? Evidence from recent U.S. airline mergers. *International Journal of Industrial Organization, 62*, 58–95.

Christoffersen, J., Plenborg, T., & Robson, M.J. (2014). Measures of strategic alliance performance, classified and assessed. *International Business Review, 23*(3), 479–489.

Contractor, F. J., & Kundu, S. K. (1998). Modal choice in a world of alliances: Analyzing organizational forms in the international hotel sector. *Journal of International Business Studies*, 29(2), 325–358.

Du, K., & Sim, N. (2016). Mergers, acquisitions, and bank efficiency: Cross-country evidence from emerging markets. *Research in International Business and Finance, 36*, 499–510. https://doi.org/10.1016/j.ribaf.2015.10.005

Elango, B., & Pattnaik, C. (2007). Building capabilities for international operations through networks: A study of Indian firms. *Journal of International Business Studies, 38*(4), 541–555. https://doi.org/10.1057/palgrave.jibs.8400280

Emrouznejad, A., Yang, G., & Amin, G. R. (2019). A novel inverse DEA model with application to allocate the CO_2 emissions quota to different regions in Chinese manufacturing industries. *Journal of the Operational Research Society, 70*(7), 1079–1090.

Gattoufi, S., Amin, G. R., & Emrouznejad, A. (2014). A new inverse DEA method for merging banks. *IMA Journal of Management Mathematics, 25*, 73–87.

Ghosh, A. (2001). Does operating performance really improve following corporate acquisitions? *Journal of Corporate Finance, 7*(2), 151–178. https://doi.org/10.1016/S0929-1199(01)00018-9

Healy, P. M., Palepu, K. G., & Rubak, R. C. (1990). Does corporate performance improve after mergers? *Journal of Financial Economics, 31*(2), 135–175. http://doi.org/10.3386/w3348

Hoelz, J.C., & Bataglia, W. (2022). Corporate reputation and strategic alliance performance. *Corporate Reputation Review, 25*, 161–175.

Ibn Boamah, M., & Amin, G.R. (2024). Partner selection in business mergers: A data envelopment analysis approach, *RAIRO-Operations Research, 58*(4), 3107–3117. https://doi.org/10.1051/ro/2024091

Kim, J.-Y., & Finkelstein, S. (2009). The effects of strategic and market complementarity on acquisition performance: evidence from the U.S. commercial banking industry. 1989–2001. *Strategic Management Journal, 30*(6), 617–646. https://doi.org/10.1002/smj.754

Lane, P. J., Salk, J. E., & Lyles, M. A. (2001). Absorptive capacity, learning, and performance in international joint ventures. *Strategic Management Journal, 22*(12), 1139–1161. https://doi.org/10.1002/smj.206

Lozano, S. (2013). Using DEA to find the best partner for a horizontal cooperation. *Computers & Industrial Engineering, 66*(2), 286–292.

Ludwig, M., Groot, W., & Van Merode, F. (2009). Hospital efficiency and transaction costs: A stochastic frontier approach. *Social Science and Medicine, 69*(1), 61–67.

Luo, Y. (2002). Contract, cooperation, and performance in international joint ventures. *Strategic Management Journal, 23*(10), 903–919.

Mishra, A., Sinha, K. K., Thirumalai, S., & Van, de V. A. (2020). Sourcing structures and the execution efficiency of information technology projects: A comparative evaluation using stochastic frontier analysis. *Journal of Operations Management, 66*(3), 281–309.

Murray, J. Y., Kotabe, M., & Zhou, J. N. (2005). Strategic alliance-based sourcing and market performance: evidence from foreign firms operating in China. *Journal of International Business Studies, 36*(2), 187–208. https://doi.org/10.1057/palgrave.jibs.8400120

Nielsen, B.B. (2007). Determining international strategic alliance performance: A multidimensional approach. *International Business Review, 16*(3), 337–361.

Rothaermel, F. T., & Deeds, D. L. (2004). Exploration and exploitation alliances in biotechnology: A system of new product development. *Strategic Management Journal, 25*(3), 201–221. https://doi.org/10.1002/smj.376

Scuotto, V., Beatrice, O., Valentina, C., Nicotra, M., Di Gioia, L., & Farina Briamonte, M. (2020). Uncovering the micro-foundations of knowledge sharing in open innovation partnerships: An intention-based perspective of technology transfer. *Technological Forecasting & Social Change*, 152–166. https://doi.org/10.1016/j.techfore.2019.119906

Sjögren, S., & Söderberg, M. (2011). Productivity of airline carriers and its relation to deregulation, privatization, and membership in strategic alliances. *Transportation Research Part E, 47*(2), 228–237. https://doi.org/10.1016/j.tre.2010.09.001

Steensma, H. K., Tihanyi, L., Lyles, M. A., & Dhanaraj, C. (2005). The evolving value of foreign partnerships in transitioning economies. *Academy of Management Journal, 48*(2), 213–235.

Su, Y.-S., Tsang, E. W. K., & Peng, M. W. (1996). How do internal capabilities and external partnerships affect innovativeness? *Asia Pacific Journal of Management, 26*(2), 309–331. https://doi.org/10.1007/s10490-008-9114-3

Sun H., Long Y., & Yuan R. (2024). Technology M&A and enterprise innovation performance–knowledge-based mediation effect. *Technology Analysis and Strategic Management, 36*(2), 365–377.

Taylor, A. (2005). An operations perspective on strategic alliance success factors: An exploratory study of alliance managers in the software industry. *International Journal of Operations & Production Management*, 25(5), 469–490.

Watson, W. E., Ponthieu, L. D., & Critelli, J. W. (1995). Team interpersonal process effectiveness in venture partnerships and its connection to perceived success. *Journal of Business Venturing*, 10(5), 393–411.

Wu, J., Zhu, Q., Cook, W. D., & Zhu, J. (2016). Best cooperative partner selection and input resource reallocation using DEA. *The Journal of the Operational Research Society, 67*(9), 1221–1237.

Yan, H., Wei, Q., & Hao, G. (2002). DEA models for resource reallocation and production input/output estimation. *European Journal of Operational Research, 136*(1), 19–31.

Yang, J., Lai, K.H., Wang, J., Rauniar, R., & Xie, H. (2015). Strategic alliance formation and the effects on the performance of manufacturing enterprises from supply chain perspective. *International Journal of Production Research*, 53(13), 3856–3870.

Zhang, W., Wang, K., Li, L., Chen, Y., & Wang, X. (2018). The impact of firms' mergers and acquisitions on their performance in emerging economies. *Technological Forecasting and Social Change, 135*, 208–216.

Zhang, Z., Deng, Y., & Tang, S.H.K. (2019). Do foreign mergers and acquisitions increase acquired firms' innovation? Evidence from Chinese manufacturing firms. *Australian Economic Review, 52*(1), 41–60.

CHAPTER 3

Uncertain Data Envelopment Analysis

Theory, Algorithms, and Applications

Casey Garner
Department of Mathematics, University of Minnesota, Minneapolis, MN, USA

Allen Holder and Nat Hurtig
Department of Mathematics, Rose-Hulman Institute of Technology, Terre Haute, IN, USA

3.1 INTRODUCTION

Data envelopment analysis (DEA) has flourished as an academic study and as an effective and pragmatic assessment technique since its inception in 1978 [1]. DEA's durable success is due to its basic but ingenious use of linear programming to make intra-comparisons among decision making units (DMUs). It is, without doubt, a straightforward, inviting, and reasoned way to use optimization for assessment and for credible decision making, which gives DEA a broad draw in academia, business, science, medicine, engineering, and the social sciences.

DEA's extensive employ has motivated numerous adaptations, and its fundamental reliance on linear programming has allowed it to harness the computational and modeling advances of this prominent study. For instance, we have discrete, stochastic, combinatorial, fuzzy, and multiple objective DEA models; we have advanced algorithms tailored to DEA; and we have leveraged sensitivity analysis and parametric programming

DOI: 10.1201/9781003345206-3

to analyze applications – all of which extend related studies in linear programming. So we have a plethora of options as we consider using DEA to discern, advise, and manage.

Our goal herein is to summarize and review how a contemporary DEA model can address one of the longest standing problems in operations research (OR), that being how to make optimal decisions under uncertainty. This question is part of OR's earliest history; indeed, one of OR's first problems was to advise transatlantic shipping strategies to reduce casualties and to limit the loss of cargo as German U-boats (randomly) patrolled the seas [2]. Moreover, the Father of Linear Programming, the preeminent George Dantzig, posited decision making under uncertainty early in his career [3]. Grappling with uncertainty continues within the OR and mathematical programming communities, resulting in several mature schemes to model, integrate, and calculate with imperfect information. The traditional tactic uses the language of probability and statistics, which moves us into the area of stochastic optimization. These studies benefit from a substantial backdrop of theory and application, but they innately rely on distributional assumptions that need to be assumed or empirically verified. Sensitivity analysis and parametric programming instead work to answer 'what if' queries from a deterministic counterpart that represents a plausible instance of an uncertain problem. Fuzzy logic replaces probability distributions with membership functions, giving whole new perspectives on likelihood and the degree of truth.

A more recent paradigm to address uncertainty is that of robust optimization, which was introduced to the OR community in the seminal works of Ben Tall and Nimrovski [4, 5], Ghaoui [6, 7], and Kouvelis and Yu [8] – see [9] as a review. Our approach herein is to adopt the modern use of robust optimization to model and incorporate imperfect data within the DEA framework. Robust optimization neither supplants nor combines other tactics, but rather, it adds a complimentary scheme to our traditional options [9]. This new perspective is, in the authors' opinion, especially appropriate for DEA because it provides a modeling framework to address the inherent uncertainty of the data on which DMUs are compared. Moreover, an uncertain DEA (uDEA) model coincides with the basic premise of DEA, which is that DMUs get to decide how they want to assess themselves against the other DMUs. The advance with uDEA is that each DMU not only gets to decide how it weighs inputs and outputs, but that it also gets to select data within a realm of uncertain options. This is a meaningful and powerful combination that extends DEA to overcome the pernicious concern that DMUs might be deemed inefficient due to dubious data, whereas a perfectly reasonable alternate data selection would have shown them to be efficient. DEA analysis critically relies on data, after all, its name is **data** envelopment analysis; however, data is innately inaccurate in almost all settings. Adopting a robust perspective on DEA straightforwardly addresses this fundamental aggravation.

Concomitant with DEA's many faceted employ, there are numerous aggregations of DEA and robust optimization, and melding these two studies has an established contemporary literature; see [10] as a modern survey as well as the recent research in [11, 12]. The exposition that follows begins by reviewing and unifying uDEA's development as presented in [13–16], giving a succinct monograph on uDEA. The resulting uDEA framework no longer lives under the computational umbrella of linear programming, and we discuss algorithms and heuristics to solve uDEA's nonconvex problems in Section 3.4. The subsequent computational model motivates more than traditional efficiency scores by imbuing an ideal of equitable efficiency among subgroups of DMUs, a concept that prompts classifying/clustering DMUs so that they are most equitable. This classification scheme is developed in Section 3.3.

We conclude by reviewing applications of uDEA in medical physics, stock evaluation, and political redistricting [13–15, 17]. We computationally interrogate the stock example to find that previous algorithmic schemes are, unfortunately, sensitive to modeling and algorithmic parameters, which prompts enhanced algorithmic design. We subsequently retool an important computational step to reduce sensitivity, and we then address a problem in the airline industry previously studied in [18–20]. This example demonstrates the efficacy of our updated algorithmic scheme on higher dimensional problems, and it suggests that uncertain clustering could be used to identify business partnerships, mergers, and possible conglomerates.

3.2 UNCERTAIN DATA ENVELOPMENT ANALYSIS

We select the input-oriented DEA model with variable returns to scale as our developmental platform from among the many DEA modeling options [21]. Nothing specific in our mathematical or experimental work depends on this selection, and our development extends straightforwardly to, e.g., output-oriented models with or without constant returns to scale. Our cognizance of uncertain data promotes segmenting DMUs into categories/clusters for the sake of analysis, and as such, we assume that DMUs are partitioned into S (indexed by s) categories. This assumption supports the search for clusters that categorize DMUs into groups most similar in their data. Classic DEA sets $S = 1$, and it subsequently compares each DMU against all other DMUs, whereas our premise of segmentation permits DMUs to be compared with only those DMUs within a category. We assume T DMUs (indexed by t), N inputs (indexed by i), and M outputs (indexed by j), with the values of the inputs and outputs being

X^s_{it} - the i-th input for DMU t in category s, and

Y^s_{jt} - the j-th output for DMU t in category s.

All of our examples assume $X^{s'}_{it} = X^{s''}_{it}$ and $Y^{s'}_{it} = Y^{s''}_{it}$ for different categories s' and s''; however, nothing mathematically changes if data differs with category membership. The data matrices for category s are X^s and Y^s, which comprise all the data if $S = 1$.

The efficiency score of DMU t' in category s under the assumption of certain data is

$$E^{t'} = \min\{c^T \eta \; : \; A^{t'} \eta \leq 0, \; B\eta = e, \; \eta \geq 0\},$$

with

$$A^{t'} = \left[\begin{array}{c|c|c} X^s & 0 & -x^{t'} \\ \hline -Y^s & y^{t'} & 0 \end{array}\right], \quad B = \left[\begin{array}{c|c|c} e^T & 0 & 0 \\ \hline 0 & 1 & 0 \end{array}\right], \quad \text{and} \quad c = \begin{pmatrix} 0 \\ \vdots \\ 0 \\ 1 \end{pmatrix}. \tag{3.1}$$

We assume DMU t' is in category s, and the column vectors $x^{t'}$ and $y^{t'}$ are the respective t' columns of X^s and Y^s. The vector of ones is e, with length being decided by the context of its use. We index the rows of A with k, which indexes from 1 to $M + N$, so the first M values correspond with the input data and the last N values correspond with the output data. The vector $\eta_{t'}$ with components

$$[\eta_{t'}]_t = \begin{cases} 1, & t \in \{t', T+1, T+2\} \\ 0, & \text{otherwise} \end{cases}$$

is always feasible, and hence, the maximum efficiency score is 1. Note that this model potentially suffers from the concept of weak efficiency, which for this model means that either a DMU could have an efficiency score less than one, whereas any increase in an output would result in a score of one, or a DMU could have an efficiency score of one, whereas any decrease in an input would result in a score less than one. A DMU satisfying either of these properties is weakly efficient.

Robust optimization's central tenant is to optimize relative to the requirement that constraints must be satisfied for all data instances within an uncertainty set, and as such, a robust optimization's solution hedges against the most restrictive cases in the uncertainty set. The optimal objective value of a robust problem is consequently no better than a deterministic counterpart whose data is selected from the uncertainty set. Note that designing a robust problem requires a model of uncertainty, and while there are many options, of which the most flexible is simple set membership, the most prevalent paradigm is row-wise uncertainty. The row-wise model assumes uncertainty per constraint, and it is the model that we employ. See [9, 22] for practical discussions of other options and a rationale as to why row-wise uncertainty is unconcerning.

The rows of the data matrices X^s and Y^s correspond with the input and output characteristics of the DMUs, and the row-wise model predicates an uncertainty set for each input and output. We, respectively, denote the i-th row of X^s and the j-th row of Y^s as X_i^s and Y_j^s, and we let $\mathcal{X}_i^s$ and $\mathcal{Y}_j^s$ be the respective input and output uncertainty sets. The uncertain model then replaces all deterministic uses of X_i^s and Y_j^s with the statements

$$\left(X_i^s \text{ for all } X_i^s \in \mathcal{X}_i^s\right) \text{ and } \left(Y_j^s \text{ for all } Y_j^s \in \mathcal{Y}_j^s\right).$$

Each row of $A^{t'}$ in (3.1) then equates with an uncertainty set from one of $\mathcal{X}_i^s$ or $\mathcal{Y}_j^s$, which results in the uDEA model.

$$\mathcal{E}^{t'}(\mathcal{U}) = \min\{c^T\eta \;:\; A_k^{t'}\eta \leq 0,\ \forall\, A_k^{t'} \in \mathcal{U}_k,\ B\eta = e,\ \eta \geq 0\},$$

with the uncertainty set of the k-th input/output characteristic being $\mathcal{U}_k$ and the uncertainty set $\mathcal{U}$ being the collection $\{\mathcal{U}_k : k = 1, 2, \ldots, N+M\}$. The value of $\mathcal{E}^{t'}(\mathcal{U})$ is the robust efficiency score for the uncertainty set $\mathcal{U}$, and we remark that $\mathcal{E}^{t'}(\mathcal{U}) = E^{t'}$ if each $\mathcal{U}_k$ is the singleton set containing the data used to calculate $E^{t'}$. Moreover, if at least one of the uncertainty sets contains more than the data used to calculate $E^{t'}$, then

$$\mathcal{E}^{t'}(\mathcal{U}) \geq E^{t'}$$

due to the deleterious effect associated with robust optimization, i.e., the minimum value is non-decreasing as more constraints are added to the problem. So the robust counterpart of a DEA problem cannot degrade the efficiency score, but it can improve it. Note that the vector $\eta_{t'}$ remains feasible, and hence, we know that $\mathcal{E}^{t'}(\mathcal{U}) \leq 1$.

DEA's salient premise is that we assess DMUs by calculating individualized formulas of efficiency for each DMU. So DEA is a competitive analysis in which DMUs are inefficient if they are outdone by other DMUs under their best possible formulas of efficiency. The uDEA model boosts this sentiment by allowing each DMU to optimize both its efficiency formula and its data selection within the uncertainty set. To see this, select $\hat{A}_k^{t'} \in \mathcal{U}_k$ so that

$$\mathcal{E}^{t'}(\mathcal{U}) = \min\{c^T\eta \, : \, \hat{A}_k^{t'}\eta \leq 0, \; \forall \, k, \; B\eta = e, \; \eta \geq 0\}.$$

The X^s and Y^s submatrices of $\hat{A}$ are then data instances in $\mathcal{X}^s$ and $\mathcal{Y}^s$ that optimize the efficiency score of DMU t' within category s, and hence, $\hat{A}$ is a data instance that DMU t' would prefer because it advances a best possible efficiency score. Note that the data instance is not necessarily unique and that others could have given the value of $\mathcal{E}^{t'}(\mathcal{U})$. Note further that the data selection is not just for DMU t' but that it is also for the other DMUs; so each DMU gets to select data from the uncertainty set for itself and for the other DMUs as it optimizes its individualized efficiency formula. The result is that uDEA hedges against every data instance in the uncertainty set as it decides the efficiency score, with all selections being optimized to make a DMU appear as efficient as possible.

The robust efficiency score satisfies a monotonicity property that prompts additional properties. We indicate that $\mathcal{U}''$ has at least the uncertainty of $\mathcal{U}'$, denoted $\mathcal{U}' \lhd \mathcal{U}''$, if $\mathcal{U}'_k \subseteq \mathcal{U}''_k$ for all k. The following result is in [13].

Theorem 3.1 *If* $\mathcal{U}' \lhd \mathcal{U}''$, *then* $\mathcal{E}^{t'}(\mathcal{U}') \leq \mathcal{E}^{t'}(\mathcal{U}'')$.

Proof If $\mathcal{U}' \lhd \mathcal{U}''$, then

$$\{\eta : \; A_k^{t'}\eta \leq 0, \; \forall \, A_k^{t'} \in \mathcal{U}''_k, \; B\eta = e, \; \eta \geq 0\}$$
$$\subseteq \{\eta : \; A_k^{t'}\eta \leq 0, \; \forall \, A_k^{t'} \in \mathcal{U}'_k, \; B\eta = e, \; \eta \geq 0\},$$

from which we have that $\mathcal{E}^{t'}(\mathcal{U}') \leq \mathcal{E}^{t'}(\mathcal{U}'')$. ■

Theorem 3.1 suggests that robust efficiency scores could improve with increased uncertainty, which advocates for more uncertainty from an inefficient DMU's perspective. The authors of [13] introduce the idea of calculating the minimum amount of uncertainty required to render a DMU efficient, which motivates the optimization problem

$$\inf\{m(\mathcal{U}) : \mathcal{E}^{t'}(\mathcal{U}) = 1\}, \tag{3.2}$$

with $m(\mathcal{U})$ being a measure of the amount of uncertainty. However, this problem is not always feasible as illustrated by an example in [13], which shows that it is possible to have a sequence of uncertainty sets whose amount of uncertainty increases to infinity, whereas the limit of the robust efficiency scores is strictly less than one. This example shows that the structure of the uncertainty set can prevent some DMUs from having a claim to efficiency no matter how large the uncertainty sets become. Such DMUs are called incapable in [13].

The common geometry associated with DEA, which for single input and output plots each DMU as a point in an input–output plane, suggests the intuition that all DMUs should be efficient with enough uncertainty by simply allowing each DMU to select data that places it on the efficient frontier of the certain problem. This intuition is misguided due to the example in [13], but it leaves us with the question of identifying structural conditions that guarantee the feasibility of (3.2). We establish such conditions in Theorem 3.2, which combines results from [15] and [16] to assure that all robust efficiency scores can become one with a sufficient amount of uncertainty. The result depends on the following property.

Property: Full-Dimensional Scaling

Assume the uncertainty sets $\mathcal{X}_i^s$ and $\mathcal{Y}_j^s$ contain closed, full-dimensional balls defined by a p-norm and centered, respectively, at $\bar{X}_i^s$ and $\bar{Y}_j^s$, i.e., assume there are full-row rank matrices R_i and R_j so that

$$\{\bar{X}_i^s + \sigma_i u^T R_i : \|u\|_p \leq 1\} \subseteq \mathcal{X}_i^s \text{ and } \{\bar{Y}_j^s + \sigma_j u^T R_j : \|u\|_p \leq 1\} \subseteq \mathcal{Y}_j^s,$$

in which the set inclusion holds for arbitrary, nonnegative scalars σ_i and σ_j – which scope the amount of uncertainty.

We denote an uncertainty set that satisfies the full-dimensional scaling property by $\mathcal{U}(\sigma)$, where σ is indexed by k. The choice of p-norm is inconsequential due to the fact that the existence of one ball implies the existence of all other p-norm balls. A straightforward consequence of the full-dimensional scaling property is that $\mathcal{U}(\sigma)$ results in a uDEA model whose certain counterpart is defined by the data $\bar{X}_i^s$ and $\bar{Y}_j^s$, an interpretation that asserts a 'robustification' of the certain model defined by $\bar{X}_i^s$ and $\bar{Y}_j^s$. There is no requirement that $\bar{X}_i^s$ and $\bar{Y}_i^s$ be (sample) means, although this is a reasonable predilection as we discuss later.

Theorem 3.2 *Suppose $\mathcal{U}(\sigma)$ satisfies the full-dimensional scaling property. Then $\mathcal{E}^{t'}(\mathcal{U}(\sigma)) = 1$ for sufficiently large components of σ.*

Proof We first argue the case with $R_i = R_j = I$ and $p = \infty$, for which

$$\{\bar{X}_i^s + \sigma_i u^T : \|u\|_\infty \leq 1\} \subseteq \mathcal{X}_i^s \text{ and } \{\bar{Y}_j^s + \sigma_j u^T : \|u\|_\infty \leq 1\} \subseteq \mathcal{Y}_j^s.$$

For the input data, set

$$\hat{\sigma}_i = \max\{\bar{X}_{it}^s : t = 1, 2, \dots, T\} - \min\{\bar{X}_{it}^s : t = 1, 2, \dots, T\}.$$

Then

$$u_t = \begin{cases} \dfrac{\max\{\bar{X}_{i\tau}^s : \tau = 1, 2, \dots T\} - \bar{X}_{it}^s}{\hat{\sigma}_i}, & \hat{\sigma}_i > 0, \\ 0, & \hat{\sigma}_i = 0, \end{cases}$$

gives a vector u so that $\|u\|_\infty \leq 1$ and so that

$$\bar{X}_i^s + \hat{\sigma}_i u = \max\{\bar{X}_{i\tau}^s : \tau = 1, 2, \dots, T\}\, e^T.$$

The result is that each DMU shares the same input data for this selection in the uncertainty set. The same argument holds for the output data upon setting

$$\hat{\sigma}_j = \max\{\bar{Y}_{jt}^s : t = 1, 2, \dots, T\} - \min\{\bar{Y}_{jt}^s : t = 1, 2, \dots, T\}.$$

So a data instance in $\mathcal{U}(\hat{\sigma})$ results in all DMUs having the same input and output data, which mandates that $\mathcal{E}^{t'}(\mathcal{U}(\hat{\sigma})) = 1$. So the theorem holds in the case of $R_i = R_j = I$ and $p = \infty$.

We now consider the case for general p and for general R_i and R_j matrices with full row rank. Let $v \in \{\bar{X}_i^s + \hat{\sigma}_i u : \|u\|_\infty \leq 1\}$. Then,

$$\|\bar{X}_i^s - v\|_p \leq \|\hat{\sigma}_i e\|_p = \hat{\sigma}_i \sqrt[p]{T}.$$

Assume

$$\sigma_i \geq \hat{\sigma}_i \sqrt[p]{T}\, \|R_i^+\|,$$

where R_i^+ is the Moore-Penrose pseudo-inverse of R_i. We have from the full row rank assumption that

$$u^T R_i = \frac{1}{\sigma_i} \left(\bar{X}_i^s - v\right)$$

has a solution, one of which is

$$u^T = \frac{1}{\sigma_i} \left(\bar{X}_i^s - v\right) R_i^+.$$

The submultiplicity of the p-norm now gives,

$$\|u\|_p \leq \frac{1}{\sigma_i} \left\|\left(\bar{X}_i^s - v\right)\right\|_p \left\|R_i^+\right\|_p \leq \frac{\hat{\sigma}_i}{\sigma_i} \sqrt[p]{T} \left\|R_i^+\right\|_p \leq 1,$$

from which we know that

$$\{\bar{X}_i^s + \hat{\sigma}_i u^T : \|u\|_\infty \leq 1\} \subseteq \{\bar{X}_i^s + \sigma_i u^T R_i : \|u\|_p \leq 1\}.$$

We conclude that

$$\begin{aligned} &\left\{\{\bar{X}_i^s + \hat{\sigma}_i u^T : \|u\|_\infty \leq 1\} : i = 1, 2, \dots, M\right\} \\ &\quad \lhd \left\{\{\bar{X}_i^s + \sigma_i u^T R_i : \|u\|_p \leq 1\} : i = 1, 2, \dots, M\right\}. \end{aligned}$$

An analogous argument gives the same property for the output data, and the result follows from Theorem 3.1. ■

Theorem 3.2 shows that the full-dimensional scaling property ensures that the optimization problem in (3.2) is feasible, and it guarantees that all DMUs have a claim to efficiency if there is a sufficient amount of uncertainty. The authors of [13] set $p = 2$, $m(\mathcal{U}) = \|\sigma\|_2$, and $R_k = I$, and they approximate a solution to the non-convex problem in (3.2) for each DMU with a first-order algorithm, which orders the DMUs relative to how much uncertainty they require to reach a robust efficiency score of one. DMUs were prostate treatments, and those that needed small amounts of uncertainty to reach an efficiency score of one were promoted as being high-quality treatments, whereas those that necessitated larger amounts of uncertainty were recommended for reconsideration (see Section 3.5).

We now show that the property of full-dimensional scaling permits an impartial treatment of weak efficiency, although this treatment is permissive and liberal because all weakly efficient DMUs agree with their efficient counterparts. We assume the full-dimensional scaling property, and we label a DMU as *efficient relative to the uncertain data* $\bar{X}$ *and* $\bar{Y}$ if

$$0 = \inf\{\|\sigma\|_p : \mathcal{E}^{t'}(\mathcal{U}(\sigma)) = 1\}, \tag{3.3}$$

with $m(\mathcal{U}(\sigma)) = \|\sigma\|_p$ in (3.2). This altered definition of efficiency includes all efficient DMUs with $\sigma = 0$, i.e., those with an efficiency score of one with the data $\bar{X}$ and $\bar{Y}$. Our input-oriented DEA model already gives an efficiency score of one to weakly efficient DMUs that become inefficient with any increase in an input, and these DMUs subsequently and automatically satisfy (3.3) with $\sigma = 0$. Moreover, weakly efficient DMUs with an efficiency score less than one but that become efficient with any increase in an output also satisfy (3.3). This fact follows because we know that setting $\sigma_{j'} = \epsilon > 0$, for some j', and all other $\sigma_k = 0$, results in $\mathcal{E}^{t'}(\mathcal{U}(\sigma)) = 1$, from which we gain (3.3) because ε is arbitrary. So all weakly efficient and efficient DMUs satisfy (3.3) in the uDEA setting under the assumption of full-dimensional scaling, which establishes the following result.

Theorem 3.3 *All efficient and weakly efficient DMUs satisfy* (3.3).

The property of full-dimensional scaling is a bit overmuch here, for we only need uncertainty in at least one output for the result to hold. The full-dimensional scaling property simply lets us select any output to make the argument.

Weak efficiency creates a mathematical nuisance because the robust efficiency score need not be continuous in σ. To see this, suppose we have three DMUs with input–output data $A(1,2)$, $B(1,1)$, and $C(2,2)$. Then DMU A is efficient, whereas DMUs B and C are weakly efficient, with the former having an efficiency score of one and the later having an efficiency score of 0.5. Supposing that the uDEA model satisfies the full rank assumption, we have for DMUs A and B that

$$\mathcal{E}^A(\mathcal{U}(\sigma)) = \mathcal{E}^B(\mathcal{U}(\sigma)) = 1, \text{ for all } \|\sigma\|_p \geq 0,$$

but for DMU C, we have

$$\mathcal{E}^C(\mathcal{U}(\sigma)) = \begin{cases} 0.5, & \|\sigma\| = 0 \\ 1, & \|\sigma\| > 0. \end{cases}$$

So our choice of developing uDEA with an input-oriented model forces a loss of continuity for weakly efficient DMUs having the property that any decrease in an output would make them inefficient. The result is that we cannot assume the existence of either

$$\min\{\|\sigma\|_p : \mathcal{E}^{t'}(\mathcal{U}(\sigma)) = 1\} \text{ or } \operatorname{argmin}\{\|\sigma\|_p : \mathcal{E}^{t'}(\mathcal{U}(\sigma)) = 1\}$$

unless we assert that there are no weakly efficient DMUs with an efficiency score less than one. That said, we are assuming uncertain data, and hence, it is not unreasonable to assume that both the minimum and the argument minimum exist, for if they did not, then we could simply make an arbitrarily small adjustment to the data to ensure their reality. The authors of [15] assert the existence of the minimum because it streamlines mathematical exposition, but we do not make such an assumption here. Our mathematical presentation is thus a slight extension of what has already been published.

Calculating the robust efficiency score equates with solving a robust optimization problem, although the form of the problem depends on the type of uncertainty sets. Robust optimization most commonly assumes polyhedral or ellipsoidal uncertainty to ensure computability; see [22] as a contemporary and practical guide. We assume ellipsoidal uncertainty in our forthcoming mathematical presentation and in our numerical exploration, so $p = 2$ in the full-dimensional scaling property. This assumption has two benefits in addition to computability, those being that $\mathcal{E}^{t'}(\mathcal{U})$ is the optimal value of a second-order cone problem (SOCP) and that the field of probability can aid model development should we want to assert stochasticity. The first benefit follows because assuming

$$\mathcal{U}_k = \{\bar{A}_k^{t'} + \sigma_k u^T R_k \ : \ \|u\| \leq 1\}$$

gives

$$\begin{aligned}
\mathcal{E}^{t'}(\mathcal{U}) &= \min\{c^T\eta : A_k^{t'}\eta \leq 0, \ \forall\, A_k^{t'} \in \mathcal{U}_k, \ B\eta = e, \ \eta \geq 0\} \\
&= \min\{c^T\eta : \left(\bar{A}_k^{t'} + \sigma_k u^T R_k\right)\eta \leq 0, \ \forall\, \|u\|_2 \leq 1, \ B\eta = e, \ \eta \geq 0\} \\
&= \min\{c^T\eta : \max\left\{\left(\bar{A}_k^{t'} + \sigma_k u^T R_k\right)\eta : \|u\| \leq 1\right\} \leq 0, \ B\eta = e, \ \eta \geq 0\} \\
&= \min\{c^T\eta : \bar{A}_k^{t'}\eta + \max\{\sigma_k u^T R_k \eta : \|u\| \leq 1\} \leq 0, \ B\eta = e, \ \eta \geq 0\} \\
&= \min\{c^T\eta : \bar{A}_k^{t'}\eta + |\sigma_k|\,\|R_k\eta\| \leq 0, \ B\eta = e, \ \eta \geq 0\}.
\end{aligned}$$

This last problem is a standard SOCP that solves efficiently with Gurobi and CPLEX.

The modeling benefit from a stochastic perspective comes from the fact that if the elements of $A_k^{t'}$ are sample means, or if they are otherwise normally distributed, and if their covariance matrix is $R_k^T R_k$, then the chance constraint

$$P(A_k^{t'}\eta \leq 0) \leq 1 - \varepsilon \quad \text{is the same as} \quad \bar{A}_k^{t'}\eta + \sigma_k \|R_k \eta\| \leq 0,$$

where $\bar{A}_k^{t'}$ is either the assumed mean or a reasonable estimate. The value of σ_k is the $1 - \varepsilon$ percentile of the distribution of $\bar{A}_k^{t'}\eta$, so large values impose high confidence, whereas low values reduce confidence. So if we assume that a DEA problem is built on average data, which is often a reasonable assumption, then we can model the covariance matrix to design interpretable R_k matrices and to aid model clarification.

While we assume ellipsoidal uncertainty and the full-dimensional scaling property in the rest of our presentation, these assumptions are not innate per se. Indeed, we could just as well assume polyhedral uncertainty that would scale to ensure a robust efficiency score of one. However, our assumptions provide a consistent and succinct standard that streamlines development and exploration.

We end this section with a few comments about what else is known about uDEA. The authors of [13] show that it is possible to design uncertainty sets so that the solutions of DEA and uDEA agree, and from this perspective, uDEA encompasses DEA as a sub-study. While this sentiment is mathematically accurate, it is impractical to replace DEA's linear programming forumlation with uDEA's non-convex paradigm. The special case of box-uncertainty, i.e., $p = \infty$ in the full-dimensional scaling property, is studied in [16], work that also demonstrates the practical use of sampling in applied problems.

3.3 UNCERTAIN CLASSIFICATION

DEA is inherently a classification process, as its basic outcome is to classify DMUs as either efficient or inefficient and maybe even as weakly efficient. So one obvious way to employ uDEA is to model uncertainty and then to use the robust efficiency score to separate DMUs into efficient and inefficient classes. However, the authors of [14, 15] have instead proposed classifying relative to the amount of uncertainty necessary to achieve efficiency, with the idea being that individual or collected groups of DMUs that require large amounts of uncertainty are more inefficient than those who need less uncertainty. This sentiment is a direct extension of DEA because efficient DMUs need no uncertainty to claim efficiency, whereas inefficient DMUs have no claim to efficiency without altered data.

The impetus to use an amount of uncertainty to decide a DMU's efficiency status promotes the idea of measuring uncertainty, and our assumed ellipsoidal model provides a straightforward scheme due to the scaling by σ_k. We set the amount of uncertainty associated with $\mathcal{U}(\sigma)$ to be $\|\sigma\|$. So the uncertainty sets collapse to singletons, i.e., $\mathcal{U}_k = \{\bar{A}_k^{t'}\}$, if there is no uncertainty. We otherwise have uncertainty with $\|\sigma\| > 0$. The question we posit is how much uncertainty does it take for an individual DMU to have a claim to efficiency? We assign DMU t' the value of

$$\inf\{\|\sigma\| \,:\, \mathcal{E}^{t'}(\mathcal{U}(\sigma)) = 1\}, \tag{3.4}$$

and we again note that the infimum cannot be replaced with a minimum without making an assumption about weak efficiency. The constraint $\mathcal{E}^{t'}(\mathcal{U}(\sigma)) = 1$ is relative to the category of DMUs containing t', which reduces to a robust version of DEA if $S = 1$. However, subdividing the DMUs into categories to make this evaluation advances how we can classify.

One of the algorithms of the next section calculates, for each t', a vector $\sigma^{t'}$ so that

$$\mathcal{E}^{t'}(\mathcal{U}(\sigma^{t'})) = 1 \;\text{ and }\; 0 \leq \|\sigma^{t'}\| - \inf\{\|\sigma\| \,:\, \mathcal{E}^{t'}(\mathcal{U}(\sigma)) = 1\} \leq \varepsilon,$$

with ε being a computational tolerance. The idea behind this calculation is that we search for the first-order conditions motivated by treating the infimum as a minimum. Calculating $\sigma^{t'}$ requires an approximate solution to a non-convex problem, so computing it is not as straightforward as calculating $\mathcal{E}^{t'}$.

One classification technique compares $\|\sigma^{t'}\|$ with $E^{t'}$ to provide a two dimensional visual aid that can aid separating the DMUs into classes. This classification process is post-uDEA, and it assumes that DMUs are benchmarked against the entire collection, i.e., that $S = 1$. A small example illustrates the technique. Consider the results of the DEA and uDEA problems with the following DMUs, with the R_i matrices being the identity for the uDEA instance.

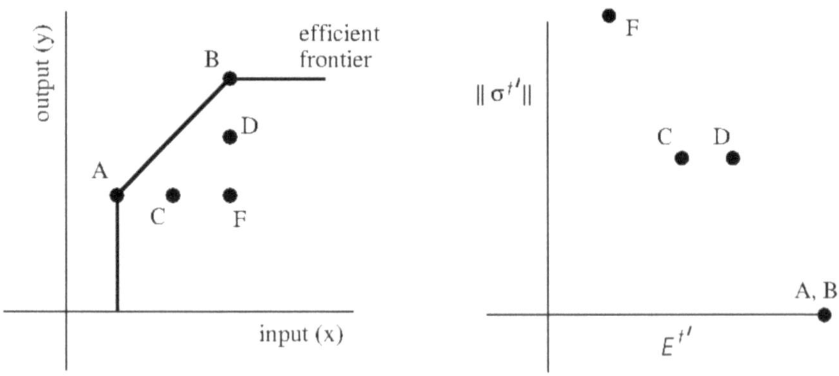

FIGURE 3.1 Small example illustrating the difference between assessing inefficient DMUs with $E^{t'}$ and $\|\sigma^{t'}\|$.

DMU	A	B	C	D	F
input (x)	1	3	2	3	3
output (y)	2	4	2	3	2
$E^{t'}$	1	1	0.50	0.67	0.33
$\|\sigma^{t'}\|$	0	0	0.55	0.55	1.15

An input–output graph, along with a plot of $E^{t'}$ versus $\|\sigma^{t'}\|$, is in Figure 3.1. The advantage of assessing efficiency relative to $\|\sigma^{t'}\|$ is that it removes the input-oriented bias imposed by our selection of calculating efficiency scores with an input-oriented model. This bias is obvious because $E^C = 0.5 < 0.67 = E^D$, whereas DMUs C and D share an identical input-to-output ratio. So these DMUs should be treated similarly in an efficiency analysis, a fact supported by them having the same $\|\sigma^{t'}\|$ value. The bias would have shifted to DMU D having a lower efficiency score if we had instead used an output-oriented model.

The important observation is that traditional DEA separates DMUs into efficient and inefficient classes, but any further distinction among the inefficient DMUs is not directly applicable with the efficiency score alone, at least not with regard to how efficiency scores are typically computed. The long-standing tactic to navigate this discrepancy is to identify targets for inefficient DMUs. This tradition assumes trustworthy data and uses DEA to identify reasonable pathways/strategies for inefficient DMUs to become efficient DMUs.

The uDEA methodology transitions the analysis of inefficient DMUs from guiding them toward improving their lot by identifying targets with assumed perfect data to identifying how uncertain the data would have to be for them to have a claim to efficiency. This data approach seems particularly apropos in the inveterate world of approximated, estimated, and simulated data, and if only a small amount of uncertainty is required to have a claim to efficiency, then it is likely reasonable for the DMU to be, for all practical purposes, efficient up to the dubiousness of the data. If a DMU instead necessitates a large amount of uncertainty, say an amount that clearly is outside the realm of standard assessments, then we have some assurance that the DMU is almost certainly inefficient. These sentiments promote the idea that we should partially order DMUs relative to the amounts of uncertainty required to claim efficiency; i.e., we should partially order DMUs by their values of $\|\sigma^{t'}\|$. Note that the small example above indicates that a linear ordering is not guaranteed because different DMUs can share a value of $\|\sigma^{t'}\|$.

A graph like the one on the right in Figure 3.1a permits the segmentation of DMUs into different categories; however, this partitioning is, seemingly, appropriate after the DEA analysis. For instance, if the goal is to separate the five DMUs into two classes, say those that are 'mostly' efficient versus those that are not, then DMUs C and D could be categorized with F or with A and B depending on how a decider appraises the value of $\|\sigma^C\| = \|\sigma^D\| = 0.55$ and maybe on how a decider weighs the values of $E^C = 0.5$ and $E^D = 0.67$. Cutoff values demarcating categories could be set a priori, but deciding clusters after-the-fact would surely be banal in many situations.

An alternative clustering perspective is to instead cluster DMUs so that their intra-data similarity with regard to efficiency is as similar as possible [14, 15]. This means that the optimization model would itself seek to partition the DMUs so that the individual categories minimized an aggregate measure of the amount of uncertainty necessitated for each DMU to claim efficiency within its category. We assume a preference for a number of categories; i.e., we assume that S is a predefined value to the classification problem. So the partitioning goal is to divide the index set of DMUs $\{1, 2, \dots, T\}$ into S nonempty sets so that the DMUs within a category are as intra-similar as possible. If we label C^s as the s-th category, then we seek individual categories that approximately calculate σ^s so that

$$\mathcal{E}^t\left(\mathcal{U}\left(\sigma^s\right)\right) = 1 \ \forall \ t \in C^s$$

and

$$\|\sigma^s\| - \varepsilon \leq \mathcal{P}^s = \inf\{\|\sigma\| \,:\, \mathcal{E}^t(\mathcal{U}(\sigma)) = 1,\ \forall\, t \in C^s\} \leq \|\sigma^s\|,$$

with $\varepsilon > 0$ again being a small tolerance necessitated by the infimum. The difference between this problem and those previously described is that σ^s now describes an amount of uncertainty that permits **all** DMUs within category C^s to have a claim to efficiency, whereas previous problems only required a single DMU to have such a claim. Note the notational differences among:

$\sigma^{t'}$, which indicates the specific DMU t' (in a category),

σ^t, which indicates a general DMU (in a category), and

σ^s, which indicates a property for all DMUs in a category s.

The value of $\mathcal{P}^s$ is called the proximity to equitable efficiency, and it is a lower bound on the amount of uncertainty required of the DMUs in category s so that all DMUs share a claim to a robust efficiency score of one. So the DMUs in category s are equitable under the uncertainty listed in σ^s because each can select a data instance in $\mathcal{U}(\sigma^s)$ so that it has an efficiency score of one, although we note that the data instance need not be the same for each DMU.

The classification problem we address seeks to identify a partition of $\{1, 2, \dots, T\}$, say $\pi = \{C^s : s = 1, 2, \dots, S\}$, that approximately solves

$$\inf_{\pi}\left\{\sum_{\pi} \mathcal{P}^s\right\} \approx \min_{\pi}\left\{\sum_{\pi} \|\sigma^s\|\right\}, \tag{3.5}$$

where the index of the summation is s over the partition π and the approximation depends on our ability to accurately estimate σ^s to make ε small. The next section introduces an algorithm to approximate $\|\sigma^s\|$, but first, there is a significant computational issue with this classification scheme, which is that the number of partitions is combinatorially explosive. Indeed, the number of different partitions is the Stirling number of the second kind:

$$\left\{ {T \atop S} \right\} = \frac{1}{S!}\sum_{q=0}^{S}(-1)^q \binom{S}{q}(S-q)^T.$$

To illustrate, if we want to partition fifty entities into five categories, then there are over 7.4×10^{32} different partitions, and if we want to partition two hundred entities into ten categories, then there are over 2.7×10^{193} partitions. We conclude that an enumerative search is impossible even with fast and accurate techniques to estimate σ^s, and hence, heuristic searches become necessary on even modest sized problems.

We say that σ^s is feasible-for-classification if $\mathcal{E}^t(\mathcal{U}(\sigma^s)) = 1$ for all $t \in C^s$. Note that $\|\sigma^s\| \geq \mathcal{P}^s$ provided that σ^s is feasible-for-classification. Moreover, we can generate a feasible-for-classification vector by first calculating σ^t for each $t \in C^s$ satisfying

$$\mathcal{E}^t(\mathcal{U}(\sigma^t)) = 1 \ \text{ and } \ 0 \leq \|\sigma^t\| - \inf\{\|\sigma\| \ : \ \mathcal{E}^t(\mathcal{U}(\sigma)) = 1\} \leq \varepsilon.$$

We then set

$$\sigma^s_k = \max_t \{\sigma^t_k\}. \tag{3.6}$$

The resulting σ^s vector is feasible-for-classification, and it bounds $\mathcal{P}^s$ within an interval.

Theorem 3.4 *Let σ^s be constructed from σ^t vectors as described in* (3.6). *Then σ^s is feasible-for-classification and*

$$\frac{\|\sigma^s\|}{\sqrt{M+N}} - \varepsilon \leq \mathcal{P}^s \leq \|\sigma^s\|.$$

Proof Notice that $\sigma^s \geq \sigma^t$ for all $t \in C^s$. So $\mathcal{U}(\sigma^s) \lhd \mathcal{U}(\sigma^t)$, and we have from Theorem 3.1 that σ^s is feasible-for-classification because

$$1 = \mathcal{E}^t(\mathcal{U}(\sigma^t)) \leq \mathcal{E}^t(\mathcal{U}(\sigma^s)), \ \ \forall\, t \in C^s.$$

Feasible-for-classification status immediately gives $\mathcal{P}^s \leq \|\sigma^s\|$.

We have for all $t \in C^s$ and all k that

$$\sigma^t_k \leq \|\sigma^t\| \leq \inf\{\|\sigma\| \ : \ \mathcal{E}^t(\mathcal{U}(\sigma^t)) = 1\} + \varepsilon \leq \mathcal{P}^s + \varepsilon.$$

So $\sigma^s_k \leq \mathcal{P}^s + \varepsilon$ for all k, from which we have

$$\|\sigma^s\| \leq \|(\mathcal{P}^s + \varepsilon)\, e\| \leq \sqrt{M+N}\,(\mathcal{P}^s + \varepsilon).$$

The result follows directly. ■

The following corollary improves the result of Theorem 3.4 in the case that $\text{argmax}_t\{\sigma_k^t\} \supseteq \{t'\}$ independent of k.

Corollary 3.5 *Suppose that t' has the property that* $\text{argmax}_t\{\sigma_k^t\} \supseteq \{t'\}$ *for each k. Then,*

$$\|\sigma^s\| - \varepsilon \leq \mathcal{P}^s \leq \|\sigma^s\|.$$

Proof Let t' be as stated. Then,

$$\begin{aligned}
\|\sigma^s\| &\leq \inf\left\{\|\sigma\| \, : \, \mathcal{E}^{t'}(\mathcal{U}(\sigma)) = 1\right\} + \varepsilon \\
&\leq \inf\{\|\sigma\| \, : \, \mathcal{E}^{t}(\mathcal{U}(\sigma)) = 1, \ \forall\, t \in C^s\} + \varepsilon \\
&= \mathcal{P}^s + \varepsilon \\
&= \|\sigma^s\| + \varepsilon.
\end{aligned}$$

■

If we assume no weakly efficient DMUs, then the infimums that define σ^s and that are used in the proof of Theorem 3.4 can be replaced with minimums, and subsequently, we can set ε to zero and obtain stronger results. Theorem 3.4 and Corollary 3.5 reduce to results in [15] in this situation.

Theorem 3.4 and Corollary 3.5 give practical mathematical certainty in support of approximating the value of $\mathcal{P}^s$. In particular, the bounds in Theorem 3.4 limit our search range, and the bounds in Corollary 3.5 essentially define $\mathcal{P}^s$ in the special case that a single DMU necessitates the largest componentwise σ^t to reach efficiency.

3.4 ALGORITHMS AND HEURISTICS

Computing the robust efficiency score $\mathcal{E}^{t'}(\mathcal{U}(\sigma))$ equates with solving an SOCP, and hence, this score is straightforwardly computed by standard solvers once the uncertainty sets are defined by the R_k matrices and the σ vector. However, calculating approximations for $\|\sigma^{t'}\|$ and $\|\sigma^s\|$, as well as searching for a partition of $\{1, 2, \ldots, T\}$ that minimizes the aggregated sum of the proximities to equitable efficiency, behooves customized algorithms. In this section we present algorithms and heuristics to

- approximate $\|\sigma^{t'}\|$ with a first-order scheme;
- estimate $\|\sigma^s\|$ with a first-order algorithm; and

- compute a partition of $\{1, 2, \ldots, T\}$, denoted by π^*, with reduced, and hopefully minimized, proximity values.

These algorithms were originally presented in [13, 15], and our goal here is to locate them in one place, to unify their presentation, and to include them for completeness as we explore algorithmic adjustments and modeling elements in the forthcoming section. Our previous mathematical presentation was careful to use infimums instead of minimums, but the numerical intent of this section motivates the perspective of minimums instead of infimums. So we assume that $\varepsilon = 0$ as we interpret results from the previous section – not because it is actually zero, but because this assumption streamlines motivation and aligns with our sentiment to approximate solutions independent of the value of ε.

We approximate gradients with finite differences if a straightforward hand calculation is unavailable, and in these cases we, respectively, denote the forward and backward differences as

$$\left[\tilde{\nabla}_+ f(x)\right]_q = \frac{f(x + \delta\, e_q) - f(x)}{\delta} \text{ and } \left[\tilde{\nabla}_- f(x)\right]_q = \frac{f(x) - f(x - \delta\, e_q)}{\delta},$$

with f being a function of the vector x and e_q being the vector of zeros except for a one in the q-th location. The approximated forward and backward gradients are, respectively, $\tilde{\nabla}_+ f(x)$ and $\tilde{\nabla}_- f(x)$. We denote the actual gradient as $\nabla f(x)$ in cases in which we can guarantee its existence and in which its calculation is clear.

The algorithm to approximate $\|\sigma^{t'}\|$ initially assumes no uncertainty, and it then iteratively increases $\sigma^{t'}$ by following the approximate gradient of $\mathcal{E}(\mathcal{U}(\sigma^{t'}))$ a step size of α. This linear approximation can deviate from the nonlinear trajectory associated with finding the **smallest** $\sigma^{t'}$ because it only considers increasing $\sigma^{t'}$. A corrective step returning the iteration to the intended trajectory first calculates a direction $\hat{d}$ solving

$$\min\{d^T \sigma^{t'} \;:\; d^T \tilde{\nabla}_+ \mathcal{E}^{t'}(\mathcal{U}(\sigma^{t'})) = 0,\ \|d\| \leq 1\},$$

which is a convex problem (linear except for a single quadratic constraint) easily solved by commercial solvers. The solution is a unit direction that minimizes the directional derivative of $\|\sigma\|$ relative to marginally maintaining the current robust efficiency score. We search this direction with the method of bisection assuming that the minimum and maximum step sizes are, respectively, 0 and 1, taking the largest step possible to ensure the current robust efficiency score up to a calculation tolerance. The process terminates once the robust efficiency score achieves its maximum value of one. The iterative process is in Algorithm 1.

Algorithm 1 Estimating $\|\sigma^{t'}\| \approx \inf\{\|\sigma\| : \mathcal{E}^{t'}(\mathcal{U}(\sigma)) = 1\}$

Set $\sigma^{t'} = 0$

if $E^{t'} < 1$ **then**

  repeat

    Update $\sigma^{t'} = \sigma^{t'} + \alpha \tilde{\nabla}_{+} \mathcal{E}^{t'}(\sigma^{t'})$

    Compute $\hat{d} \in \text{argmin}\{d^T \sigma^{t'} : d^T \tilde{\nabla}_{+} \mathcal{E}^{t'}(\mathcal{U}(\sigma^{t'})) = 0, \|d\| \leq 1\}$

    Compute $\hat{\beta} \in \min\{1, \text{argmax}\{\beta : \mathcal{E}^{t'}(\mathcal{U}(\sigma^{t'} + \beta \hat{d})) \geq \mathcal{E}^{t'}(\mathcal{U}(\sigma^{t'})) - \varepsilon\}\}$

    Update $\sigma^{t'} = \sigma^{t'} + \hat{\beta}\hat{d}$

  until $\mathcal{E}(\mathcal{U}(\sigma^{t'})) \geq 1$

end if

return Both $\sigma^{t'}$ and $\|\sigma^{t'}\|$

The algorithm to estimate $\|\sigma^s\|$ starts with a feasible-for-classification σ^s and then incrementally decreases σ^s while maintaining feasible-for-classification status. The improving direction, $\hat{d}$, for each iteration solves

$$\min\{d^T \sigma^s : d^T \tilde{\nabla}\_\Gamma(\sigma^s) \geq 0, \|d\| \leq 1\}, \tag{3.7}$$

with σ^s being the incumbent solution and

$$\Gamma(\sigma^s) = \sum_{t \in C^s} \mathcal{E}^t(\mathcal{U}(\sigma^s)).$$

The maximum value of $\Gamma(\sigma^s)$ is $|C^s|$, which is only possible if σ^s is feasible-for-classification. So a solution to the optimization problem in (3.7) results in a unit direction that minimizes the directional derivative of $\|\sigma\|$ as it marginally maintains feasible-for-classification status. The step size, $\hat{\alpha}$, is bounded by

$$0 \leq \hat{\alpha} \leq -\hat{d}^T \sigma^s - \sqrt{\left(\hat{d}^T \sigma^s\right)^2 - \|\sigma^s\|^2 \left(1 - \frac{1}{\sqrt{M+N}}\right)} = B(\sigma^s, \hat{d}),$$

for which the upper bound is calculated directly from Theorem 3.4 with $\varepsilon = 0$. The calculation scheme is in Algorithm 2.

Algorithm 2 Estimating $\|\sigma^s\| \approx \mathcal{P}^s$

```
Input σ^t for all t ∈ C^s
Set σ^s_k = max_{t∈C^s} {σ^t_k}
if σ^s ≠ σ^t for all t ∈ C^s then
  Improvment = true
  repeat
    Compute d̂ ∈ argmin{d^T σ^s : d^T ∇̃_Γ(σ^s) ≥ 0, ‖d‖ ≤ 1}
    if d̂^T σ^s < 0 then
      Compute α̂ ∈ argmax{α : Γ(σ^s + α d̂) ≥ |C^s| − ε, 0 ≤ α ≤ B(σ^s, d̂)}
      Update σ^s = σ^s + α̂ d̂
    else
      Improvment = false
    end if
  until Improvement = false
end if
return Both σ^s and ‖σ^s‖
```

The general classification problem suffers from combinatorial explosion as already noted, which essentially necessitates a heuristic search instead of explicit enumerate. The authors of [14, 15] suggest the use of an adopted P-median problem to initialize a partition π, which is then iteratively improved with a neighborhood search until no improvement is identified. The adapted P-median problem is

$$\left.\begin{array}{rrcl}\min & \sum_{t'\,t''} \left|\,\|\sigma^{t'}\| - \|\sigma^{t''}\|\,\right| \zeta_{t't''} & & \\ \text{subject to} & \sum_{t''} \zeta_{t't''} & = & 1,\ \forall\, t' \\ & \sum_{t'} \zeta_{t't'} & = & S, \\ & \zeta_{t''t''} & \geq & \zeta_{t't''},\ \forall t', t'', t' \neq t'' \\ & \sum_{a} \omega_{t''a} & \leq & 1,\ \forall\, t'' \\ & \sum_{t'} \zeta_{t't''} & = & \sum_{a} p'_a \omega_{t''k},\ \forall t'' \\ & \sum_{t''} \omega_{t''a} & = & \left|\,\{p_q : p_q = p'_a\}\,\right|,\ \forall a \\ & \zeta_{t't''} & \in & \{0,1\},\ \forall t', t'' \\ & \omega_{t'a} & \in & \{0,1\},\ \forall t'', a. \end{array}\right\} \tag{3.8}$$

The problem assumes that $(p'_1, p'_2, \dots, p'_{S'})$ is the tuple of unique elements of p. This problem is solved for each vector $p = (p_1, p_2, \dots, p_S)$ so that

$$\sum_{s=1}^{S} p_s = S \text{ and } p_s \geq 2,\ \forall\, s.$$

The resulting partition for each p is

$$\begin{aligned}\pi^p &= \{C^s : s = 1, 2, \dots S\} \\ &= \{\{t' : \zeta_{t't''} = 1,\ t'' = 1, 2, \dots, T\} \ :\ s = 1, 2, \dots, S\},\end{aligned}$$

with s indexing the collection of t' so that $\zeta_{t't'} = 1$. The first incumbent partition, π, is then selected so that

$$\pi \in \operatorname*{argmin}_{p} \left\{ \sum_{\pi^p} \|\sigma^s\| \right\}.$$

The search to minimize the sum of the proximity measures continues with a neighborhood search, with the neighborhood of π being,

$$\begin{aligned}\mathcal{N}(\pi) = \Big\{ \big(\pi \backslash \big(\{C^{s'}\} \textstyle\bigcup \{C^{s''}\}\big)\big) \textstyle\bigcup \big(\{\hat{C}^{s'}\} \textstyle\bigcup \{\hat{C}^{s''}\}\big) : \\ C^{s'} \in \pi,\ t' \in C^{s'},\ \hat{C}^{s'} = C^{s'} \backslash \{t'\},\ \hat{C}^{s''} = C^{s''} \cup \{t'\} \Big\}.\end{aligned}$$

So the neighborhood of π contains all partitions that remove an object from one of its categories and places it in another. Notice that $\mathcal{N}(\pi)$ does not contain π, and hence, the neighborhood is a deleted neighborhood. If one of the new partitions has a lower sum of proximity measures, then we accept the one with the lowest possible sum as the new incumbent. We otherwise cease the search and declare the incumbent to be our (estimate of) the optimal solution (see Algorithm 3). The heuristic search is embarrassingly parallel, and the **parfor** syntax indicates parallel calculations.

Algorithm 3 Approximating Optimal Partitions

parfor each $t' \in \{1, 2, \dots, T\}$ **do**
  Calculate $\sigma^{t'}$ with Algorithm 1
end for
parfor each $p = (p_1, p_2, \dots, p_S)$ with each $p_s \geq 2$ **do**
  Calculate partition π with adopted P-median problem (3.8)
  Calculate $\sum_{\pi} \|\sigma^s\|$ for π with Algorithm 2
end for
Select $\pi^* \in \operatorname{argmin}\left\{\sum_{\pi} \|\sigma^s\|\right\}$ {initial incumbent solution}
repeat
  parfor each $\pi \in \mathcal{N}(\pi^*)$ **do**
    Calculate $\sum_{\pi} \|\sigma^s\|$
  end for
  if min $\left\{\sum_{\pi} \|\sigma^s\| : \pi \in \mathcal{N}(\pi^*)\right\} < \sum_{\pi^*} \|\sigma^s\|$ **then**
    Select $\pi^* \in \operatorname{argmin}\left\{\sum_{\pi} \|\sigma^s\| : \pi \in \mathcal{N}(\pi^*)\right\}$
  end if
until min $\left\{\sum_{\pi} \|\sigma^s\| : \pi \in \mathcal{N}(\pi^*)\right\} \geq \sum_{\pi^*} \|\sigma^s\|$

All algorithms are coded in python, with optimization models formulated in Pyomo [23] and solved with Gurobi. Calculating each of $\sigma^{t'}$, σ^s, and π requires thousands upon thousands of SOCP solves to repeatedly estimate gradients and to calculate proximity measures, and while SOCP solvers are much improved, they still lack the unflappable trustworthiness associated with other problem domains like linear programming. We have learned that Gurobi works well in our experimental studies as long as we set convergence and computational tolerances to their most strict settings, i.e., so that computations are conducted under the most stringent of assurances. Default settings regularly work, but they have not proved unflappable. Tightening tolerances reduces speed, which only exacerbates what are already computationally burdensome calculations.

The embarrassingly parallel aspect of the search for an optimal π is a redeeming feature, and we make use of a large number of nodes in the examples below. We have also added several practical speed-ups. For instance, we record the value of $\mathcal{P}^s$ so that we do not have to recalculate it should it appear in a future calculation. This scheme proves especially useful when classifying DMUs into several groups because the neighborhood search only updates two of the groups. The combination of distributed parallel computing along with non-repetitive calculations has reduced calculation times on small examples, say 30–45 DMUs with single inputs and outputs, down to minutes from days. These efficiency gains promote larger and more diverse applications, and we subsequently address a problem in the next section with 28 DMUs, three inputs, and two outputs.

3.5 APPLICATIONS AND COMPUTATIONAL IMPROVEMENTS

We introduce three applications of uDEA from the literature, those being the analysis of radiotherapy treatments [13, 15–17], the segmentation of stocks into performant tiers [15], and the evaluation of political redistricting maps [14]. We review these applications and make a special study of the stock example to identify sensitivities and computational shortcomings, the latter of which prompts algorithmic and modeling improvements. We then solve a problem in the aviation industry that clusters airlines into peer groups. This last application has multiple inputs and outputs, and it's solution buttresses the veracity of our computational platform. The airline example also advances the idea that uncertain clusters can identify industrial peer groups among which cooperation could be beneficial.

3.5.1 Assessing Radiotherapy Treatments

The original application of uDEA analyzed 42 prostate cases to identify treatments that would have benefited from continued planning [13]; see [17] as the original DEA praxis for this problem. Radiotherapy is a common cancer treatment that focuses ionizing beams of radiation on a cancerous target, and the primary goal of the planning process is to select beams, as well as exposure levels for those beams, so that dose accrues to an aggregate tumoricidal level on the cancerous region while sparing nearby structures. Each treatment is tailored to the individual needs of a patient, taking into account the stage of cancer, the type of cancer, and the unique physiological and anatomical features of a patient.

Treatment design is a tedious trial and error process. Images are delineated to identify structures, beam pathways are selected, and exposure durations are calculated. Optimization aids each of these processes, but still the process is far from automated; see [24, 25] as reviews. Experts instead create reasonable treatments based on experience and medical standards. Each treatment is assessed on several metrics, and initial designs routinely fail acceptability and require adjustment and revision until they qualify for delivery. There is, importantly, no unifying assessment to indicate if a treatment would improve with continued design, and it is this problem that is addressed in [13]. The authors consider 42 prostate treatments delivered to patients in New Zealand, and they use uDEA to identify cases that would likely benefit from continued design.

The DEA framework is natural in the evaluation of radiotherapy treatments due to the competing goals of achieving a high dose on the target and a low dose on the surrounding structures. We use the dose received by 95% of the planning treatment volume (PTV) as the output metric, denoted by D95, and the rectal generalized equivalent uniform dose (gEUD) as the input metric. These assessments are common for prostate cases; a high D95 value means a high dose to the cancerous region and a low gEUD value means a low dose to the rectum.

The authors of [13] calculate the efficiency score and the value of $\|\sigma^t\|$ for each treatment under the assumption of a single category, and they plot these values against each other to find that they do not correlate as might be expected (see Figure 3.2a). In particular, several cases have low efficiency scores and low values of $\|\sigma^t\|$, indicating that traditional DEA would have flagged them for reconsideration, whereas uDEA would have found them acceptable. Importantly, almost all aspects of treatment planning are influenced by a myriad of hand-cured processes, making the input and output data highly uncertain. So a low value of $\|\sigma^t\|$ almost surely indicates a high-quality treatment, at least as compared against its comparative cohort. The authors of [16] continue this study under the assumption of

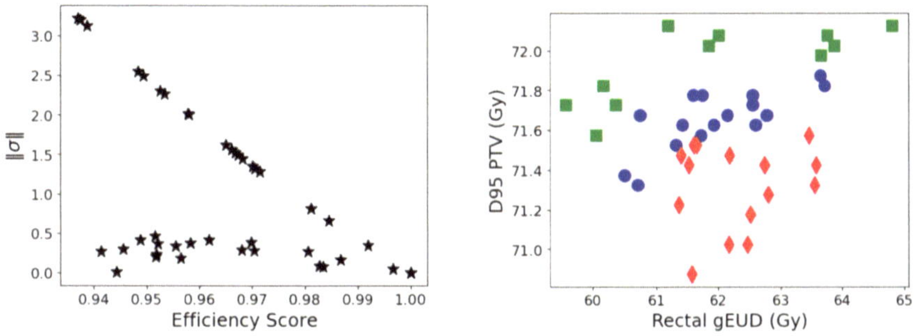

(a) Assessment of Prostate Cases

(b) Classified Prostate Cases

FIGURE 3.2 Applications of uDEA in the assessment of prostate cases.

box-uncertainty, a case that nicely lends itself to mathematical analysis and random sampling.

The work in [15] advances the analysis by classifying the prostate cases into three categories, those being a collection of exemplar treatments, a collection of clinical standards, and a collection suggested for improvement. This strategy has an advantage over the previous studies because it does not require a clinician to interpret uncertainty to decide clinical acceptability. Plotting efficiency scores against $\|\sigma^t\|$ is beneficial, but it mandates that someone establish a rule, or maybe just a rule-of-thumb, to distinguish clinical standards from treatments that would likely improve with continued design. The classification process removes this awkwardness and lets the optimization model decide what is best for the clustering goal. Figure 3.2b shows the three clusters, with the green squares being exemplar treatments, the blue circles being clinical standards, and the red diamonds being treatments suggested for continued improvement.

3.5.2 Political Redistricting

The work in [14] uses uDEA to assist the political redistricting process, which is mandated every decade following the census to realign voting districts with demographic and population changes. Redistricting is replete with political influence and is further complicated by numerous laws and procedures at the federal and state levels. The struggle to fairly redistrict has lead to numerous Supreme Court cases, and it is regularly in the news.

The work in [14] does not address the creation of political districts per se, but rather, it uses uDEA to assess candidate maps suggested by FiveThiryEight's *The Atlas of Redistricting*[1]. The input metric is wasted

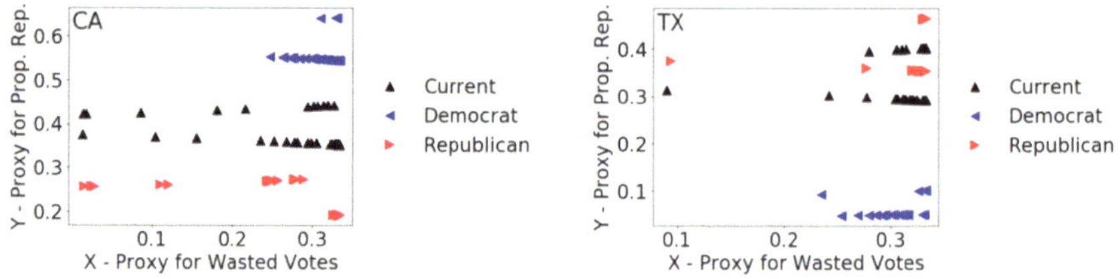

(a) California District Maps (b) Texas District Maps

FIGURE 3.3 Assessing the current, Democrat, and Republican maps of California and Texas.

votes, which should be as small as possible, and the output metric is proportional representation, which should be as large as possible. Voting districts are DMUs, and a state map is preferred if it has a lower proximity to equitable efficiency.

Figure 3.3 illustrates the problem for three of the maps in California and Texas, the former of which has 53 districts and the latter of which has 36. The current California map, shown as the black triangles in Figure 3.3a, has a proximity to equitable efficiency of 0.1361. This value is a proxy for how much an election would need to deviate from historical benchmarks for each of the 53 districts to be efficient with regard to wasted votes and proportional representation. The proximity values for the Democrat and Republican maps are, respectively, 0.0954 and 0.0986, both of which indicate that these maps would give efficient districts with smaller deviations from historical expectations. The conclusion is that the Democrat and Republican maps would be improvements over the current map, with a slight preference for the Democrat map. The analysis for Texas is similar, but the proximity values are 0.1222 (Current), 0.0969 (Democrat), and 0.1505 (Republican). Complete results for all 50 states and all eight maps are in [14].

3.5.3 Classifying Stocks

We now classify the thirty stocks in the Dow Jones Industrial average, a problem originally found in [15]. The goal is to classify the stocks into three performant tiers. We calculate the annual return of each stock ending on the last trading day of each month in 2021. The input characteristic of each stock is its semi-deviation, which is the standard deviation of its downside return, and the output characteristics is its average annual return. The DEA interpretation is that efficient stocks should dominate the others relative to these risk-reward proxies, and the classification problem seeks to categorize the stocks into co-dominate tiers.

We extend the previous analysis of this problem on several fronts. All previous examples set the R_i matrices to the identity, which from a stochastic perspective assumes independence of the row-data and the normality of the constraint bodies. We extend previous studies in two ways:

1. We explore algorithmic sensitivity,
2. We model R_i matrices other than the identity.

The first study shows that the algorithms in Section 3.4 are sensitive to differences in SOCP solvers, a result that is somewhat unsurprising because the authors of [13] and [15] have already noted complications due to a loss of convexity and due to numerical instabilities. The second study advances our understanding of if, and if so how, the model of uncertainty alters analysis. We show that uDEA is sensitive to small changes in the uncertainty model defined by R_i, and hence, modeling the R_i matrices is important and should be done assiduously. These first two studies challenge interpretability and reproducibility, and they suggest the need for an enhanced model to limit sensitivity. We introduce a novel dominance relation and alter the calculation of σ^s in agreement with this goal.

We select the stock problem for computational experimentation for two reasons: first, the availability of historical data prompts the reasoned employ of correlation matrices for each R_i, and second, the two-dimensional model solves quickly with our recent code improvements. These two qualities permit us to efficiently evaluate algorithmic sensitivity to both underlying SOCP solves and to modeling adaptations of the R_i matrices.

We first note that calculating the proximity to equitable efficiency depends on numerous SOCP solves, and while standard solvers such as Gurobi efficiently compute solutions, these commercial algorithms continue to develop. Moreover, SOCP solvers do not yet have the reliability associated with other convex problems like linear and quadratic programming. In particular, we have found that different versions of Gurobi, which we use repeatedly to calculate the proximity to equitable efficiency, render different initial, intermediate, and final clusters. For instance, the authors of [15] use the classification process in Section 3.4 with Gurobi 9.5 to produce the results in Figure 3.4, whereas the same

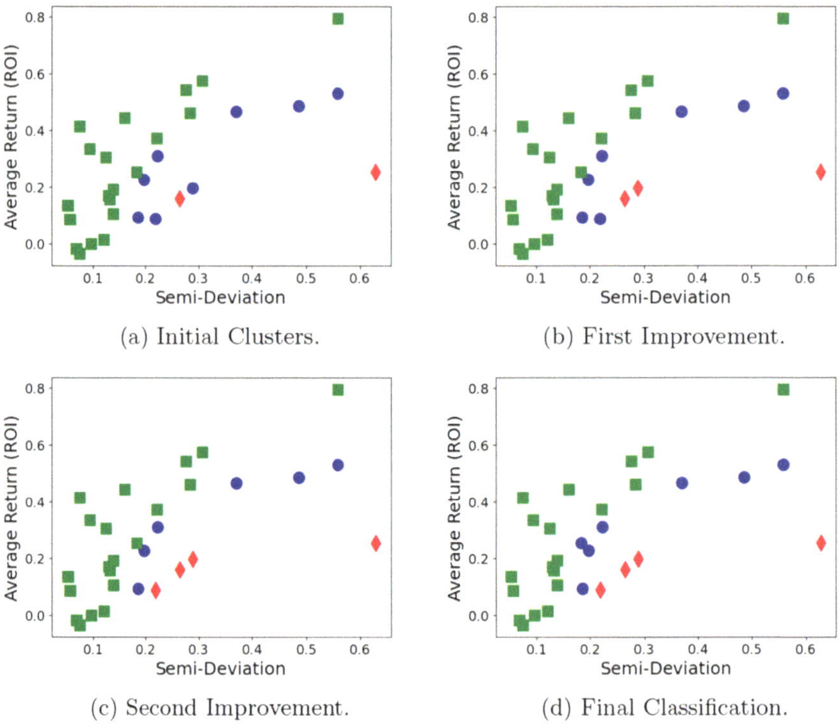

FIGURE 3.4 Results from Algorithm 3 with Gurobi 9.5 from [15].

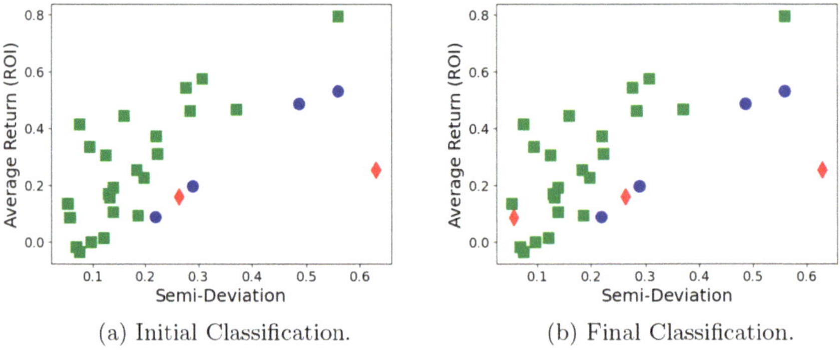

FIGURE 3.5 Results from Algorithm 2 with all settings and data from [15] with Gurobi 11.

code run with the same settings, on the same machine, and with the same data, but with Gurobi 11, produces the results in Figure 3.5. The original outcome with Gurobi 9.5 reasonably segments the stocks into clusters that adhere to our sensibilities of the risk-reward trade-off. However, the contemporary outcome with Gurobi 11 counters our expectation by, for instance, intermixing the cluster of red diamonds among the other two classifications. The proximity measures of the cluster of red diamonds and the cluster of blue dots are zero, so the algorithm is acting reasonably from the perspective of minimizing the sum of the proximity measures, but the result is less than comforting against the intent to divide the stocks into clusters of similar economic efficiencies.

Notice also that the initial classifications, as well as the number of neighborhood search iterations, differ between the two versions of Gurobi. This means that the results of the p-median problem are different, a discrepancy caused by the objective coefficients in (3.8) varying between the two versions of Gurobi. Indeed, we find small changes in the calculation of σ^t, and subsequently in σ^s, if the only difference in Algorithm 1 is the version of Gurobi, which submits that this first-order algorithm is sensitive to small changes resulting from the SOCP solver. The number of neighborhood search iterations being different is unsurprising with different initial classifications; however, we have seeded Algorithm 3 with the same initial classification in Figure 3.4 and found that Gurobi 11 deviates from the iterates in both Figures 3.4 and 3.5. So we again find sensitivity to the SOCP solver in Algorithm 3.

We next investigate sensitivity to the R_i matrices. If we replace the R_i matrices with correlation matrices, which we estimate from our 12 months of data in 2021, then the results differ from the identity matrices; see Figure 3.6 for the results with the correlation matrices and Figure 3.5 with identities. All algorithmic settings, as well as the same version of Gurobi, were identical between these two outcomes. While the resulting categories are similar, their differences note that incorporating standard stochastic assumptions alters the outcomes. We conclude that modeling uncertainty is important and that it should be considered relative to the application and the availability of data.

The statistical inference of replacing an identity matrix with a correlation matrix is reasonable depending on statistical assumptions, and the fact that the resulting classifications differ is almost anticipated. However, this particular numerical experiment does not adequately evaluate sensitivity to small changes in the components of the R_i matrices. We tested such sensitivity by randomly sampling a matrix E_i and setting

$$R_i = I + E_i.$$

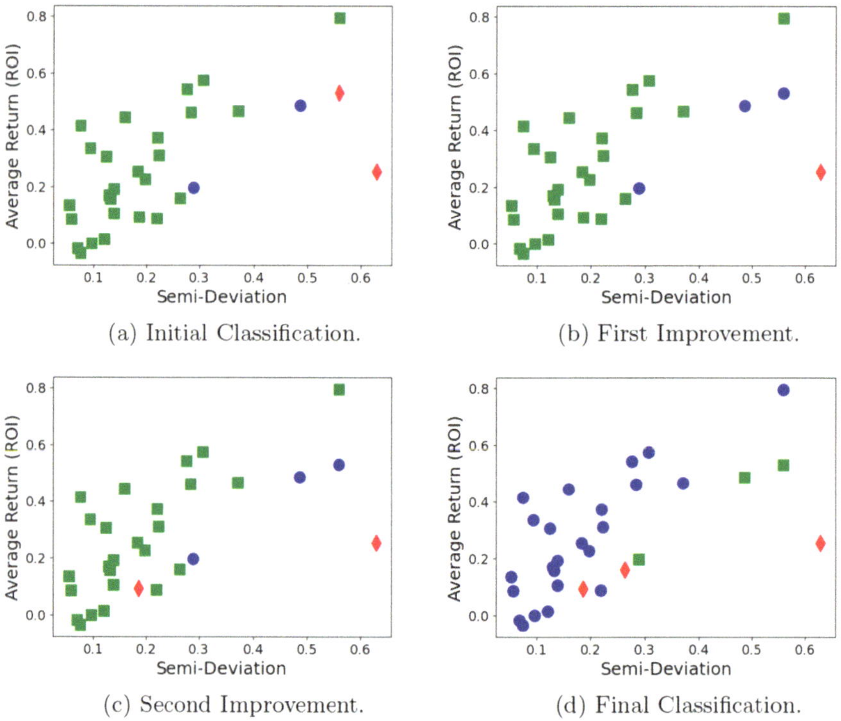

FIGURE 3.6 The classification process with correlation.

Each element of E_i is independently drawn from a normal distribution with a mean of 0 and standard deviation of s, i.e., $[S_i]_{ab} \sim N(0, s^2)$ (note that s is not a sample deviation, but we refrain from using σ due to its previous use as a vector of scalars for the uncertainty sets). These sampled R_i matrices should result in classifications similar to the final classification in Figure 3.5 if the algorithmic scheme was stable to small perturbations in R_i. However, Figure 3.7 demonstrates that we have varying outcomes even with s being the relatively low value of 0.1, and these variations not only disagree with the one from the identity matrix, but they also vary with each other. We also conducted a wider study by setting s to 0.1, 0.2, ..., 0.9, and for each value we created five classifications by randomly sampling R_i matrices. Several different classifications were rendered for each value of s, and there were numerous variations over the entire experiment. We conclude that Algorithm 3 is sensitive to small perturbations in the R_i matrices.

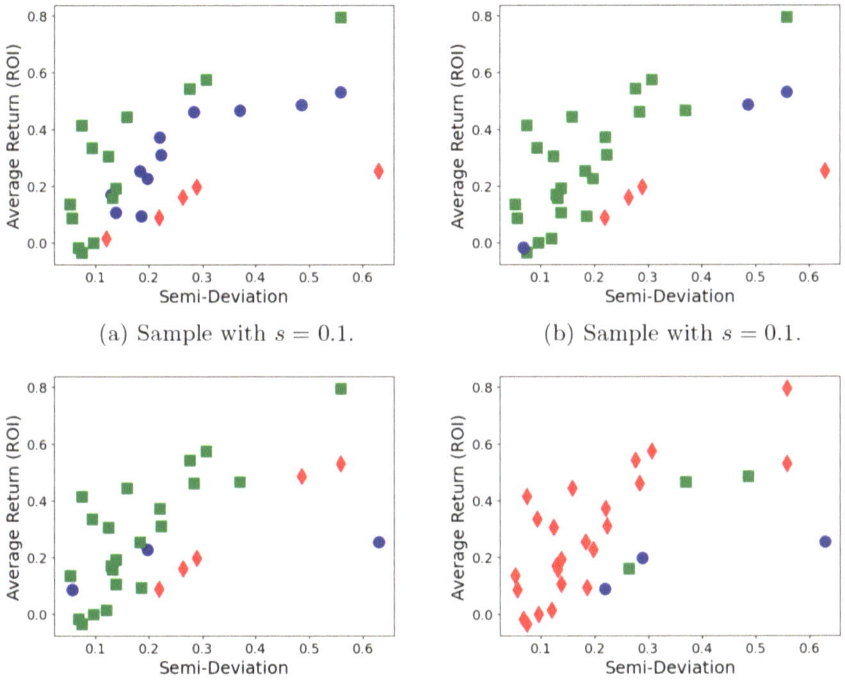

(a) Sample with $s = 0.1$.

(b) Sample with $s = 0.1$.

(c) Sample with $s = 0.1$.

(d) Sample with $s = 0.1$.

FIGURE 3.7 Sampled Final Classifications with $s = 0.1$.

3.5.3.1 Conclusions from Our Sensitivity Study

Our recent advances in reducing the calculation time have thankfully permitted wide-scale testing of a previously studied, low-dimensional example, and we have explored different models of uncertainty and different SOCP solvers. These computational investigations overwhelming point to a clustering scheme that is sensitive to small numerical perturbations due to either variations in the SOCP solver or due to small adjustments in the R_i matrices. The conclusion is that we have a mathematically sound classification scheme that has found promise in previous studies, but one that is difficult to predict or repeat with small changes in algorithmic parameters, with variations in modeling uncertainty, or with updated SOCP solvers.

Part of the problem is the fact that the local, neighborhood search converges to a local minimum, a fact illustrated by several of the final classifications in Figures 3.5–3.7, many of which have two small clusters of sizes three or four. The proximity measure of almost any cluster with three or four DMUs is zero, and hence, the neighborhood search tends to swap DMUs toward having two categories with a small number of DMUs, and it does so regardless of 'overlapping' other categories. For instance, the far-left red diamond of the final classification in Figure 3.5 joins the other red diamonds to give a proximity value of zero for this cluster, creating an 'overlap' with the large cluster of green squares. This DMU moves from the cluster of green squares to the cluster of red diamonds because it reduces the total proximity of the green squares (by about 0.026) while maintaining the zero proximity of the red diamonds. The magnitude of the change is small, and the fact that this DMU is selected for a swap indicates that it plays a role in deciding the proximity value of the cluster of green squares in the initial classification. We comment that there are two close competitors for the DMU that moves from the green squares to the red diamonds, both of which would decrease the total proximity by about 0.025. These competitors are the two, nearly weakly efficient, green squares below the one that swaps to the red diamonds. So the local solution identified by our neighborhood search algorithm is numerically close to several others. This fact suggests that small changes could lead to different outcomes.

3.5.3.2 Modeling Adaptations and Algorithmic Improvements

The entirety of uDEA, as well as our classification process, relies on calculating σ^t, and small deviations in this calculation propagate through all subsequent algorithms. We have found in our numerical study of the Dow Jones stocks that iterative updates are often decided by small margins, and hence, small variations in the calculation of σ^t can lead to different iterates and to different classifications. Small deviations in the calculation of σ^t can be the result of adjustments to algorithmic tolerances, including updates to commercial solvers, to minor algorithmic parameter changes, and to small modeling perturbations in the R_i matrices. This fact motivates improvement in our modeling and algorithmic processes to increase stability and to guide the heuristic search toward meaningful classifications. We introduce adaptions in support of this goal.

Both adaptations address the issue of imposing an order on the clusters. The issue is visible in Figure 3.7, for which subparts (a)–(c) somewhat order the clusters from the green squares in the upper left, to the blue circles in the middle, and to the red diamonds in the lower right. However, subpart (d) swaps this labeling and has, for instance, red diamonds in the upper left. The swap in symbols is nothing more than our algorithm returning clusters in a different order, so while we inherently perceive an order to aid interpretation, no order is imposed or required by the classification process.

All classifications in Figure 3.7 show clusters that 'overlap,' a result that counters intuition. Part of the issue is that our mathematical and heuristic development only partly accounts for the natural expectation that clusters should segment the DMUs akin to the concept of efficiency and that they should do so without intersection or overlap. The result in Figure 3.4 with Gurobi 9.5 illustrates the expectation. The p-median initialization nearly satisfies the intent, and the subsequent neighborhood search terminates with a classification that nicely tiers the DMUs. The classification in Figure 3.5 with Gurobi 11 instead leaves us with a cluster of red diamonds that intersects the other two. However, in all cases the proximity values of the two clusters toward the bottom right are (near) zero, which means that the large cluster to the upper left is driving the iterative process. The algorithm is essentially seeking two clusters to remove from the entire collection that will reduce the proximity value as much as possible, and hence, fabricating small clusters with DMUs from the bottom right makes sense, especially if these clusters have individual proximity values near zero. The real issue is that we can add DMUs to the small clusters toward the bottom right while keeping their proximity values at or near zero and while decreasing the proximity value of the larger cluster, creating overlapping categories that decrease the total proximity measure.

We adjust the calculation of (3.4) to help guide our heuristic search toward intuitive solutions and to help reduce sensitivity. We specifically assume a minimum, as we have already noted in Section 3.4, and adjust (3.4) to

$$\min\left\{\|\sigma\| : \mathcal{E}^{t'}\left(\mathcal{U}(\sigma)\right) = 1\right\},$$

with the constraint $\mathcal{E}^{t'}\left(\mathcal{U}(\sigma)\right) = 1$ being relative to the cluster containing t' ***and to the clusters deemed lower tiered***. Consider the clusters in Figure 3.7c to illustrate. Let C^1 be the cluster of green squares, C^2 be the cluster of red diamonds, and C^3 be the cluster of blue circles. If we order the clusters $(1, 2, 3)$, then $t' \in C^1$ now means that $\sigma^{t'}$ permits enough uncertainty so that DMU t' is robustly efficient compared with the entire collection of DMUs, i.e., against $C^1 \cup C^2 \cup C^3$. Nothing substantially changes with regard to the DMUs in C^1 because the overlapping blue circles in C^3 only play a minor role in deciding $\sigma^{t'}$ if $t' \in C^1$. However, there is a significant difference for the DMUs in C^2 because the red diamonds must now be robustly efficient against the red diamonds in C^2 and the blue circles in C^3. So if $t' \in C^2$, then $\sigma^{t'}$ must now give enough uncertainty to guarantee the robust efficiency of DMU t' among the combined collection of $C^2 \cup C^3$. This new calculation results in $\sigma^{t'}$ no longer being zero if t' indexes any of the three red diamonds in the lower left of C^2, rendering this classification inviable due to it's higher proximity value. Nothing changes with regard to the third cluster because it is last in the $(1, 2, 3)$ preference.

The algorithmic adaptation to (3.4) increases the total proximity of classifications whose categories overlap or extend through others, making non-overlapping classifications more attractive due to their lower total proximity. This adaptation depends on a preferential ordering of the clusters, and we lengthen our computational process by searching all orderings of the initial clusters from the p-median problem. To illustrate, suppose that $p = (5, 22, 3)$ in the adapted p-median problem and that the respective clusters are C^1, C^2, and C^3. We then calculate the total proximity for every permutation of $(1, 2, 3)$ to decide the order. Suppose the permutation with the lowest total proximity is $(2, 1, 3)$. Then, the following criteria hold:

- The DMUs in C^2 are robustly efficient against the DMUs in $C^2 \cup C^1 \cup C^3$.
- The DMUs in C^1 are robustly efficient against the DMUs in $C^1 \cup C^3$.
- The DMUs in C^3 are robustly efficient against the DMUs in C^3.

If $p = (5, 22, 3)$ with order $(2, 1, 3)$ results in the lowest possible option over all p vectors, then C^1, C^2, and C^3 are the initial classifications and ordered as C^2, then C^1, and then C^3. The order remains intact as the classification proceeds with the neighborhood search, so all subsequent calculations of $\sigma^{t'}$ are relative to the order established after the p-median problem.

The assumption of an ordering streamlines the neighborhood search process because we only need to consider swaps between neighboring clusters. So if we order the clusters as C^2, C^1, and C^3, then we only swap DMUs between C^2 and C^1 and between C^1 and C^3. So the neighborhood search iterations are more efficient with an ordering. Our numerical exploration shows that this computational savings does not outweigh the computational burden of considering all permutations to decide an order, but still it is a reduction in calculation. The algorithmic adaption lengthens

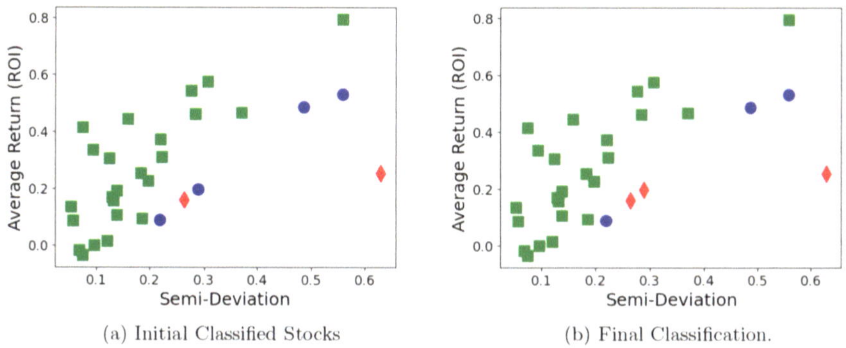

FIGURE 3.8 The classification of the stocks with an updated model that asserts an order on the clusters.

calculation time from 12 to 15 minutes to 25 to 30 minutes on the stock problem. Results with each R_i being an identity matrix and with the updated imposition of ordering the categories are in Figure 3.8. Notice most importantly that the clusters no longer overlap, indicating that the ordering scheme improves the classification intent.

3.5.4 Identifying Peer Groups in the Airline Industry

Our concluding example is from the airline industry, and we consider the problem of identifying peer groups among 28 airlines. The original data comes from [19], but it was also studied with DEA in [18, 20]. We use three inputs, those being the number of staff, the gallons of fuel, and the number of aircraft, and we assume two outputs, those being the number of passenger-kilometers and the amount of freight. Our primary goal is to demonstrate the ability of our computational platform to solve classification problems with multiple inputs and outputs. Indeed, solving large, high-dimensional problems prompts computational and verification challenges, the latter of which is due to the fact that we lose two-dimensional visual acuity.

We assume that each R_i is an identity matrix, and we divide the 28 airlines into three peer groups, each with a claim to robust efficiency within its group. The problem solves, somewhat surprisingly, almost as quickly as the single input and output problems, and we know that the algorithm produces an optimal classification because the total proximity is zero. Indeed, the algorithm terminates after three neighborhood search iterations with three categories in which the DMUs are efficient within their clusters. Note that the solution is optimal independent of the R_i matrices,

TABLE 3.1 Optimal airline peer groups

Cluster Order	Airlines
1	JAL, Quantas, Saudia, SIA, AUA, Lufthansa, TAP Air Portugal, Swissair, America West, American, Northwest, Pan American, United
2	Nippon Airways, Cathay Pacific, Malaysia Airlines, British Airways, Finnair, SAS, Canadian, Continental, Delta, TWA
3	Garuda, Iberia, Air Canada, Eastern, USAIR

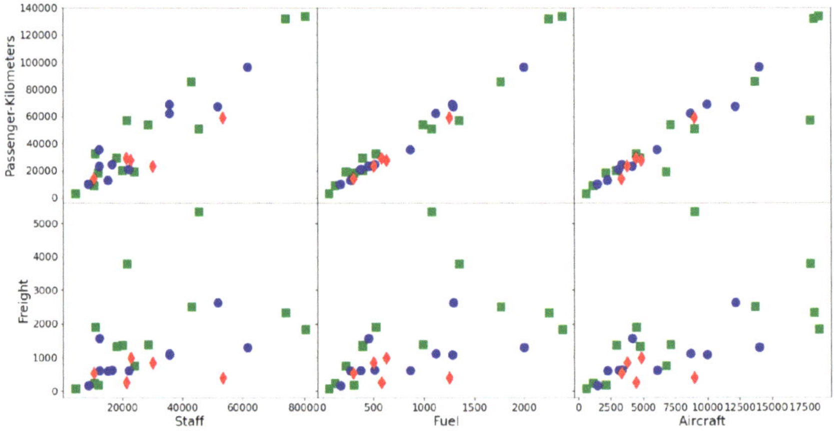

FIGURE 3.9 Airline peer groups with outputs as rows and inputs as columns. The green squares are Category 1, the red diamonds are Category 2, and the blue circles are Category 3.

and hence, modeling uncertainty is moot in this example. We had not expected this outcome, but the reality that zero proximity is possible, and the performance of our algorithm to identify such a solution, testifies to the efficacy of our classification technique. Although we did not expect a classification to be possible without uncertainty, such an outcome is reasonable due to the high number of inputs and outputs and due to the relatively small number of DMUs per category: the first category had 13 DMUs, the second had 10, and the third had 5. The peer groups are listed in Table 3.1, and two-dimensional projections of the DMUs onto their input and output data are in Figure 3.9.

The recent work in [26–28] adapts DEA to identify benefits of mergers and partnerships, and the airline clusters suggest similar industrial amalgamations. The airlines in the first cluster do not have an obvious reason to combine operations relative to inputs, outputs, or both because they are efficient, and hence, these companies already out-compete their advisories. However, companies in the second and third clusters operate with similar efficiencies, and they are robustly intra-efficient; i.e., they are robustly efficient within their segmented cohorts. A plausible rationale is that they share similar operational and market perspectives as they excel in different aspects within their cluster. Operational and market similarity could encourage combined efforts, for instance, mergers, partnerships, and acquisitions, that would move the resulting cooperative entity into the first tier. For instance, maybe Finnair and SAS could partner their resources so that both would gain tier one status, or maybe USAIR and Air Canada could merge to create a new, more efficient entity. The fact that each pair is within the same competitive cohort means that they are data-similar, and hence, they likely share concomitant goals and could have similar tendencies toward cooperation. Indeed, the clusters could suggest large-scale conglomerates in which many companies would cooperate to reach tier one status.

3.6 CONCLUDING REMARKS

Perfect data is banally suspect because flawed information is incontrovertible, but robust optimization gives an ability to model and leverage uncertainty that can circumvent DEA's intrinsic reliance on accurate data. The uDEA extension of DEA reviewed herein provides a contemporary scheme to account for uncertainty. The uDEA model rests on the idea that DMUs should be able to decide their own weights *and* their own data from a realm of uncertainty. The uDEA framework also leverages uncertainty to posit the question of how much uncertainty is required to make a DMU, or a collection of DMUs, equitably efficient. This last quality principally distinguishes uDEA from previous models of DEA created to handle imperfect information, e.g., stochastic DEA and fuzzy DEA, as these paradigms lack the natural concept of equity. This observation does not diminish the employ of previous DEA extensions, but it shows that uncertainty adds to our interpretive repertoire. The goal of each application should pair with a technique that supports it. If expected efficiency is the goal, then stochastic DEA is perfect, but if we want to know how little data needs to change to make everyone efficient, then uDEA is the correct tactic.

The four example applications demonstrate a latitude of use, and uDEA can be applied per DMU, as shown in the original radiotherapy example, or it can cluster DMUs into equitable categories, as shown in all examples. The computational burden of uDEA is significant compared with DEA, but the added analysis of uDEA seems worth the effort in some cases.

The future of uDEA depends on continued improvement, and we conclude by discussing what the next steps might be. Our current code has improved to the point that we can efficiently solve reasonably sized uDEA problems, but several investigations remain. In particular, we need to reduce the algorithm's sensitivity to minor parametric perturbations and to small numerical discrepancies, say from the underlying solver. This pursuit will likely require modeling and computational adjustments.

Modelers would benefit from templates of uncertainty, and a wide-scale study of the effects of uncertainty on uDEA would help identify best practices. The stock example demonstrates that modeling uncertainty is important, but this sole example does not give a backdrop to identify trends and to hone standards. We need to collect a large test-bank of applications and to vet them against one another within the ethos or uDEA.

Using uDEA to identify constructive and profitable business mergers and partnerships is a worthy continuation, and in particular, melding the models in [26–28] with uDEA and its clustering prowess seems auspicious. An obvious first application would be to use uDEA clusters to identify beneficial cooperative agreements, and a second, albeit less obvious, application would be to combine and retool uDEA and inverse DEA models for partnerships and mergers. A third continuation would be to use uDEA clustering to spot large industrial amalgamations that might benefit from the creation of a conglomerate of shared and directed interests.

Lastly, uDEA would benefit from software development that would seamlessly integrate with spreadsheets and other DEA platforms. This would allow managers to use uDEA with a few clicks instead of hiring experts to run code and write reports. Interpretive tools seem plausible, and our hope is to work toward easy-to-use software.

REFERENCES

[1] A. Charnes, W.W. Cooper, and E. Rhodes. Measuring the efficiency of decision making units. *European Journal of Operational Research*, 2(6):429–444, 1978.

[2] Charles Kittel. The nature and development of operations research. *Science*, 105(2719):150–153, 1947.

[3] George B. Dantzig. Linear programming under uncertainty. *Management Science*, 1(3-4):197–206, 1955.

[4] A. Ben-Tal and A. Nemirovski. Robust truss topology design via semidefinite programming. *SIAM Journal on Optimization*, 7(4):991–1016, 1997.

[5] A. Ben-Tal and A. Nemirovski. Robust convex optimization. *Mathematics of Operations Research*, 23(4):769–805, 1998.

[6] Laurent El Ghaoui and Hervé Lebret. Robust solutions to least-squares problems with uncertain data. *SIAM Journal on Matrix Analysis and Applications*, 18(4):1035–1064, 1997.

[7] Laurent El Ghaoui, Francois Oustry, and Hervé Lebret. Robust solutions to uncertain semidefinite programs. *SIAM Journal on Optimization*, 9(1):33–52, 1998.

[8] Panos Kouvelis and Gang Yu. *Robust Discrete Optimization: Past Successes and Future Challenges*, pages 333–356. Springer US, Boston, MA, 1997.

[9] Aharon Ben-Tal, Laurent El Ghaoui, and Arkadi Nemirovski. *Robust optimization*, volume 28. Princeton university press, 2009.

[10] P. Peykani, E. Mohammadi, R. F. Saen, S. J. Sadjadi, and M. Rostamy-Malkhalifeh. Data envelopment analysis and robust optimization: A review. *Expert Systems*, 37(4):e12534, 2020.

[11] Maziar Salahi, Mehdi Toloo, and Narges Torabi. A new robust optimization approach to common weights formulation in dea. *Journal of the Operational Research Society*, 72(6):1390–1402, 2021.

[12] Mehdi Toloo, Emmanuel Kwasi Mensah, and Maziar Salahi. Robust optimization and its duality in data envelopment analysis. *Omega*, 108:102583, 2022.

[13] M. Ehrgott, A. Holder, and O. Nohadani. Uncertain Data Envelopment Analysis. *European Journal of Operational Research*, 268:231–242, 2018.

[14] C. Garner and A. Holder. Comparing voting districts with uncertain data envelopment analysis. *Optimization Letters*, 2023. https://doi.org/10.1007/s11590-023-02046-0.

[15] C. Garner and A. Holder. Classifying with uncertain data envelopment analysis. *EURO Journal on Computational Optimization*, 12:100090, 2024.

[16] E. Stubington, M. Ehrgott, and O. Nohadani. Uncertain Data Envelopment Analysis: Box Uncertainty. preprint, 2020.

[17] K. Lin, J. Simpson, G. Sasso, A. Raith, and M. Ehrgott. Quality assessment for VMAT prostate radiotherapy planning based on data envelopment analysis. *Physics in Medicine and Biology*, 58(16):5753, 2013.

[18] Juan Aparicio, Juan F. Monge, and Nuria Ramón. A new measure of technical efficiency in data envelopment analysis based on the maximization of hypervolumes: Benchmarking, properties and computational aspects. *European Journal of Operational Research*, 293(1):263–275, 2021.

[19] Tim Coelli, Emili Grifell-Tatjé, and Sergio Perelman. Capacity utilisation and profitability: A decomposition of short-run profit efficiency. *International Journal of Production Economics*, 79(3):261–278, 2002.

[20] Nuria Ramón, José L. Ruiz, and Inmaculada Sirvent. Cross-benchmarking for performance evaluation: Looking across best practices of different peer groups using dea. *Omega*, 92:102169, 2020.

[21] W. Cooper, L. Seiford, and K. Tone. *Data Envelopment Analysis: A Comprehensive Text with Models, Applications, References and DEA-solver Software.* Springer, New York, 2007.

[22] B. Gorissen, İ. Yanıkoğlu, and D. den Hertog. A practical guide to robust optimization. *Omega*, 53:124–137, 2015.

[23] W. Hart, C. Laird, J.-P. Watson, D. Woodruff, G. Hackebeil, B. Nicholson, and J. Siirola. *Pyomo - Optmization Modeling in Python*. Springer, 2017.

[24] M. Ehrgott and A. Holder. Operations research methods for optimization in radiation oncology. *Journal of Radiation Oncology Informatics*, 6(1):1–41, Oct. 2017.

[25] A. Holder. Radiotherapy treatment design and linear programming. In M.L. Brandeau, F. Sainfort, and W.P. Pierskalla, editors, *Operations Research and Health Care*, pages 741–774. Kluwer Academic Publishers, Norwell MA, 2004.

[26] Gholam R. Amin, Saeed Al-Muharrami, and Mehdi Toloo. A combined goal programming and inverse dea method for target setting in mergers. *Expert Systems with Applications*, 115:412–417, 2019.

[27] Gholam R. Amin and Mustapha Ibn Boamah. Modeling business partnerships: A data envelopment analysis approach. *European Journal of Operational Research*, 305(1):329–337, 2023.

[28] Gholam R. Amin and Amar Oukil. Flexible target setting in mergers using inverse data envelopment analysis. *International Journal of Operational Research*, 35(3):301–317, 2019.

CHAPTER 4

Planning Large-Scale Partnerships in the Hotel Industry

An Inverse DEA Perspective

Amar Oukil

College of Economics & Political Science, Sultan Qaboos University, Muscat, Oman

4.1 INTRODUCTION

The global hotel industry remains fiercely competitive, with firms constantly vying for expansion and market dominance. Amid this intensifying landscape, growth via mergers and acquisitions (M&As) has become a prominent strategy (Pohlman, 2017). High-profile deals such as Marriott International's acquisition of Starwood Hotels & Resorts Worldwide for US$13.6 billion, and Accor's acquisition of Fairmont Hotels & Resorts for US$2.9 billion, have marked significant milestones in the sector. However, despite these landmark transactions, analysts argue that the hotel industry remains highly fragmented, with no dominant player commanding a substantial portion of the global market (Zhang et al., 2020). This fragmented structure is further complicated by the regional diversification of hotel brands and varying levels of market maturity, necessitating continual consolidation efforts to achieve meaningful competitive advantage.

DOI: 10.1201/9781003345206-4

While M&A offers firms opportunities to expand their market presence, the sector's inherent fragmentation continues to challenge operators in their quest to secure a decisive global foothold. The likelihood of increased M&A activity within the hospitality industry is undeniable, particularly given the potential for consolidating a diverse array of operational frameworks and corporate cultures. While this integration is not without its challenges, M&As are widely regarded as a more efficient path to growth compared to organic expansion, which typically focuses on scaling outputs and increasing sales at the individual hotel level (Khairy, 2019).

From a practical standpoint, the necessary scope of output expansion can be precisely determined by assessing a hotel's efficiency relative to industry benchmarks. Frontier efficiency methodologies, including stochastic frontier analysis (SFA) and data envelopment analysis (DEA), offer robust support for such evaluations. Unlike SFA, which requires predefined functional forms (Greene, 2008), DEA (Emrouznejad et al., 2022) operates as a non-parametric technique, free from assumptions about data distributions (Sow et al., 2016). Furthermore, DEA's ability to assess the efficiency of decision making units (DMUs) that operate with multiple inputs and outputs makes it ideally suited to complex environments like the hospitality sector, where diverse performance metrics must be integrated into decision making processes.

The application of DEA spans broad business areas. The most recent applications include power generation (Eguchi et al., 2021), production scheduling (Oukil et al., 2022a), open field agriculture (Zalaghi et al., 2021), carbon emission reduction (Yang et al., 2020), material handling selection (Hassan & Oukil, 2021; Oukil et al., 2022a, 2022b, 2022c), greenhouse production (Al-Mezeini et al., 2020), performance of airlines (Ngo & Tsui, 2022), water quality assessment (Oukil et al., 2022c), irrigation systems (Kamiyama et al., 2021), manufacturing systems (Oukil, 2020), banking (Kamarudin et al., 2019; Moghaddas et al., 2022), ranking investment firms (Peykani et al., 2021), stock markets (Amin & Oukil, 2019b), regional transport sustainability (Tian et al., 2020), faculty academic performance (Oral et al., 2014, 2015), to mention just a few areas.

In the hotel industry, the most prevailing research stream is concerned with the application of DEA to estimate efficiency at the hotel firm level (e.g., Kim & Chung, 2022; Dolasinski et al., 2019; Sáez-Fernández et al., 2020; Lado-Sestayo & Fernández-Castro, 2019; Higuerey et al., 2020; Kularatne et al., 2019; Ang et al., 2018; Mariani & Visani, 2019; Tan & Despotis, 2021; Dobrovič et al., 2021; Karakitsiou et al., 2020; El Alaoui et al., 2023; Martins et al., 2021; Nguyen & Nguyen, 2019; Zaki, 2019;

Frančeškin & Bojnec, 2022a; Singh et al., 2022; Guo et al., 2019; Li et al., 2021; Bire, 2020; Oukil et al., 2016). The DEA Malmquist index has also been used to measure the hotels' productivity change over the time horizon (e.g., Fang et al., 2020; Tzeremes, 2020; Tzeremes & Tzeremes, 2021; Frančeškin & Bojnec, 2022b; Kim et al., 2021; Peypoch et al., 2021). More advanced DEA models have been developed to handle technological heterogeneity among hotels and other structural forms (e.g., Yu & Chen, 2020a, 2020b; Chen, 2018; Lee et al., 2019; Chiu & Lin, 2018; Deng et al., 2020).

Interestingly, despite the vast number of DEA-based studies focused on hotel industry efficiency, there is a notable lack of research explicitly examining the efficiency of hotel mergers and acquisitions (M&As). This study aims to address that gap, among others. We view M&A as a critical strategic decision that requires thorough consideration before any agreements are finalized, as it could represent either a step forward or backward for the future of the hotel business. Therefore, it is crucial to develop decision making tools that take a forward-looking approach, rather than simply relying on retrospective analysis, to guide decision makers in the M&A planning process.

Adopting a prospective view of M&A means equipping DMs with the ability to answer key questions, such as: (1) What level of efficiency gains can be expected from a potential hotel merger? (2) Is the proposed merger a worthwhile endeavor?

Several DEA models have been developed to estimate potential merger gains (e.g., Xiao *et al.* 2017; Shi et al., 2017; Li et al., 2018; Nguyen & Pham, 2020; Xie et al., 2020; Chiu et al., 2021; Ray & Sethia, 2022; Khoveyni & Eslami, 2022; Al Tamimi et al., 2022). However, these models primarily concentrate on the post-merger phase, where the DM's focus is limited to realizing cost and profit efficiency gains based on the existing levels of inputs and outputs. In contrast, the approach we propose seeks to guide DMs at an earlier stage—namely, the pre-merger phase—by providing insight into the specific levels of inputs and outputs necessary for the prospective merger to meet a predefined efficiency target. Such forward-looking decisions can be effectively supported through the application of inverse DEA techniques (Wei et al., 2000; Yan et al., 2002), which allow for the calculation of required input or output adjustments to achieve the desired performance benchmarks prior to the merger being finalized.

Conventional DEA is designed to evaluate the efficiency of DMUs by analyzing their known input and output data. Conversely, inverse DEA (IDEA) shifts the focus toward determining the optimal input and/or output levels required to achieve a specified efficiency target. In other words, rather than assessing a hotel's efficiency based on its current inputs and outputs, the IDEA approach assumes that these values are initially unknown and works to identify the precise input–output configuration needed to meet a predetermined efficiency benchmark.

Except for a few studies in the agriculture production (Oukil, 2023), the majority of known applications of IDEA to M&A are limited to the banking sector, with the work of Gattoufi et al. (2014), Amin et al. (2017b), Gerami et al. (2021), Amin et al. (2019), Amin and Al-Muharrami (2018), Soltanifar et al. (2023), Ghobadi (2021), Amin et al. (2017a), Kamyab et al. (2019), Amin and Ibn Boamah (2021), Amin and Ibn Boamah (2020), Oukil et al. (2024), and Soltanifar et al. (2022). Accordingly, the present study is the first application of IDEA to mergers in the hotel sector, where M&As are major strategic decisions. Using the IDEA model of Gattoufi et al. (2014), which is slightly modified in Amin and Ibn Boamah (2020), we will estimate the optimal resource gains for prospective merging hotels with reference to a preset efficiency target.

Knowing that partnership, which underpins M&A, constitutes the ultimate mirror for potential growth, competition dynamics, and prospective gains (Huhtilainen et al., 2022), we develop a procedure that will help DMs to select the best partners for merger based on potential merger gains. Partner selection, based on input and output distribution, is another major business area where DEA models have been developed (see, e.g., Lozano, 2013; Amin & Ibn Boamah, 2023, Wu et al., 2016; Zhu et al., 2017; Wang et al., 2018; Lin et al., 2020; Zhu et al., 2021; Oukil, 2024).

While existing studies typically focus on evaluating isolated pairs of DMUs, this study adopts a more comprehensive approach by addressing the question: What are the best partners for prospective hotel mergers that would enable the entire hotel industry—represented hypothetically by a selected sample of hotels—to achieve maximum collective performance? In other words, assuming each hotel in the sample is a candidate for merger, the goal is to identify the optimal pairings among them.

To answer this question, we developed a methodology that operates across four stages: (1) The IDEA model is applied to assess the potential gains from pairwise hotel consolidations. (2) The hotels with productive post-merger potential—those likely to generate real gains—are identified. (3) These productive post-merger hotels are then ranked according to their expected outcomes. (4) Finally, the best prospective merger plan for the entire hotel sample is established by identifying the optimal merger partners.

To the best of the author's knowledge, no previous studies have specifically tackled hotel M&A from a DEA efficiency perspective to develop an optimal merger plan for a sample of hotels. Therefore, the contributions of this chapter to the tourism literature are threefold: (1) It is the first DEA study dedicated to hotel M&A. (2) It introduces the concept of inverse DEA to the tourism literature for the first time. (3) It provides a comprehensive analysis of the challenge of identifying the best merger partners across the entire hotel industry.

The relevance of this methodology is demonstrated through an evaluation of potential merger gains among 58 hotels in the Sultanate of Oman, leading to the formulation of the best prospective merger plan.

The remainder of the chapter is structured as follows. Sections 4.2 and 4.3 present the methodological foundation, including DEA concepts and the IDEA model used for the merger process. Section 4.4 introduces the case study and discusses the results. Section 4.5 outlines the approach to selecting the best merger partners, followed by a set of managerial implications, before concluding remarks in the final section.

4.2 HOTEL EFFICIENCY EVALUATION

Let K be the number of hotels to be evaluated, with each hotel T_k represented with I inputs x_{ik} and J outputs y_{ik}, where $i=1,\ldots,I$ and $j=1,\ldots,J$ The following Banker, Charnes and Cooper (BCC) (Banker et al., 1984) input-oriented model is used to estimate the efficiency score e_o^* of $T_o=\left(x_o\ y_o\right)$ and set targets for inefficient hotels.

$$
(\text{BCC})\quad
\begin{aligned}
e_o^* &= \min \pi \\
\text{s.t.} \sum_{k=1}^{K} \alpha_k x_{ik} &\le \pi x_{io} && i=1,\ldots,I \\
\sum_{k=1}^{K} \alpha_k y_{jk} &\ge y_{jo} && j=1,\ldots,J \\
\sum_{k=1}^{K} \alpha_k &= 1 \\
\alpha_k &\ge 0 && k=1,\ldots,K
\end{aligned}
$$

The optimal efficiency score e_o^* of T_o represents the minimum inputs' reduction that is required to achieve full efficiency for given outputs' levels. Thus, T_o is efficient if $e_o^*=1$, otherwise, it is inefficient, i.e., it is not using its inputs optimally.

The intensity value α_k measures the contribution of each hotel T_k toward the efficiency evaluation of T_o. An optimum $\alpha_k^*>0$ indicates that T_k is a potential benchmark for T_o (Soltani et al., 2021).

In order to achieve full efficiency, an inefficient hotel T_o can decrease its ith input with an amount s_{io}^* while keeping its outputs unchanged, where

$$s_{io}^* = e_o^* x_{io} - \sum_{k=1}^{K} \alpha_k^* x_{ik} \qquad i = 1,\ldots,I$$

s_{io}^* is the slack value of the ith input. T_o is strongly efficient if $s_{io}^* = 0$ for all inputs $i = 1,\ldots,I$. Otherwise, T_o is weakly efficient if $e_o^* = 1$ and $s_{io}^* > 0$ for at least one input $i = 1,\ldots,I$ (Oukil et al., 2021). In other words, s_{io}^* represents the savings of input i that an inefficient or a weakly efficient hotel T_o is required to achieve if it is willing to upgrade to efficient.

4.3 INVERSE DEA MODEL FOR MERGER

We assume, for modeling simplicity, that two pre-merger hotels, T_A and T_B, are merging into a larger hotel M_m. Meanwhile, this assumption does not exclude the fact that the proposed methodology applies to the merger of more than two hotels.

Let $\bar{\pi}$ be the efficiency target set *a priori* for the post-merger hotel M_m. Under a DEA input orientation setting, hotel M_m keeps the entire outputs of the merging hotels T_A and T_B, while consuming the minimum levels of their current inputs with, as objective, achieving a preset efficiency target $\bar{\pi}$.

Let ∂_{iA} and ∂_{iB} denote the minimum levels of ith input associated to hotels T_A and T_B, respectively. Hence, the level of ith input for hotel M_m is $\partial_{im} = \partial_{iA} + \partial_{iB}$, which is expected to be less or equal to $x_{im} = x_{iA} + x_{iB}$, while the level of its jth output remains $y_{jm} = y_{jA} + y_{jB}$, for $i = 1,\ldots,I$ and $j = 1,\ldots,J$. The problem, as stated, can be formulated using the following IDEA model (Amin & Ibn Boamah, 2020):

$$\min \sum_{i=1}^{I} \left(\partial_{iA} + \partial_{iB}\right)$$

$$s.t.$$

$$(\text{IDEA}) \quad \begin{aligned} & \sum_{k \in P} \alpha_k x_{ik} + \alpha_m x_{im} - \left(\partial_{iA} + \partial_{iB}\right) \times \bar{\pi} \le 0 & i = 1,\ldots,I \\ & \sum_{k \in P} \alpha_k y_{jk} + \alpha_m y_{jm} \ge y_{jm} & j = 1,\ldots,J \\ & \sum_{k \in P} \alpha_k + \alpha_m = 1 & \\ & 0 \le \partial_{iA} \le x_{iA}, 0 \le \partial_{iB} \le x_{iB} & i = 1,\ldots,I \\ & \alpha_k \ge 0, k \in P, \alpha_m \ge 0 & \end{aligned}$$

Model (IDEA) is an inverse DEA model, because, here, the efficiency score $\overline{\pi}$ is known and the objective is to find the minimum required levels of corresponding inputs, i.e., the values of the $2I$ variables ∂_{iA} and ∂_{iB}, $i = 1,\ldots,I$.

With P denoting the set of peers available for the evaluation of the post-merger hotel M_m, P may contain one of the merging hotels, T_A and T_B, or none of them, depending on the form of merger. If the merger is a *survival*, only the acquiring hotel will continue to operate under its former name (Amin & Oukil, 2019a). Else, if the merger is a *consolidation*, the merging hotels will be fusing into a new entity with a new name, such as M_m. The objective of our study being specifically boosting hotels' efficiency through consolidation, none of the two hotels will be considered in P. The reader is referred to Oukil et al. (2022a) for a detailed discussion of possible merger forms.

Regardless of the merger's form, the intensity variables α_k are now restricted to $k \in P$, which includes α_m, the intensity variable associated to M_m. If $\alpha_m^* = 1$ and $\overline{\pi} = 1$ are the only solutions of model (IDEA), the merger will be a *major consolidation*, where the post-merger hotel M_m uses the totality of the inputs, i.e., $\partial_{im} = x_{im}$, no input gains are expected from the merger.

4.4 APPLICATION AND PRACTICAL SCOPE

Utilizing the hotel database from the Department of Statistics and Geographic Information at the Ministry of Tourism (Oman), data on four outputs and four inputs were collected for the annual operations of 58 hotels. The outputs include *Annual Revenue*, *Number of Guests*, *Number of Nights*, and *Occupancy Rate*. The inputs comprise *Number of Beds*, *Number of Rooms*, *Number of Employees*, and *Employee Salaries*. Table 4.1 provides a statistical summary of the dataset for these hotels.

Without loss of generality, we assume that the sample of 58 hotels represents the whole Omani hotel sector.

4.4.1 Hotels' Performance Evaluation

Using IBM-ILOG CPLEX version 12.6, BCC models are solved to compute the optimal efficiency scores e_k^* for all hotels T_k, $k = 1,\ldots,58$. The results are summarized in Table 4.2.

The average efficiency score across the 58 hotels is 0.789, with 20 hotels (34.48%) classified as strongly efficient, meaning they do not require input reductions. The remaining hotels are deemed inefficient and must benchmark against these strongly efficient hotels to improve their performance, as highlighted by Oukil et al. (2021). For an inefficient hotel to achieve

TABLE 4.1 Statistical Summary of the Hotels' Dataset

Variables	Unit	Mean	SD	Min.	Max.
Output					
Annual revenue	$ /year	6,911,989	13,731,079	3,004	78,795,452
Number of guests		17,864	20,273	597	96,877
Number of nights		23,882	28,091	669	147,084
Occupancy rate	%	55	24	2	87
Inputs					
Number of beds		135	138	23	937
Number of rooms		91	97	13	640
Number of employees		111	191	4	1,193
Salary of employees	$ /year	400,603	845,686	5,520	4,504,122

TABLE 4.2 DEA Efficiency Results for 58 Hotels

Hotel	e_k^*
T_{01}	0.73
T_{02}	0.865
T_{03}	1
T_{04}	1
T_{05}	0.678
T_{06}	0.819
T_{07}	0.743
T_{08}	0.974
T_{09}	0.88
T_{10}	0.506
T_{11}	0.903
T_{12}	0.742
T_{13}	0.393
T_{14}	1
T_{15}	1
T_{16}	1
T_{17}	0.801
T_{18}	0.49
T_{19}	0.463
T_{20}	0.509
T_{21}	0.515
T_{22}	0.46
T_{23}	0.771
T_{24}	0.704

(Continued)

TABLE 4.2 (Continued)

Hotel	e_k^*
T_{25}	1
T_{26}	1
T_{27}	1
T_{28}	1
T_{29}	0.935
T_{30}	1
T_{31}	1
T_{32}	1
T_{33}	0.747
T_{34}	0.231
T_{35}	0.771
T_{36}	0.374
T_{37}	1
T_{38}	1
T_{39}	0.847
T_{40}	1
T_{41}	1
T_{42}	0.671
T_{43}	0.746
T_{44}	0.647
T_{45}	0.898
T_{46}	0.625
T_{47}	0.535
T_{48}	0.318
T_{49}	0.774
T_{50}	0.503
T_{51}	0.885
T_{52}	0.967
T_{53}	0.766
T_{54}	1
T_{55}	0.594
T_{56}	1
T_{57}	1
T_{58}	1

efficiency, it must reduce its inputs by the slack values while maintaining its output levels. The pattern of input savings required for the entire sample of hotels is detailed in Table 4.3.

The analysis reveals that the input savings rates vary from 21.41% for employee salaries to 24.51% for the number of beds. Notably, these savings

TABLE 4.3 Input Savings for 58 Hotels

Input	Present Inputs	Target Inputs	Input Savings	Savings Rate (%)
Number of beds	7,856	5,930	1,926	24.51
Number of rooms	5,289	4,018	1,271	24.03
Number of employees	6,418	4,990	1,428	22.25
Salary of employees ($)	23,234,970	18,261,066	4,973,904	21.41

apply exclusively to the inefficient hotels as a prerequisite for achieving an efficient status. In the subsequent section, we will demonstrate that within an M&A framework, potential savings can still be realized even when both merging hotels are already strongly efficient.

4.4.2 Mergers of Hotels

Using the sample of hotels, the IDEA model is solved for each potential pairwise hotel merger $M_m = (T_A, T_B)$, $A \neq B$, with efficiency target $\bar{\pi} = 1$. With K = 58 hotels, 1,653 pairwise mergers M_m are possible. The optima of IDEA reveal that 83.24% of the mergers are major consolidations, where the post-merger hotels M_m retain the entire inputs of the merging hotels T_A and T_B, i.e., $\partial_{iA} = x_{iA}$, $\partial_{iB} = x_{iB}$ for all $i = 1,\ldots,I$ and $\alpha_m^* = 1$. In other words, no input gains are expected from these mergers, which is not surprising with the high levels of efficiency noted in the sample. The remaining 277 potential post-merger hotels are potentially productive.

Definition 1

A merger $M_m = (T_A, T_B)$, $A \neq B$, is productive if and only if $\alpha_m^* = 0$ and $\partial_{iA} < x_{iA}$ or $\partial_{iB} < x_{iB}$ for at least one input i $(i = 1,\ldots,I)$.

Given the large number of productive mergers, Table 4.4 shows only ten pairs of merging hotels for illustrative purposes.

Consider, for instance, the consolidation of hotels T_{14} and T_{57}, which are both strongly efficient, i.e., $e_{14}^* = e_{57}^* = 1$. With $\bar{\pi} = 1$ as the efficiency target for the merger, the minimum levels of inputs required to create the new merged hotel M_{10} are $\partial_{114}^* = 36$ and $\partial_{157}^* = 23$ beds, $\partial_{214}^* = 30$ and $\partial_{257}^* = 9$ rooms, $\partial_{314}^* = 44$ and $\partial_{357}^* = 0$ employees, and $\partial_{414}^* = \$116,481$ and $\partial_{457}^* = \$0$ salary of employees. With $\partial_{357}^* = 0$ and $\partial_{457}^* = 0$, it appears that the contribution of the employees of hotel T_{57} to the merger M_{10} is not needed, i.e., the new business can be run with only the employees of hotel T_{14}.

TABLE 4.4 Potential Inputs for a Sample of Ten Productive Post-Merger Hotels

Hotels			Number of beds		Number of rooms		Number of employees		Salary of employees	
M_m	T_A	T_B	∂^*_{1A}	∂^*_{1B}	∂^*_{2A}	∂^*_{2B}	∂^*_{3A}	∂^*_{3B}	∂^*_{4A}	∂^*_{4B}
M_{01}	T_{37}	T_{57}	30	15	15	4	4	1	5,520	6,54
M_{02}	T_{38}	T_{41}	26	0	23	0	4	4	7,680	1,797
M_{03}	T_{38}	T_{54}	27	0	21	0	4	3	7,680	1,494
M_{04}	T_{38}	T_{56}	27	0	22	0	4	4	7,680	2,486
M_{05}	T_{38}	T_{57}	30	0	17	0	4	1	7,581	0
M_{06}	T_{38}	T_{58}	60	93	41	39	4	64	7,680	173,969
M_{07}	T_{41}	T_{57}	69	0	26	0	6	1	7,125	915
M_{08}	T_{54}	T_{57}	48	23	17	9	5	2	7,928	0
M_{09}	T_{56}	T_{57}	28	23	20	5	5	2	8,210	496
M_{10}	T_{14}	T_{57}	36	23	30	9	44	0	116,481	0

Definition 2

Given a merger $M_m = (T_A, T_B)$, $A \neq B$, the potential gains of T_A and T_B for input i are $\gamma_{iA} = x_{iA} - \partial^*_{iA}$ and $\gamma_{iB} = x_{iB} - \partial^*_{iB}$, respectively.

Noting that the pre-merger inputs of hotels T_{14} and T_{57} are $x_{114} = 36$, $x_{157} = 23$, $x_{214} = 30$, $x_{257} = 18$, $x_{314} = 63$, $x_{357} = 6$, $x_{414} = \$136{,}176$, and $x_{457} = \$8{,}322$, the potential savings consist of $\gamma_{114} = x_{114} - \partial^*_{114} = 36 - 36 = 0$ and $\gamma_{157} = x_{157} - \partial^*_{157} = 23 - 23 = 0$ beds, $\gamma_{214} = x_{214} - \partial^*_{214} = 30 - 30 = 0$ and $\gamma_{257} = x_{257} - \partial^*_{257} = 18 - 9 = 9$ rooms, $\gamma_{314} = x_{314} - \partial^*_{314} = 63 - 44 = 19$ and $\gamma_{357} = x_{357} - \partial^*_{357} = 6 - 0 = 6$ employees, and $\gamma_{414} = x_{414} - \partial^*_{414} = 136{,}176 - 116{,}481 = \$19{,}695$ and $\gamma_{457} = x_{457} - \partial^*_{457} = 8{,}322 - 0 = \$8{,}322$ salaries. Interestingly, hotels T_{14} and T_{57} have been classified as strongly efficient, indicating that neither hotel requires any input reductions when operating independently. However, after merging, there remains potential for further savings. One of the most significant findings from this IDEA application is that, among the productive post-merger combinations, 30 pairs (10.83%) involve exclusively strongly efficient hotels. This suggests that even highly efficient hotels can benefit from mergers by realizing additional efficiencies.

Definition 3

The cumulative potential gain of merger $M_m = (T_A, T_B)$, $A \neq B$, for input i is $g_{im} = \gamma_{iA} + \gamma_{iB}$. and the corresponding proportion of savings is $g_{im} / x_{im} \times 100\%$.

Following the merger M_{10}, the cumulative savings are substantial: $g_{14} = \gamma_{114} + \gamma_{157} = 0$ beds, $g_{24} = \gamma_{214} + \gamma_{257} = 9$ rooms, $g_{34} = \gamma_{314} + \gamma_{357} = 25$ employees, and $g_{44} = \gamma_{414} + \gamma_{457} = \$28{,}017$ in salaries, reflecting savings of at least 17.78%, 35.76%, and 19.39%, respectively, on the total number of rooms, the number of employees, and salaries for hotels T_{14} and T_{57}. This demonstrates that, even though these hotels are individually efficient in their resource use, significant additional savings can still be realized through the merger. Table 4.5 offers a comprehensive summary of the cumulative savings and their proportions for the selected set of ten post-merger hotels.

Although Table 4.5 provides results for only ten post-merger hotels, the rows labeled "Avg" and "Max" represent the average and maximum proportions of gains observed across the entire sample of 277 potential productive post-merger hotels. The average gain proportions range from 47.05% to 52.79%, while the maximum values reach as high as 82.81% (M_{22}) for the number of beds, 82.44% (M_{179}) for the number of rooms, 87.91% (M_{179}) for the number of employees, and 86.75% (M_{183}) for employee salaries. This indicates that merging the strongly efficient hotels T_{38} and T_{41} into M_{02} could achieve efficiency target $\bar{\pi} = 1$ by utilizing only a fraction of the combined inputs, with as little as 17.19% of the beds needed. Figure 4.1 illustrates

TABLE 4.5 Cumulative Gains for a Sample of Ten Post-Merger Hotels With Related Proportions

Hotels			Number of beds		Number of rooms		Number of employees		Salary of employees	
M_m	T_A	T_B	g_{1m}	%	g_{2m}	%	g_{3m}	%	g_{4m}	%
M_{01}	T_{37}	T_{57}	8	15.60	14	43.87	5	52.69	7,668	55.39
M_{02}	T_{38}	T_{41}	124	82.81	45	66.46	2	24.92	5,328	35.99
M_{03}	T_{38}	T_{54}	81	74.90	37	63.61	2	23.54	8,166	47.09
M_{04}	T_{38}	T_{56}	61	69.79	39	63.14	1	15.87	5,724	36.02
M_{05}	T_{38}	T_{57}	53	63.94	42	71.63	5	49.30	8,421	52.62
M_{06}	T_{38}	T_{58}	9	5.56	22	21.15	15	18.16	52,298	22.35
M_{07}	T_{41}	T_{57}	44	39.31	19	42.20	5	44.51	7,407	47.95
M_{08}	T_{54}	T_{57}	0	0.00	9	25.41	4	40.21	10,054	55.91
M_{09}	T_{56}	T_{57}	0	0.00	13	34.17	4	35.25	7,826	47.34
M_{10}	T_{14}	T_{57}	0	0.00	9	17.78	25	35.76	28,017	19.39
	Averages			48.91		47.05		52.79		49.66
	Maxima			82.81		82.44		87.91		86.75

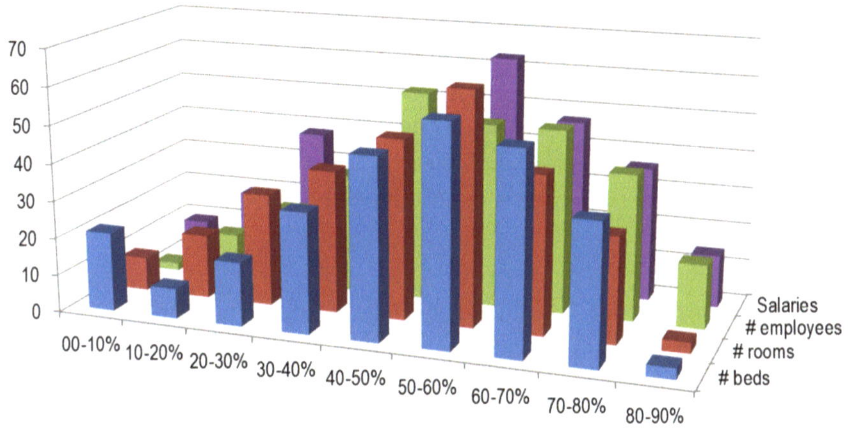

FIGURE 4.1 Frequency distributions of the input gains' proportions for post-merger hotels.

the frequency distributions of these gain proportions for the post-merger hotels.

Figure 4.1 demonstrates that the gain proportions for all categories closely follow a normal distribution, as evidenced by the median values being very similar to the means: 52.00% for the number of beds, 49.41% for the number of rooms, 53.21% for the number of employees, and 53.68% for salaries.

Additionally, it is important to note that most of the 277 potential post-merger hotel pairs have common partners. For example, four mergers listed in Table 4.4 involve hotel T_{38} paired with five different hotels. Since a hotel can only merge with one partner, it falls to the DMU to choose the optimal match from the list of potential candidates. This process involves ranking the alternative post-merger options and selecting the best set of non-overlapping pairs, where each pair consists of distinct hotels T_A and T_B.

4.5 SELECTING THE BEST PARTNERS OF A MERGER

A crucial step in devising the optimal merger strategy entails selecting the best set of post-merger hotel pairs to achieve two key objectives: (1) maximizing the collective performance of the merger process and (2) ensuring that there are no overlaps among the selected post-merger pairs. Given that each productive post-merger hotel is characterized by four cumulative gains—namely, the number of beds, the number of rooms, the number of employees, and employee salaries—assessing the collective performance of these mergers is inherently a multi-criteria decision making challenge.

To address this, we first employ a DEA model to rank the post-merger hotels according to their relative efficiency. Following this, a greedy heuristic is applied to identify the optimal set of non-overlapping pairs of post-merger hotels. This approach ensures that each pair contributes maximally to the overall performance while avoiding redundancy. The methodology for these processes is detailed in the subsequent sections.

4.5.1 Ranking the Post-Merger Hotels

The cumulative gains g_{im} of inputs $i\ (i=1,\ldots,4)$ are plausible post-merger outcomes for each pair of hotels $M_m=(T_A,T_B)\ (m=1,\ldots,277)$. It is evident that the higher the value of g_{im}, the more attractive is M_m. As such, g_{im} abides by the dictum "more is better," and, hence, each gain g_{im} can be treated as an output of the post-merger production technology. Subsequently, each post-merger hotel M_m can be regarded as a DMU that produces the cumulative expense gains $g_{im}\ (i=1,\ldots,4)$ as outputs, and DEA could be a suitable approach for ranking the post-merger hotels M_m. The following Best Partner Model (BPM) is used.

$$
\text{(BPM)}\quad
\begin{aligned}
& E_o^* = \max \psi \\
& \text{s.t.} \\
& \quad \sum_{m=1}^{277} \beta_m g_{im} \ge \psi\, g_{io} && i=1,\ldots,4 \\
& \sum_{m=1}^{277} \beta_m = 1 \\
& \beta_m \ge 0 && m=1,\ldots,277
\end{aligned}
$$

BPM is a variable returns to scale (VRS) output-oriented DEA model for the evaluation of post-merger hotel M_o, defined with two outputs and no inputs. β_m is the intensity variable associated with the post-merger hotel M_m. Hence, M_o is efficient if $E_o^*=1$, otherwise, it is inefficient ($E_o^*>1$).

Even with $M=277$ post-merger hotels and four outputs ($I=4$) only, the model (BPM) does not enable full ranking. In order to discriminate among the three efficient post-merger hotels that BPM produces, we use the associated benchmarking powers P_o, as shown in Table 4.6, where P_o is the number of times the efficient post-merger hotel M_o appears in the reference sets of inefficient post-merger hotels.

TABLE 4.6 Ranking of the Efficient Post-Merger Hotels

M_o	T_A	T_B	P_o	Rank
M_{99}	T_{01}	T_{48}	95	3
M_{184}	T_{34}	T_{48}	178	2
M_{257}	T_{48}	T_{50}	217	1

Based on the benchmarking powers, there is no doubt that M_{257} is the leading post-merger hotel, followed by M_{184} and M_{99}. The remaining 274 inefficient post-merger hotels are ranked based on the corresponding scores E_k^*.

4.5.2 Building a Merger Plan

Let $\Re$ denote the list of ranked post-merger hotels, where $\Re = \{M^{(r)} = \left(T_A^{(r)}, T_B^{(r)}\right) \mid r = 1,\ldots,277\}$, and let $\Im$ denote the list of hotels that have already been paired.

The following greedy heuristic is implemented to select the list $\aleph$ of post-merger hotels that are best suited for the merger plan.

Iteration 0: $\aleph = \varphi,\ \Im = \varphi$

$r = 1$

Iteration r: while $\left(r \leq 277\right)$ do {

If $(S_A^{(r)} \notin \Im$ and $T_B^{(r)} \notin \Im)$ {

$$\aleph \leftarrow \aleph \cup \left\{M^{(r)} = \left(T_A^{(r)}, T_B^{(r)}\right)\right\}$$

$$\Im \leftarrow \Im \cup \left\{T_A^{(r)}, T_B^{(r)}\right\}$$

$$\Re \leftarrow \Re \setminus \left\{M^{(r)}\right\}$$

}

$r \leftarrow r + 1$

}

Thus, M_{257}, the lead post-merger hotel, is selected in Iteration r = 1, i.e., $\aleph = \left\{M_{257}\right\} \Rightarrow \Im = \left\{T_{48}, T_{50}\right\}$. Although M_{184} ranks in the second position, it is discarded from the list $\aleph$ in Iteration r=2 because it is overlapping on T_{48} with M_{257}. Similarly, M_{99} is discarded in Iteration r = 3. In Iteration r = 4, M_{96} enters $\aleph$,, i.e., $\aleph = \left\{M_{257}, M_{96}\right\} \Rightarrow \Im = \left\{T_{48}, T_{50}, T_{01}, T_{44}\right\}$.

TABLE 4.7 Merger Plan With Pairs of Hotels and Corresponding Gains

M_m	T_A	T_B	Number of Beds	Number of Rooms	Number of Employees	Salary of Employees
M_{257}	T_{48}	T_{50}	217	97	190	348,692
M_{96}	T_{01}	T_{44}	144	87	235	3,468,440
M_{179}	T_{34}	T_{36}	149	146	72	145,907
M_{222}	T_{42}	T_{55}	187	62	53	58,497
M_{174}	T_{33}	T_{46}	153	23	97	166,505
M_{78}	T_{38}	T_{47}	112	94	30	116,308
M_{272}	T_{51}	T_{57}	0	8	173	389,066
M_{128}	T_{13}	T_{49}	81	80	94	197,755
M_{109}	T_{10}	T_{27}	75	64	41	106,626
M_{192}	T_{35}	T_{45}	73	40	14	10,446
M_{139}	T_{20}	T_{37}	66	42	21	30,242
Gains			1,258	745	1,020	5,038,484
Proportions			47.36%	43.93%	53.99%	64.55%

The above algorithm is run again for $r = 5,\ldots,277$ based on the rank succession of the remaining 274 pairs of hotels. The final list of post-merger hotels $\aleph$, involves only 11 pairs of hotels, as presented in Table 4.7.

Thus, 36 hotels have not been paired due to the fact that all pairwise mergers among these hotels are unproductive.

4.6 MANAGERIAL IMPLICATIONS

In light of this study's findings, potential managerial implications can be south, including the following:

Strategic partnership formation: The study provides hotel managers with data-driven insights to identify optimal merger partners. By understanding the post-merger potential for productivity gains, hotels can target strategic partnerships that maximize collective performance, leading to a stronger competitive edge in the market.

Cost optimization in mergers: Even for hotels operating efficiently, the results demonstrate the potential for further savings post-merger, particularly in areas like staffing and resource utilization. Managers should therefore consider M&A not only as a growth strategy but also as an opportunity for operational cost reduction.

Data-driven merger decisions: The use of DEA and IDEA models offers a robust, evidence-based framework for merger decisions, encouraging a shift from intuition-based M&A decisions to ones grounded in performance analytics. Managers can rely on quantitative tools to predict the merger's success before committing to the transaction.

Customized merger plans: The methodology highlights that while some hotels are likely to benefit significantly from mergers, others may not. This reinforces the need for tailored merger strategies that avoid a one-size-fits-all approach. Managers should focus on selecting the right partners rather than merging for expansion's sake.

Policy advocacy for M&A: Hotel industry policymakers can use these findings to advocate for M&A as a viable growth and efficiency-enhancement strategy. By supporting policies that encourage mergers, the industry could achieve a more consolidated and efficient market structure.

Addressing methodological challenges: For larger hotel chains or portfolios, computational complexity in analyzing all merger possibilities may arise. Future investment in advanced analytics or hybrid models integrating both partner selection and evaluation could streamline the merger planning process, making it more accessible for larger operations.

Adapting to contextual realities: Not all business environments are conducive to formalized mergers. Managers should explore horizontal cooperation strategies, such as alliances or joint ventures, where full mergers might not be feasible but efficiency gains can still be realized through collaborative operations.

Fostering a culture of efficiency: Beyond mergers, this study highlights the importance of ongoing efficiency evaluations. Managers can adopt a culture of continuous improvement, using models like DEA regularly to benchmark performance and identify potential improvements, even outside of M&A contexts.

4.7 CONCLUDING REMARKS

In this chapter, we presented a hybrid methodology based on DEA to develop the optimal merger strategy for the hotel industry, focusing on the potential performance improvements of individual hotels post-merger. The aim of this approach is to determine the best merger partners that, once combined, would enable the entire sector—represented by a selected sample of hotels—to achieve maximum collective performance. Specifically, assuming that each hotel is interested in merging with another hotel from the sample, all possible pairwise combinations are assessed using an IDEA model. Only those mergers that show the potential for productivity gains are further considered. In the next phase, a hybrid process combining a DEA model with a greedy heuristic is employed to select the most productive merger pairs, ensuring that the overall performance of the final group of merged hotels is optimized.

To evaluate the practical applicability of the proposed methodology, we conducted a case study involving 58 hotels. Using the IDEA approach, we analyzed 1,653 possible pairwise mergers, identifying 277 as potentially productive—meaning these post-merger combinations were likely to result in performance gains.

One of the most striking findings at this stage is the discovery that, among the productive post-merger combinations, some involve only highly efficient hotels—hotels that, individually, were not expected to need any cost reductions based on their production possibility sets. Nevertheless, after merging, these pairs still show potential for savings. A detailed analysis of these mergers revealed that the potential savings could reach as high as 82.81% in the number of beds, 82.44% in the number of rooms, 87.91% in the number of employees, and 86.75% in employee salaries. In the context of hotel mergers and acquisitions, these savings proportions offer compelling evidence of the potential gains mergers can bring.

A crucial step in formulating the optimal merger strategy involves selecting the best non-overlapping pairs from the set of productive post-merger hotels. Each productive post-merger hotel is characterized by four cumulative performance gains—the number of beds, the number of rooms, the number of employees, and employee salaries—making this a multi-criteria decision problem. To address this, a DEA model is used as a ranking tool. While the ranking provided by the DEA model helps clearly differentiate between the productive post-merger hotels, a greedy heuristic was developed to identify the best non-overlapping hotel pairs. This process resulted in a final list of 11 productive post-merger hotel pairs, meaning that 22 out of the 58 hotels were matched with their optimal merger partner. The remaining 36 hotels, while free to merge with each other, are unlikely to form productive consolidations.

These findings are strategically significant for informing policies that encourage hotel mergers as a viable option for achieving collective gains. However, the proposed methodology has limitations from both methodological and practical standpoints. From a methodological perspective, using a standard DEA model to rank productive post-merger hotels may not provide sufficient discrimination, particularly when dealing with more complex sets of inputs and outputs. To address this, more sophisticated ranking methods could be employed, such as DEA cross-efficiency, which would allow for incorporating decision makers' preferences into the evaluation process.

On the practical side, conducting a comprehensive analysis of all possible pairwise consolidations can become computationally expensive as the number of hotels increases. Future research could explore the development of models that integrate both the merger evaluation and optimal partner selection stages within a unified framework.

Additionally, the concept of a "merger plan" may not be feasible or realistic in certain business contexts. Future studies might focus on exploring different settings for horizontal cooperation and corporate restructuring strategies where designing large-scale merger plans could be both conceptually sound and practically applicable.

REFERENCES

Al Tamimi, H.A.H., Duqi, A., Kanas, A., & Zervopoulos, P.D. (2022). Directional distance function DEA estimators for evaluating efficiency gains from possible mergers and acquisitions. *Journal of the Operational Research Society*, *73*(6), 1240–1257.

Al-Mezeini, N.K., Oukil, A., & Al-Ismaili, A.M. (2020). Investigating the efficiency of greenhouse production in Oman: A two-stage approach based on Data Envelopment Analysis and double bootstrapping. *Journal of Cleaner Production*, *247*, 119160.

Amin, G.R., & Al-Muharrami, S. (2018). A new inverse data envelopment analysis model for mergers with negative data. *IMA Journal of Management Mathematics*, *29*(2), 137–149.

Amin, G.R., Al-Muharrami, S., & Toloo, M. (2019). A combined goal programming and inverse DEA method for target setting in mergers. *Expert Systems with Applications*, *115*, 412–417.

Amin, G.R., Emrouznejad, A., & Gattoufi, S. (2017a). Minor and major consolidations in inverse DEA: Definition and determination. *Computers & Industrial Engineering*, *103*, 193–200.

Amin, G. R., Emrouznejad, A., & Gattoufi, S. (2017b). Modelling generalized firms' restructuring using inverse DEA. *Journal of Productivity Analysis*, *48*(1), 51–61.

Amin, G.R., & Ibn Boamah, M. (2020). A new inverse DEA cost efficiency model for estimating potential merger gains: a case of Canadian banks. *Annals of Operations Research*, 295, 21–36.

Amin, G.R., & Ibn Boamah, M. (2021). A two-stage inverse data envelopment analysis approach for estimating potential merger gains in the US banking sector. *Managerial and Decision Economics*, *42*(6), 1454–1465.

Amin, G.R., & Ibn Boamah, M. (2023). Modeling business partnerships: A data envelopment analysis approach. *European Journal of Operational Research*, *305*(1), 329–337.

Amin, G.R. Oukil, A. (2019a). Flexible target setting in mergers using inverse data envelopment analysis. *International Journal of Operational Research*, 35(3), 301–317.

Amin, G.R., & Oukil, A. (2019b). Gangless cross-evaluation in DEA: an application to stock selection. *RAIRO-Operations Research*, *53*(2), 645–655.

Ang, S., Chen, M., & Yang, F. (2018). Group cross-efficiency evaluation in data envelopment analysis: An application to Taiwan hotels. *Computers & Industrial Engineering*, 125, 190–199.

Banker, R.D., Charnes, A., & Cooper, W.W. (1984). Some models for estimating technical and scale inefficiencies in data envelopment analysis. *Management Science*, 9, 1078–1092.

Bire, R.B. (2020). Efficiency Analysis of Hotels Listed in the Indonesian Stock Exchange Market. *TOURISM: Jurnal Travel, Hospitality, Culture, Destination, and MICE*, *3*(1), 1–6.

Chen, L.F. (2018). An improved three-stage DEA as business intelligence for benchmarking luxury hotels with environmental effect. In *Proceedings of the 10th International Conference on Computer Modeling and Simulation* (pp. 72–76).

Chiu, S.H., & Lin, T.Y. (2018). Performance evaluation of Taiwanese international tourist hotels: Evidence from a modified NDEA model with ICA technique. *Technological and Economic Development of Economy*, *24*(4), 1560–1580.

Chiu, Y.H., Lin, T.Y., Chang, T.H., Lin, Y.N., & Chiu, S.Y. (2021). Prevaluating efficiency gains from potential mergers and acquisitions in the financial industry with the Resample Past–Present–Future data envelopment analysis approach. *Managerial and Decision Economics*, *42*(2), 369–384.

Deng, Z., Gao, Y., Liang, B., & Morrison, A.M. (2020). Efficiency evaluation of hotel operations in Mainland China based on the super efficiency SBM model. *Tourism Economics*, *26*(2), 276–298.

Dobrovič, J., Čabinová, V., Gallo, P., Partlová, P., Váchal, J., Balogová, B., & Orgonáš, J. (2021). Application of the DEA Model in Tourism SMEs: An Empirical Study from Slovakia in the Context of Business Sustainability. *Sustainability*, *13*(13), 7422.

Dolasinski, M. J., Roberts, C., & Zheng, T. (2019). Measuring hotel channel mix: A DEA-BSC model. *Journal of Hospitality & Tourism Research*, *43*(2), 188–209.

Eguchi, S., Takayabu, H., & Lin, C. (2021). Sources of inefficient power generation by coal-fired thermal power plants in China: A metafrontier DEA decomposition approach. *Renewable and Sustainable Energy Reviews*, *138*, 110562.

El Alaoui, M.H., Ibrahim, M.D., Daneshvar, S., Alola, U.V., & Alola, A.A. (2023). A two-stage data envelopment analysis approach to productivity, efficiency and their sustainability in the hotel industry of Tunisia. *Quality & Quantity*, 57, 955–972.

Emrouznejad, A., Yang, G., Khoveyni, M., & Michali M. (2022). Data envelopment analysis: Recent developments and challenges. In S. Salhi & J. Boylan (Ed.) *The Palgrave handbook of operations research* (pp. 307–350). Palgrave Macmillan, Springer Nature Switzerland.

Fang, D., Hsueh, M.C., & Chiu, C.R. (2020). An investigation of international hotel productivity considering quasifixed inputs and negative outputs. *Managerial & Decision Economics*, *41*(3), 380–388.

Frančeškin, J., & Bojnec, Š. (2022a). Economic efficiency of coastal hotel companies. *Economic Research-Ekonomska Istraživanja*, *35*(1), 4425–4436.

Frančeškin, J., & Bojnec, Š. (2022b). Total factor productivity of the Slovenian hotel companies. *Economic Research-Ekonomska Istraživanja*, 36(1), 1–13.

Gattoufi, S. Amin, G.R. Emrouznejad, A. (2014). A new inverse DEA method for merging banks. *IMA Journal of Management Mathematics*, 25, 73–87.

Gerami, J., Mozaffari, M.R., Wanke, P.F., & Correa, H.L. (2021). A generalized inverse DEA model for firm restructuring based on value efficiency. *IMA Journal of Management Mathematics*, dpab043, https://doi.org/10.1093/imaman/dpab043

Ghobadi, S. (2021). Merging decision making units with interval data. *RAIRO-Operations Research*, *55*, S1605–S1631.

Greene, W. 2008. The econometric approach to efficiency analysis. In H.O. Fried, C.A.K. Lovell, & S.S. Schmidt (Eds.), *The measurement of productive efficiency and productivity growth*, Oxford University Press.

Guo, X., Wei, W., Li, Y., & Wang, L.Y. (2019). A study of different types of air pollutants on the efficiency of china's hotel industry. *International Journal of Environmental Research and Public Health*, *16*(22), 4319.

Hassan, M., & Oukil, A. (2021). Design of efficient systems of commercial material handling equipment for supply chain and logistics facilities using DEA. *International Journal of Logistics Systems & Management*, *39*(2), 241–272.

Higuerey, A., Viñan-Merecí, C., Malo-Montoya, Z., & Martínez-Fernández, V.A. (2020). Data Envelopment Analysis (DEA) for Measuring the efficiency of the hotel industry in Ecuador. *Sustainability*, *12*(4), 1590.

Huhtilainen, M., Saastamoinen, J., & Suhonen, N. (2022). Determinants of mergers and acquisitions among Finnish cooperative and savings banks. *Journal of Banking Regulation*, *23*(3), 339–349.

Kamarudin, F., Sufian, F., Nassir, A.M., Anwar, N.A.M., & Hussain, H.I. (2019). Bank efficiency in Malaysia a DEA approach. *Journal of Central Banking Theory and Practice*, *8*(1), 133–162.

Kamiyama, H., Kefi, M., & Kashiwagi, K. (2021). Irrigation water use efficiency in olive trees in Kairouan, Tunisia. *Asian Journal of Agriculture and Rural Development*, *11*(3), 255–261.

Kamyab, P., Mozaffari, M. R., & Gerami, J. (2019). Determining minor and major consolidations in network inverse data envelopment analysis. *Journal of Mathematical Extension*, *14*, 195–223.

Karakitsiou, A., Kourgiantakis, M., Mavrommati, A., & Migdalas, A. (2020). Regional efficiency evaluation by input-oriented data envelopment analysis of hotel and restaurant sector. *Operational Research*, 20(4), 2041–2058.

Khairy, H.A. (2019). How organizational merger affects employees' engagement and job security in the hotel industry? A pre-post analysis. *International Journal of Heritage, Tourism and Hospitality*, *13*(2), 245–262.

Khoveyni, M., & Eslami, R. (2022). Merging two-stage series network structures: A DEA-based approach. *OR Spectrum*, *44*(1), 273–302.

Kim, C., & Chung, K. (2022). Measuring customer satisfaction and hotel efficiency analysis: An approach based on data envelopment analysis. *Cornell Hospitality Quarterly*, *63*(2), 257–266.

Kim, C., Kang, H.J., Chung, K., & Choi, K. (2021). COVID-19 and hotel productivity changes: An empirical analysis using malmquist productivity index. *Service Science*, *13*(4), 243–257.

Kularatne, T., Wilson, C., Månsson, J., Hoang, V., & Lee, B. (2019). Do environmentally sustainable practices make hotels more efficient? A study of major hotels in Sri Lanka. *Tourism Management*, *71*, 213–225.

Lado-Sestayo, R., & Fernández-Castro, Á.S. (2019). The impact of tourist destination on hotel efficiency: A data envelopment analysis approach. *European Journal of Operational Research*, *272*(2), 674–686.

Lee, Y.L., Kuo, S.H., Jiang, M.Y., & Li, Y. (2019). Evaluating the performances of Taiwan's international tourist hotels: Applying the directional distance function and meta-frontier approach. *Sustainability*, *11*(20), 5773.

Li, F., Liang, L., Li, Y., & Emrouznejad, A. (2018). An alternative approach to decompose the potential gains from mergers. *Journal of the Operational Research Society*, *69*(11), 1793–1802.

Li, J., Li, X., & Chen, C. (2021). The CO2 emission efficiency of China's hotel industry under the double carbon objectives and homestay growth. *Energies*, *14*(24), 8228.

Lin, Y., Wang, Y.M., & Shi, H.L. (2020). Mergers and acquisitions matching for performance improvement: a DEA-based approach. *Economic research-Ekonomska istraživanja*, *33*(1), 3545–3561.

Lozano, S. (2013). Using DEA to find the best partner for a horizontal cooperation. *Computers & Industrial Engineering*, *66*(2), 286–292.

Mariani, M.M., & Visani, F. (2019). Embedding eWOM into efficiency DEA modelling: An application to the hospitality sector. *International Journal of Hospitality Management*, *80*, 1–12.

Martins, C., Vaz, C.B., & Alves, J.M.A. (2021). Financial performance assessment of branded and non-branded hotel companies. Analysis of the Portuguese case. *International Journal of Contemporary Hospitality Management*, 33(10), 3134–3156.

Moghaddas, Z. Oukil, A. Vaez-Ghasemi, M. (2022). Global multi-period performance evaluation: A new model and a new productivity index. *RAIRO Operations Research*, 56 (3), 1503–1521.

Ngo, T., & Tsui, K.W.H. (2022). Estimating the confidence intervals for DEA efficiency scores of Asia-Pacific airlines. *Operational Research*, *22*(4), 3411–3434.

Nguyen, N.T., & Nguyen, L.X.T. (2019). Applying DEA model to measure the efficiency of hospitality sector: The case of Vietnam. *International Journal of Analysis and Applications*, *17*(6), 994–1018.

Nguyen, P.A., & Pham, L.D. (2020). Non-parametric analysis of bank merger gains: The case of Vietnam. *Cogent Business & Management*, *7*(1), 1823582.

Oral, M., Amin, G.R., & Oukil, A. (2015). Cross-efficiency in DEA: A maximum resonated appreciative model. *Measurement*, *63*, 159–167.

Oral, M., Oukil, A., Malouin, J.-L., & Kettani, O. (2014). The appreciative democratic voice of DEA: A case of faculty academic performance evaluation. *Socio-Economic Planning Sciences*, 48, 20–28.

Oukil, A. (2020). Exploiting value system multiplicity and preference voting for robust ranking. *Omega, International Journal of Management Science*, 94, 102048. https://doi.org/10.1016/j.omega.2019.03.006

Oukil, A. (2023). Investigating prospective gains from mergers in the agricultural sector through Inverse DEA. *IMA Journal of Management Mathematics*, 34(3), 465–490.

Oukil, A. (2024). Devising the best prospective merger plan for a banking sector through a hybrid DEA-based methodology: An inverse DEA perspective. In Boubaker, S. & Ngo, T. (Eds.) *Handbook on data envelopment analysis in business, finance, and accounting: Recent trends and developments*, 273–306. World Scientific publishing Co.

Oukil, A., Channouf, N., & Al-Zidi, A. (2016). Performance evaluation of the hotel industry in an emerging tourism destination: Case of Oman. *Journal of Hospitality & Tourism Management*, 29, 60–68.

Oukil, A. El-Bouri, A. Emrouznejad, A. (2022a). Energy-aware job scheduling in a multi-objective production environment–An integrated DEA-OWA model. *Computers & Industrial Engineering*, 168, 108065.

Oukil, A. Kennedy, R.E. Al-Hajri, A. & Soltani, A.A. (2024). Unveiling the potential of hotel mergers: A hybrid DEA approach for optimizing sector-wide performance in the hospitality industry. *International Journal of Hospitality Management*, 116, 103620.

Oukil, A. Nourani, A. Soltani, A.A. & Benchikh, A. (2022b). Using inverse data envelopment analysis to investigate potential impact of mergers on energy use optimization - Application in agricultural production. *Journal of Cleaner Production*, 381(1), 135199.

Oukil, A., Soltani, A.A., Boutaghane, H., Abdalla, O., Bermad, A., Hasbaia, M., & Boulassel, M.R. (2021). A Surrogate Water Quality Index to assess groundwater using a unified DEA-OWA framework. *Environmental Science and Pollution Research*, 28, 56658–56685.

Oukil, A., Soltani, A.A., Zeroual, S., Boutaghane, H., Abdalla, O., Bermad, A., & Boulassel, M.R. (2022c). A DEA cross-efficiency inclusive methodology for assessing water quality: A composite water quality index. *Journal of Hydrology*, *612*, 128123.

Peykani, P., Mohammadi, E., & Emrouznejad, A. (2021). An adjustable fuzzy chance-constrained network DEA approach with application to ranking investment firms. *Expert systems with applications*, *166*, 113938.

Peypoch, N., Song, Y., & Zhang, L. (2021). The nature of technological change in the Chinese hotel sector. *Journal of Hospitality & Tourism Research*, *45*(1), 151–170.

Pohlman, M. (2017). Mergers and acquisitions in hospitality expected to accelerate. *Hotel Management*. https://www.hotelmanagement.net/transactions/mergers-and-acquisitions-hospitality-expected-to-accelerate

Ray, S.C., & Sethia, S. (2022). Nonparametric measurement of potential gains from mergers: an additive decomposition and application to Indian bank mergers. *Journal of Productivity Analysis*, *57*(2), 115–130.

Sáez-Fernández, F.J., Jiménez-Hernández, I., & Ostos-Rey, M.D.S. (2020). Seasonality and efficiency of the hotel industry in the balearic Islands: Implications for economic and environmental sustainability. *Sustainability*, *12*(9), 3506.

Shi, X., Li, Y., Emrouznejad, A., Xie, J., & Liang, L. (2017). Estimation of potential gains from bank mergers: A novel two-stage cost efficiency DEA model. *Journal of the Operational Research Society*, *68*(9), 1045–1055.

Singh, R., Charan, P., & Chattopadhyay, M. (2022). Evaluating the hotel industry performance using efficiency and effectiveness measures. *International Journal of Hospitality & Tourism Administration*, *23*(2), 408–431.

Soltani, A.A., Oukil, A., Boutaghane, H., Bermad, A., & Boulassel, M.R. (2021). A new methodology for assessing water quality, based on data envelopment analysis: Application to Algerian dams. *Ecological Indicators*, 121, 106952.

Soltanifar, M., Ghiyasi, M., Emrouznejad, A., & Sharafie, H.A. (2022) Novel model for merger analysis and target setting: A CSW-inverse DEA approach. *Available at SSRN 4115552.*

Soltanifar, M., Ghiyasi, M., & Sharafi, H. (2023). Inverse DEA-R models for merger analysis with negative data. *IMA Journal of Management Mathematics*, 34(3), 491–510.

Sow, O. Oukil, A. Ndiaye, B.M. & Marcos, A. (2016). Efficiency analysis of public transportation subunits using data envelopment analysis (DEA) and bootstrap: Dakar Dem Dikk case study. *Journal of Mathematics Research*, 8, 114–127.

Tan, Y., & Despotis, D. (2021). Investigation of efficiency in the UK hotel industry: A network data envelopment analysis approach. *International Journal of Contemporary Hospitality Management*, 33(3), 1080–1104.

Tian, N., Tang, S., Che, A., & Wu, P. (2020). Measuring regional transport sustainability using super-efficiency SBM-DEA with weighting preference. *Journal of Cleaner Production*, *242*, 118474.

Tzeremes, N.G. (2020). Robust Malmquist productivity measurement: evidence from Spanish hotel industry during the Great Recession. *International Journal of Productivity and Performance Management*, *70*(2), 408–426.

Tzeremes, P., & Tzeremes, N.G. (2021). Productivity in the hotel industry: an order-α Malmquist productivity indicator. *Journal of Hospitality & Tourism Research*, *45*(1), 133–150.

Wang, C.N., Nguyen, X.T., Le, T.D., & Hsueh, M.H. (2018). A partner selection approach for strategic alliance in the global aerospace and defense industry. *Journal of Air Transport Management*, *69*, 190–204.

Wei, Q., Zhang, J., & Zhang, X. (2000). An inverse DEA model for inputs/outputs estimate. *European Journal of Operational Research*, *121*(1), 151–163.

Wu, J., Zhu, Q., Cook, W.D., & Zhu, J. (2016). Best cooperative partner selection and input resource reallocation using DEA. *Journal of the Operational Research Society*, *67*(9), 1221–1237.

Xiao, S., Li, Y., Emrouznejad, A., Xie, J., & Liang, L. (2017). Estimation of potential gains from bank mergers: A novel two-stage cost efficiency DEA model. *Journal of the Operational Research Society*, *68*(9), 1045–1055.

Xie, J., Zhu, X., & Liang, L. (2020). A multiplicative method for estimating the potential gains from two-stage production system mergers. *Annals of Operations Research*, *288*(1), 475–493.

Yan, H., Wei, Q., & Hao, G. (2002). DEA models for resource reallocation and production input/output estimation. *European Journal of Operational Research*, *136*(1), 19–31.

Yang, M., Hou, Y., Ji, Q., & Zhang, D. (2020). Assessment and optimization of provincial CO2 emission reduction scheme in China: An improved ZSG-DEA approach. *Energy Economics*, *91*, 104931.

Yu, M.M., & Chen, L.H. (2020a). A meta-frontier network data envelopment analysis approach for the measurement of technological bias with network production structure. *Annals of Operations Research*, *287*(1), 495–514.

Yu, M.M., & Chen, L.H. (2020b). Evaluation of efficiency and technological bias of tourist hotels by a meta-frontier DEA model. *Journal of the Operational Research Society*, *71*(5), 718–732.

Zaki, K.G. (2019). Using the mixed methods research to model the hotel performance measurement in Egypt: An example from a hotel chain. *Journal of Global Business Insights*, *4*(1), 18–33.

Zalaghi, A.H., Lotfalian Dehkordi, A., Abedi, A., & Taki, M. (2021). Applying data envelopment analysis (DEA) to improve energy efficiency of apple fruit, focusing on cumulative energy demand. *Energy Equipment and Systems*, *9*(1), 37–52.

Zhang, F., Xiao, Q., Law, R., & Lee, S. (2020). Mergers and acquisitions in the hotel industry: A comprehensive review. *International Journal of Hospitality Management*, *91*, 102418.

Zhu, Q., Li, X., Li, F., & Amirteimoori, A. (2021). Data-driven approach to find the best partner for merger and acquisitions in banking industry. *Industrial Management & Data Systems*, *121*(4), 879–893.

Zhu, Q., Wu, J., Chu, J., Amirteimoori, A., & Sun, J. (2017). DEA-based models for best partner selection for merger. *RAIRO-operations Research*, *51*(4), 1345–1357.

CHAPTER 5

Centralized DEA Model for Resource Allocation Among Internal Business Partnerships

Ming-Miin Yu
Department of Transportation Science, National Taiwan Ocean University, Taiwan

Bo Hsiao
National Yunlin University of Science and Technology, Taiwan

Kok Fong See
Economics Programme, School of Distance Education, University Sains Malaysia, Malaysia

5.1 INTRODUCTION

In today's competitive business landscape, the quest for organizational efficiency is paramount. Efficient intraorganizational units not only optimize resource utilization but also contribute to improved performance and competitive advantage. Internal business partnerships refer to collaborative relationships and cooperative efforts between different units, departments, divisions, or subsidiaries within the same organization.

DOI: 10.1201/9781003345206-5

These internal partnerships aim to optimize the organization's overall performance by leveraging the strengths, resources, and expertise of various internal entities and play a pivotal role in driving innovation, expanding market reach, and achieving strategic objectives within an organization. A centralized resource allocation setting to improve efficiency and effectiveness in allocating resources across various units or departments within an organization or system is important (Asmild et al., 2009). However, managing resources efficiently within these partnerships presents a significant challenge. The concept of centralization in resource allocation among internal business partnerships has garnered increasing attention as organizations seek to streamline operations and maximize synergies. Centralized resource allocation offers the potential for optimizing resource utilization, mitigating redundancy, and enhancing overall partnership efficiency (Lozano and Villa, 2005). However, implementing centralized resource allocation mechanisms requires a comprehensive understanding of the diverse needs, capabilities, and objectives of each partner, as well as the complex interdependencies within the partnership network.

Motivated by the need for robust resource allocation frameworks tailored to the unique dynamics of internal business partnerships, this chapter explores the application of the data envelopment analysis (DEA) model. DEA has emerged as a powerful methodology for evaluating and improving the efficiency of decision making units (Charnes et al., 1978), making it well suited for addressing the complexities of resource allocation in collaborative settings (Lozano and Villa, 2004). This chapter delves into the centralized DEA (CDEA) approach as a powerful tool for enhancing intra-organizational efficiency.

The CDEA model has undergone significant development over the years. DEA was initially introduced by Charnes, Cooper, and Rhodes in the late 1970s as a nonparametric method for evaluating the relative efficiency of decision making units (DMUs) within an organization (Charnes et al., 1978). In the early stages of development, DEA focused primarily on evaluating the efficiency of decentralized decision making units. However, as the model evolved, researchers recognized the need for a centralized approach to address scenarios where decisions are made at both the individual and aggregate levels within an organization (Fang, 2015).

Over the years, various modifications and extensions to the CDEA model have been proposed to accommodate centralized decision making structures. These developments include the introduction of centralized target setting for internal business partnerships using CDEA, the perspective of centralized management, a common set of weights (Hatami-Marbini et al., 2015), resource allocation for multiple periods (Xiong et al., 2022), how DEA can be employed to assess the efficiency of firms in using emission permits and to determine optimal reallocation strategies to enhance overall environmental performance within the cap and trade framework (Momeni et al., 2019), examining the potential gains from splits and mergers, analyzing the impact of centralized resource reallocation (Contreras and Lozano, 2022), and evaluating the efficiency of resource allocation within each subset while considering the degree of centralization across different subsets (Zhu et al., 2023), among others. The methods adopted in the aforementioned studies are regarded as narrow CDEA approaches.

The broad CDEA approach encompasses several distinct models, with three primary types being hierarchical DEA models, network DEA models, and dynamic DEA models. Hierarchical DEA models organize institutions or units into multiple levels or tiers and assess the efficiency of each level. For example, a company may have multiple departments, each containing several teams. Hierarchical DEA models can be used to evaluate the efficiency of departments and teams at different levels (Yu et al., 2022). Network DEA models consider the interrelationships and influences among different units. They can be used to assess a network composed of multiple interdependent units, such as supply chains or social networks. Network DEA models allow for the evaluation of the interactions between units, providing a more comprehensive assessment of the entire system's efficiency (Zhu and Cook, 2007). Dynamic DEA models consider time factors and are typically used to evaluate how the efficiency of an organization or system changes over time. These models can be used to analyze trends, changes, and fluctuations in efficiency, providing insights for forecasting and improvement (Tone and Tsutsui, 2010).

The commonality of both broad and narrow CDEA methods lies in their utilization of DEA as a tool for assessing efficiency while also taking into account the specific characteristics and requirements of the organization or system in various aspects. Overall, the development of the CDEA model has been driven by the need to provide decision makers with robust tools for assessing and improving the efficiency of centralized decision making structures across various domains, including banks (Varmaz et al., 2013; Zhou, et al., 2013; Zhou et al., 2017; Zhu et al., 2023), supermarkets (Fang, 2013), transportation (Yu et al., 2013; Yu and Chen, 2016; Hsiao and Chen, 2022; Cesaroni, 2020; Lozano et al., 2011), universities (Xiong et al., 2022; Contreras and Lozano, 2022; Lozano and Contreras, 2022), education (Davutyan et al., 2010; López-Torres and Prior, 2016), public administration (Arocena et al., 2022), power plants (Toloo et al., 2022; Zhang et al., 2024), and emission allowances within the cap and trade system (Lozano et al., 2009; Feng, et al., 2015; Momeni et al., 2019; Yang et al., 2022). This chapter introduces only the narrow CDEA approach.

The motivation behind this chapter stems from the recognition of the gap between existing efficiency evaluation methods and the evolving organizational landscape. By focusing on the CDEA approach, we aim to bridge this gap and provide organizations with a comprehensive framework for optimizing their internal operations. Furthermore, this chapter aims to contribute to the academic discourse by exploring the application of the CDEA in the literature. By revealing its effectiveness in capturing the complexities between intraorganizational business partnerships, we anticipate paving the way for further research and advancements in the field of organizational efficiency assessment. Through this exploration, we aim to provide practitioners and scholars with valuable insights into enhancing intraorganizational efficiency using the CDEA approach.

The remaining sections of the chapter are organized as follows: Section 5.2 offers a basic understanding of DEA and the CDEA, followed by a review of the CDEA methodology and its applications in Section 5.3. To further explore CDEA research, Section 5.4 uses bibliometric analysis to describe publication trends, authorship, affiliations, publication outlets, and citation networks. Section 5.5 provides important insights into future challenges and opportunities in applying the CDEA model. Concluding remarks are summarized in the final section.

5.2 UNDERSTANDING DEA

5.2.1 Basic Principles of DEA and Its Application in Evaluating Organizational Efficiency

DEA is a powerful quantitative technique used to evaluate the relative efficiency of DMUs within an organization. Since the development of DEA by Charnes et al. (1978), it has been applied to problems in many areas, and several theoretical additions have been made. At its core, DEA compares the input–output relationships of multiple DMUs to identify those that are operating at maximum efficiency and those that could improve their performance.

The basic principles of DEA revolve around the concept of efficiency measurement without the need for a predefined functional form or explicit assumptions about the underlying production process. Instead, DEA serves as a nonparametric model for evaluating the performance of a particular entity in comparison to others that share similar characteristics and objectives (Mardani et al., 2017; Koronakos et al., 2019). Within DEA, every DMU is depicted as a set of inputs and outputs. Inputs denote the resources utilized by the DMU, whereas outputs signify the products or services generated. The efficiency of each DMU is gauged by contrasting its output–input ratio with that of other DMUs within the dataset.

DEA offers several advantages in evaluating organizational efficiency. First, it does not require explicit specifications of production functions, making it suitable for complex and diverse organizational settings where the production processes may not be well understood. Second, DEA can handle multiple inputs and outputs simultaneously, providing a comprehensive assessment of efficiency across different dimensions. Finally, DEA allows for the identification of best practices by benchmarking against the most efficient DMUs, offering valuable insights for performance improvement (Thomas et al., 1998).

In general, DEA proves to be an invaluable tool for organizations aiming to increase efficiency by identifying opportunities for optimizing resources and refining processes across different departments or units. Its adaptability, nonparametric design, and capacity to manage numerous inputs and outputs make it the favored option for evaluating efficiency in a wide array of organizational settings.

5.2.2 Limitations of DEA Compared With Those of CDEA

Nevertheless, the DMUs under assessment might belong to a centralized decision maker, who could prioritize optimizing the operations of all units globally. Specifically, in the context of input orientation, the central decision maker might aim to collectively decrease the total inputs of all units simultaneously rather than independently reducing the inputs of individual DMUs (Fang, 2015).

In recent years, the utilization of CDEA for resource allocation in centralized environments has been increasing. Numerous papers have explored DEA applications for resource allocation in such settings (Asmild et al., 2009; Fang, 2013; Sharma and Yu, 2010). Importantly, these studies typically assume that each subdecision making unit is expected to achieve overall organizational efficiency following resource allocation. This differs from the logic of traditional DEA methods, which seek only individual optimal efficiency.

Compared with CDEA, traditional DEA assumes that DMUs operate independent of each other, a premise that may not necessarily reflect the real-world situation, where DMUs may be interconnected or influenced by centralized decision making. Additionally, DEA lacks explicit consideration of centralized resource allocation strategies, a deficiency addressed by the CDEA, which integrates centralized decision making to optimize resource allocation across multiple DMUs. Furthermore, while DEA evaluates DMU efficiency solely on the basis of their current input–output relationships, without considering potential adjustments in resource allocation within the centralized optimization, CDEA promotes resource allocation optimization among DMUs to increase overall organizational efficiency. Moreover, DEA may encounter difficulties in modeling complex interdependencies and feedback loops among DMUs, whereas the CDEA, with its centralized decision making approach, is better equipped to capture and manage these interdependent relationships. Finally, in decentralized decision making environments where DEA is frequently applied, inefficiencies may arise due to suboptimal resource allocation, a challenge that the CDEA addresses through its centralized resource allocation optimization framework. Overall, while DEA is a powerful tool for assessing the relative efficiency of DMUs, the CDEA offers advantages in scenarios where centralized decision making and resource allocation are significant factors influencing performance.

5.3 REVIEW OF METHODS AND EXTENSIONS IN THE CDEA

CDEA empowers decision makers by providing a systematic approach to evaluate the relative efficiency of DMUs, identify areas for improvement, and optimize resource allocation. This tool is designed to help organizations overcome the complexities of centralized decision making, enhancing overall performance. By employing the CDEA, central management can identify best practices, inefficiencies, and areas for enhancement within the partnership network. To ensure accessibility, this chapter summarizes the 14 various types of CDEA. While detailed explanations may not be possible owing to limited space, interested readers can rest assured that these studies are readily available for a deeper understanding of the formulation.

5.3.1 Basic CDEA Models

Lozano and Villa (2004) developed a CDEA-based approach for resource allocation and target setting. The suggested method can be executed in two stages, adhering to the standard DEA protocol. Initially, the focus is on maximizing the expansion of the aggregated output. Upon identifying the optimal expansion of the aggregated output, the emphasis subsequently shifts to maximizing the slack of the aggregated input. Additionally, Lozano and Villa (2005) introduced two CDEA models, referred to as Models (5.1) and (5.2), aimed at reducing total input consumption and enhancing output production, respectively. The model is summarized as follows. Let j, $r = 1,\ldots,n$ be indexes for DMUs; $i = 1,2,\ldots,m$, be an index for inputs; $o = 1,2,\ldots,O$ be an index for output; x_{ij} be the amount of input i consumed by DMU j; y_{oj} be the quantity of output o produced by DMU j; θ be radial contraction of the total input vector; s_i be slack along the input dimension i; t_o be additional increase along the output dimension o; $\lambda_{1r},\lambda_{2r},\ldots,\lambda_{nr}$ be vector for projecting DMU r.

$$
\begin{aligned}
&\min \theta \\
&\text{s.t.} \\
&\qquad \sum_{r=1}^{n}\sum_{j=1}^{n}\lambda_{jr}x_{ij} \le \theta\sum_{j=1}^{n}x_{ij}; i=1,\ldots,m; \\
&\qquad \sum_{r=1}^{n}\sum_{j=1}^{n}\lambda_{jr}y_{oj} \ge \sum_{j=1}^{n}y_{oj}; o=1,\ldots,O; \\
&\qquad \sum_{j=1}^{n}\lambda_{jr} = 1; r=1,\ldots,n; \\
&\qquad \lambda_{jr} \ge 0; \theta \text{ is free.}
\end{aligned}
\tag{5.1}
$$

$$
\begin{aligned}
&\max \sum_{i=1}^{m}s_i + \sum_{o=1}^{O}t_o \\
&s.t., \\
&\quad \sum_{r=1}^{n}\sum_{j=1}^{n}\lambda_{jr}x_{ij} = \theta^{*}\sum_{j=1}^{n}x_{ij} - s_i; i=1,\ldots,m; \\
&\quad \sum_{r=1}^{n}\sum_{j=1}^{n}\lambda_{jr}y_{oj} = \sum_{j=1}^{n}y_{oj} + t_o; o=1,\ldots,O; \\
&\quad \sum_{j=1}^{n}\lambda_{jr} = 1; r=1,\ldots,n; \\
&\quad \lambda_{jr}, s_i, t_o \ge 0; i=1,\ldots,m; o=1,\ldots,O, j=1,\ldots,n
\end{aligned}
\tag{5.2}
$$

These models differ from traditional DEA models in two key ways. First, they project all DMUs simultaneously. Second, Models (5.1) and (5.2) permit a contraction in total input consumption without any reduction in output production.

5.3.2 MOLP CDEA Model

Under the conditions provided by the CDEA, managers can use reallocation strategies to address specific challenges and opportunities within different sub-DMUs. Korhonen and Syrjänen (2004) demonstrated this practical method by creating an interactive formal strategy that utilizes DEA and multiple-objective linear programming (MOLP) to resolve a resource allocation issue within a centralized decision making framework. Their work, as shown in Model (5.3), serves as a valuable reference, highlighting the relevance and usefulness of our research for a k-th observed DMU.

$$
\begin{aligned}
&\max t_{1k}\\
&\max t_{2k}\\
&\ldots\\
&\max t_{ok}\\
&s.t.\\
&\qquad \sum_{j=1}^{n} \lambda_j y_{oj} = y_{ok} + t_{ok}; o = 1,\ldots,O;\\
&\qquad \sum_{j=1}^{n} \lambda_j x_{ij} = x_{ik} - s_{ik}; i = 1,\ldots,m;\\
&\qquad t_{ok} \le \min\left(\frac{s_{ik}}{x_{ik}} \mid i = 1,\ldots,m\right) y_{ok}; o = 1,\ldots,O;\\
&\qquad s_{ik}, t_{ok} \ge 0; o = 1,\ldots,O; i = 1,\ldots,m;\\
&\qquad \lambda_j \ge 0; j = 1,\ldots,n
\end{aligned}
\tag{5.3}
$$

This assumption implies that a DMU can adjust its outputs to achieve the smallest ratio of $\frac{s_{ij}}{x_{ij}}$. The method was demonstrated to optimize the total output and distribute available resources across 25 supermarkets in Finland.

5.3.3 CDEA Model That Considers Only Inefficiently Adjusted DMUs

Asmild et al. (2009) revised the centralized models of Lozano and Villa (2004, 2005) to focus on adjusting inefficient DMUs (set I). They utilized their model (Model (5.4)) for resource allocation in a public service organization.

$$
\begin{aligned}
&\min \theta \\
&\text{s.t.} \\
&\qquad \sum_{j\in I}\sum_{e\in E}\lambda_{je}x_{ek} \le \theta \sum_{j\in I} x_{ij}; i = 1,\ldots,m; \\
&\qquad \sum_{j\in I}\sum_{e\in E}\lambda_{ie}y_{eo} \ge \sum_{j\in I} y_{jo}; o = 1,\ldots,O; \\
&\qquad \sum_{e\in E}\lambda_{ie} = 1; i = 1,\ldots,m; \\
&\qquad \lambda_{ie} \ge 0, \forall i \in I; \forall e \in E
\end{aligned}
\tag{5.4}
$$

where E means only considering the efficient set, and I means only considering the inefficient set.

5.3.4 Non-Radial CDEA

Hosseinzadeh Lotfi et al. (2010) introduced an innovative CDEA model, improving upon the Russell model, to either increase non-radial total output production or decrease non-radial total input consumption, as shown in Model (5.5).

$$
\begin{aligned}
&\min \tau = \frac{\frac{1}{m}\sum_{i=1}^{m}\theta_i}{\frac{1}{O}\sum_{o=1}^{O}\psi_o} \\
&\text{s.t.} \\
&\qquad \sum_{r=1}^{n}\sum_{j=1}^{n}\lambda_{jr}x_{ij} \le \theta_i \sum_{j=1}^{n} x_{ij}; i = 1,\ldots,m; \\
&\qquad \sum_{r=1}^{n}\sum_{j=1}^{n}\lambda_{jr}y_{oj} \ge \psi_o \sum_{j=1}^{n} y_{oj}; o = 1,\ldots,O; \\
&\qquad \sum_{j=1}^{n}\lambda_{jr} = 1; r = 1,\ldots.,n; \\
&\qquad \lambda_{jr} \ge 0; \theta \text{ is free}; \\
&\qquad \theta_i \le 1; i = 1,\ldots.,m; \\
&\qquad \psi_o \ge 1; o = 1,\ldots,O
\end{aligned}
\tag{5.5}
$$

5.3.5 Non-radial CDEA Extension Model

Using the CDEA for centralized target setting in business partnerships offers a systematic, data-driven approach to boost intraorganizational efficiency. This method enables organizations to optimize performance, allocate resources effectively, and promote sustainable growth within their partnership network. To achieve this, Lozano et al. (2011) proposed a non-radial CDEA model (Model (5.6)) to set the total output target for the Spanish port agency.

$$\max \frac{1}{O}\sum_{o=1}^{O}\gamma_o$$

$$s.t$$

$$\begin{aligned}
&\sum_{j=1}^{n}\lambda_{ij}x_{ij} \le \widehat{x}_{ir}; r=1,\ldots,n; i=1,\ldots,m;\\
&\widehat{x}_{ir} = x_{ir} + s_{ir}^{+}; \forall i \notin NA; r=1,\ldots,n;\\
&\widehat{x}_{ir} = x_{ir} + s_{ir}^{+} - s_{ir}^{-}; \forall i \notin NA; r=1,\ldots,n;\\
&z_i = \sum_{r=1}^{n} s_{ir}^{+}; \forall i \in NA;\\
&z_i = \sum_{r=1}^{n} s_{ir}^{+} - \sum_{r=1}^{n} s_{ir}^{-}; \forall i \notin NA;\\
&\sum_{i\in I^{realloc}} \alpha_i \sum_{r=1}^{n} s_{ir}^{-} + \sum_{i=1}^{m} c_i z_i \le B; \qquad (5.6)\\
&\sum_{j=1}^{n}\lambda_{jr}y_{oj} \ge \widehat{y}_{or}; o=1,\ldots,O; r=1,\ldots,n;\\
&\widehat{y}_{or} \ge y_{or}; o=1,\ldots,O; r=1,\ldots,n;\\
&\sum_{r=1}^{n}\widehat{y}_{or} \ge \gamma_o \sum_{j=1}^{n}\lambda_{jr}y_{oj}; o=1,\ldots,O;\\
&\sum_{j=1}^{n}\lambda_{jr} = 1; r=1,\ldots,n;\\
&\lambda_{jr} \ge 0; j=1,\ldots,n; r=1,\ldots,n;\\
&s_{ir}^{+} \ge 0; \forall j \forall r, s_{ir}^{-} \ge 0; \forall i \notin NA, r=1,\ldots,n;\\
&z_i \ge 0, \forall i, \widehat{x}_{ir} \ge 0, \forall i \forall r; \widehat{y}_{or} \ge 0, o=1,\ldots,O; r=1,\ldots,n
\end{aligned}$$

where c_i denotes the unit cost of input i; B denotes the available budget for the acquisition of additional inputs; NA denotes the nonadjustable input variables; $\widehat{x}_{ir}$ represents the target for input i of unit r; $\widehat{y}_{or}$ represents the target for output o of unit r; γ_o represents the percentage increase in the total amount of output o; α_i represents the unit cost of reallocating input i; and z_i represents the total amount of input i to be acquired.

5.3.6 Stochastic CDEA Model

In some cases, resources may not be easy to identify. However, CDEA offers a compelling solution for addressing the complexities introduced by stochastic inputs and outputs within a centralized decision making framework. Unlike traditional DEA models that assume dcterministic inputs and outputs, CDEA incorporates stochastic elements into the analysis, providing a more robust framework for decision making under uncertainty. This approach enables organizations to evaluate the efficiency of DMUs while accounting for variability in inputs and outputs, thereby offering a more accurate representation of performance. The benefits are significant, allowing for better-informed resource allocation decisions, improved risk management strategies, and enhanced overall efficiency in the face of stochasticity. For example, Hosseinzadeh Lotfi et al. (2012) presented a stochastic CDEA model to allocate resources under stochastic inputs and outputs. Assume that $\tilde{x}_j = \left(\tilde{x}_{1j},\ldots,\tilde{x}_{mj}\right)^T$ and $\tilde{y}_j = \left(\tilde{y}_{1j},\ldots,\tilde{y}_{oj}\right)^T$ are the stochastic input and output vectors, respectively. Additionally, let $x_j = \left(x_{1j},\ldots,x_{mj}\right)^T$ and $y_j = \left(y_{1j},\ldots,y_{oj}\right)^T$ be the mean input and output vectors, respectively. The probabilistic constraint from the CDEA model with stochastic data is Model (5.6).

$$
\begin{aligned}
&\min \theta \\
&s.t.\, p\left\{\sum_{j=1}^{n}\sum_{k=1}^{n}\lambda_{jk}\tilde{x}_{ij} \le \theta\sum_{j=1}^{n}\tilde{x}_{ij}\right\} \ge 1-\alpha; i = 1,\ldots,m; \\
&p\left\{\sum_{j=1}^{n}\sum_{k=1}^{n}\lambda_{jk}\tilde{y}_{oj} \ge \sum_{j=1}^{n}\tilde{y}_{oj}\right\} \ge 1-\alpha; o = 1,\ldots,O; \\
&\sum_{j=1}^{n}\lambda_{jk} = 1; k = 1,\ldots,n; \\
&\lambda_{jk} \ge 0, k = 1,\ldots,n; j = 1,\ldots,n
\end{aligned}
\tag{5.6}
$$

where in Model (5.6), p is the probability and α is the level of error between 0 and 1, which is a predetermined number.

5.3.7 Generalized CDEA Model

The development of a generalized CDEA framework is crucial for addressing the diverse and complex nature of decision making environments. Its adaptability enables organizations to tailor the analysis to their specific context, capturing the unique nuances in their performance evaluation. By developing a generalized CDEA approach, organizations can overcome the limitations of traditional models and make more informed decisions to improve efficiency and productivity. Fang (2013) expanded the models from Lozano and Villa (2004) and Asmild et al. (2009) to develop a generalized CDEA model for resource allocation.

$$\begin{aligned}
&\min \theta \\
&s.t. \sum_{j=1}^{n}\sum_{r=1}^{n} \lambda_{jr} x_{ji} \leq \theta \sum_{j=1}^{n} b_j x_{ji}; r = 1,\ldots,n; i \notin NA; \\
&\sum_{r=1}^{n} \lambda_{jr} x_{ji} \leq x_{ji}; j = 1,\ldots,n; i \in NA; \\
&\sum_{j=1}^{n}\sum_{r=1}^{n} \lambda_{jr} y_{jo} \geq \sum_{j=1}^{n} b_j y_{oj}; o \notin NT; \\
&\sum_{r=1}^{n} \lambda_{jr} y_{jo} \geq y_{jo}; j = 1,\ldots,n; o \in NT; \\
&\sum_{j=1}^{n} \lambda_{jr} = b_r; r = 1,\ldots,n; \\
&\lambda_{jr} \geq 0, r = 1,\ldots,n; j = 1,\ldots,n
\end{aligned} \tag{5.7}$$

In Model (5.7), $b_j \in [0,1]$ are user-provided constraints for DMU *j*. The central management decision-maker will decide which DMUs to consider in the overall optimization by setting $b_j = 1$ accounting for different contexts; otherwise, $b_j = 0$. The second constraint implies that any decrease in input is allowed for nonadjustable input variables. NT denotes the nontransferable output variables, and a further constraint implies that only an output increase is allowed for the nontransferable output variables.

5.3.8 Multiphase CDEA Model

Adopting a multiterm approach encompassing long-term, middle-term, and short-term perspectives in CDEA is crucial for comprehensively addressing the dynamics and complexities of decision making environments. By integrating long-term, middle-term, and short-term perspectives into the CDEA, organizations can develop a holistic understanding of their performance dynamics, implement targeted interventions across different time horizons, and continuously improve efficiency and competitiveness in dynamic decision making environments. Following the above concepts, other policies may exist for various phases. Yu et al. (2013) proposed CDEA models based on Russell's measure and the maximization of total output production to reallocate human resources across 18 Taiwanese airports under short-term, medium-term, and long-term policies, including only the long-term cases in their study. We included only long-term cases; short- and medium-term cases can be found in the study of Yu et al. (2013).

$$
\begin{aligned}
&\max \frac{1}{O}\sum_{o=1}^{O}\theta_{io} \\
&s.t. \\
&\qquad \sum_{r=1}^{n}\sum_{j=1}^{n}\lambda_{jr}x_{ij} \le \sum_{r=1}^{n}x_{ir};i \notin NA; \\
&\qquad \sum_{j=1}^{n}\lambda_{jr}x_{ij} \le x_{ir};i \in NA;r = 1,\ldots,n; \\
&\qquad \sum_{r=1}^{n}\sum_{j=1}^{n}\lambda_{jr}y_{oj} \ge \theta_o\sum_{r=1}^{n}y_{or};o \notin NT; \\
&\qquad \sum_{j=1}^{n}\lambda_{jr}y_{oj} \ge y_{oj};o \in NT;j = 1,\ldots,n; \\
&\qquad \sum_{j=1}^{n}\lambda_{jr} = 1;r = 1,\ldots,n; \\
&\qquad \lambda_{jr} \ge 0, j = 1,\ldots,n;r = 1,\ldots,n
\end{aligned}
\tag{5.8}
$$

Long-term perspectives in the CDEA enable the identification of systemic inefficiencies, guiding structural changes and strategic initiatives to enhance overall efficiency and competitiveness in the long run. The middle-term perspective bridges the gap between long-term strategic goals and short-term operational considerations. It enables organizations to evaluate the effectiveness of mid-range strategies, resource allocation decisions, and operational adjustments in achieving long-term objectives. By incorporating middle-term perspectives, CDEA facilitates the identification of operational bottlenecks, process improvements, and resource reallocation opportunities to optimize efficiency and performance. Short-term considerations focus on immediate operational concerns, such as day-to-day management, resource utilization, and performance monitoring. Adopting a short-term perspective in the CDEA allows organizations to identify and address operational inefficiencies quickly, adapt to changing market conditions, and implement timely corrective actions to improve efficiency and productivity in the short term. It provides actionable insights for operational decision making, ensuring alignment with broader strategic objectives and long-term performance goals.

$$
\begin{aligned}
&\max \sum_{r=1}^{n} \left(s_{ir}^{+} - s_{ir}^{-} \right) \\
&s.t. \\
&\quad \sum_{r=1}^{n} \sum_{j=1}^{n} \lambda_{jr} x_{ij} = \sum_{r=1}^{n} \left(x_{ir} + s_{ir}^{+} - s_{ir}^{-} \right); i \notin NA; \\
&\quad \sum_{j=1}^{n} \lambda_{jr} x_{ij} = x_{ir} + s_{ir}^{+} - s_{ir}^{-}; r = 1, \ldots, n; i \notin NA; \\
&\quad \sum_{r=1}^{n} s_{ir}^{+} \leq \sum_{r=1}^{n} s_{ir}^{-}; i \notin NA; \\
&\quad \sum_{j=1}^{n} \lambda_{jr} x_{ij} \leq x_{ir}; i \in NA; \\
&\quad \sum_{r=1}^{n} \sum_{j=1}^{n} \lambda_{jr} y_{oj} = \theta_{or}^{*} \sum_{r=1}^{n} y_{or}; o \notin \mathrm{NT}; \\
&\quad \sum_{j=1}^{n} \lambda_{jr} y_{oj} \geq y_{or}; o \in \mathrm{NT}; \\
&\quad \sum_{j=1}^{n} \lambda_{jr} = 1; r = 1, \ldots, n; \\
&\quad \lambda_{jr} \geq 0, j = 1, \ldots, n; r = 1, \ldots, n
\end{aligned}
\tag{5.9}
$$

5.3.9 Directional Distance Function CDEA

The development of a distance function version of the CDEA enhances flexibility, enables the assessment of non-radial efficiency measures, accommodates directional inputs and outputs, improves robustness, and facilitates policy analysis, making it a valuable tool for efficiency assessment in decision making contexts. Tohidi et al. (2014) introduced a new CDEA model using a directional distance function to address undesirable outputs $\left(b_{tj};t=1,\ldots,T;j=1,\ldots,n\right)$ in Model (5.10).

$$
\begin{aligned}
&\max \theta \\
&\text{s.t.} \\
&\quad \sum_{r=1}^{n}\sum_{j=1}^{n}\lambda_{jr}x_{ij} \le (1-\theta)\sum_{j=1}^{n}x_{ij};i=1,\ldots,m; \\
&\quad \sum_{r=1}^{n}\sum_{j=1}^{n}\lambda_{jr}y_{oj} \ge (1+\theta)\sum_{j=1}^{n}y_{oj};o=1,\ldots,O; \\
&\quad \sum_{r=1}^{n}\sum_{j=1}^{n}\lambda_{jr}b_{tj} \le (1-\theta)\sum_{j=1}^{n}b_{tj};t=1,\ldots,T; \\
&\quad \sum_{j=1}^{n}\lambda_{jr}=1;r=1,\ldots,n; \\
&\quad \lambda_{jr}\ge 0,r=1,\ldots,n,j=1,\ldots,n
\end{aligned}
\tag{5.10}
$$

5.3.10 Revenue-Centric CDEA

The development of a CDEA model based on revenue efficiency for resource allocation provides a revenue-focused perspective, comprehensive performance evaluation, strategic resource allocation, competitive benchmarking, and alignment with organizational goals. For instance, Feng et al. (2015) developed a CDEA model based on revenue efficiency for resource allocation in a centralized decision making environment (Model (5.11)).

$$
\begin{aligned}
&\max \psi = \sum_{r=1}^{n}\sum_{o=1}^{O}p_{or}y_{or}+\varepsilon\sum_{i\notin NA}\sum_{r=1}^{n}s_{ir} \\
&s.t. \\
&\quad \sum_{j=1}^{n}\lambda_{jr}x_{ij}\le x_{ir};r=1,\ldots,n;i\in NA; \\
&\quad \sum_{j=1}^{n}\lambda_{jr}x_{ij}+s_{ir}=x_{ir};r=1,\ldots,n;i\notin NA; \\
&\quad \sum_{j=1}^{n}\lambda_{jr}y_{oj}=y_{or};r=1,\ldots,n;o=1,\ldots,O; \\
&\quad \sum_{j=1}^{n}\lambda_{jr}=1;r=1,\ldots,n; \\
&\quad \lambda_{jr}\ge 0;r=1,\ldots,n;j=1,\ldots,n
\end{aligned}
\tag{5.11}
$$

where p_{or} denotes the price of output r, s_{ir} indicates the slack of input i for DMU r, and ε is a non-Archimedean element smaller than any positive real number. The first two constraints of Model (5.11) indicate that the reallocation does not change the original production possibility set.

The objective function of Model (5.12) seeks to find the minimum reallocation cost between DMUs.

$$
\begin{aligned}
&\min \Delta = \sum_{i\in NA}\sum_{r=1}^{n} c_{ir}^{-}\gamma_{ir}^{-} + \sum_{i\notin NA}\sum_{k=1}^{n} c_{ir}^{+}\gamma_{ir}^{+} \\
&s.t. \\
&\quad \sum_{j=1}^{n} \lambda_{jr} x_{ij} \le x_{ir}; r = 1,\ldots,n; i \in NA; \\
&\quad \sum_{j=1}^{n} \lambda_{jr} x_{ij} + s_{ir} = x_{ir}; r = 1,\ldots,n; i \notin NA; \\
&\quad \sum_{j=1}^{n} \lambda_{jr} y_{oj} = y_{ro}; r = 1,\ldots,n; o = 1,\ldots,O; \\
&\quad \sum_{r=1}^{n}\sum_{o=1}^{O} p_{or} y_{ro} + \varepsilon \sum_{i\notin NA}\sum_{r=1}^{n} s_{ir} = \psi^{*}; \\
&\quad x_{ir} + \gamma_{ir}^{-} + \gamma_{ir}^{+} = x_{ir}; r = 1,\ldots,n; i \notin NA; \\
&\quad \sum_{r=1}^{n} \lambda_{jr} = 1; r = 1,\ldots,n; \\
&\quad \lambda_{jr} \ge 0; r = 1,\ldots,n; j = 1,\ldots,n; \\
&\quad \gamma_{ir}^{-} \ge 0; r = 1,\ldots,n; i \notin NA; \\
&\quad \gamma_{ir}^{+} \ge 0; r = 1,\ldots,n; i \notin NA
\end{aligned}
\tag{5.12}
$$

where ψ^{*} is the optimal value for Model (5.11); γ_{ir}^{-} and γ_{ir}^{+} denote the negative deviation and positive deviation from the current level of input i for DMU r, respectively; and c_{ir}^{-} and c_{ir}^{+} denote the costs of moving one unit of input i from DMU r and into DMU r, respectively.

5.3.11 CDEA Model Considering the Optimum Number of DMUs

The need for the CDEA to accommodate nontransferable inputs and environmental factors is paramount because of its significant impact on organizational efficiency assessment and resource allocation. Nontransferable inputs, such as managerial skills or location-specific resources, cannot be easily transferred between DMUs, making their evaluation challenging. By incorporating these inputs into CDEA models, organizations can accurately assess the efficiency of DMUs and allocate resources more effectively. Additionally, environmental factors, such as regulatory constraints or market conditions, influence organizational performance but are often beyond managerial control. CDEA's ability to account for ecological factors enables organizations to evaluate efficiency within external influences, leading to more informed decision making and strategic planning. López-Torres and Prior (2016) introduced a CDEA model to manage nontransferable inputs and environmental factors for reallocating human resources in a public school in Spain, demonstrating its practical effectiveness (Model (5.13)).

$$\begin{aligned}
&\min \theta \\
&s.t. \\
&\qquad \sum_{j=1}^{n} \lambda_j x_{ij} \le \theta \sum_{j=1}^{n} x_{ij}; i = 1,\ldots,m; \\
&\qquad \sum_{j=1}^{n} \lambda_j x_{ij} \le \sum_{j=1}^{n} x_{ij}; i \in NA; \\
&\qquad \sum_{j=1}^{n} \lambda_j y_{oj} \ge \sum_{j=1}^{n} y_{oj}; o = 1,\ldots,O; \\
&\qquad \sum_{j=1}^{n} \lambda_j = n^*; \\
&\qquad \lambda_j \ge 0; \theta \; is \; free.
\end{aligned} \tag{5.13}$$

In Model (5.13), n^* denotes the optimum number of DMUs that the network should have to be more efficient. Once we obtain this n^*, we can decide how to reallocate the resources among the number of DMUs.

5.3.12 Network CDEA Model

Network CDEA is necessary in complex decision making environments characterized by interdependent units and network structures. This approach offers a holistic perspective on organizational performance and resource allocation by considering the interactions and dependencies among DMUs within a centralized network. The network CDEA enables organizations to identify synergies, inefficiencies, and optimal resource allocation strategies across interconnected units, improving overall efficiency and productivity. Its ability to capture the interplay between DMUs in a centralized network allows for more accurate performance evaluations and strategic decision making. By leveraging network CDEA, organizations can uncover hidden inefficiencies, exploit synergies, and optimize resource utilization across the entire network, ultimately enhancing competitiveness and achieving sustainable growth.

Yu and Chen (2016) proposed a network CDEA model considering undesirable outputs to allocate resources across 14 Asian shipping routes operated by a Taiwanese container shipping company. To minimize organizational resistance, they presented minor and major adjustment policies. The minor policy maintained the original levels of input resources, whereas the central policy could reduce total input resources. The container shipping company's operational structure, which is based on common inputs shared among routes, is depicted in Figure 5.1. Model (5.14) corresponds to the first phase, and Model (5.15) corresponds to the second.

Yu and Chen (2016) sought maximal increases in desirable outputs and maximal decreases in undesirable outputs in the first phase, with

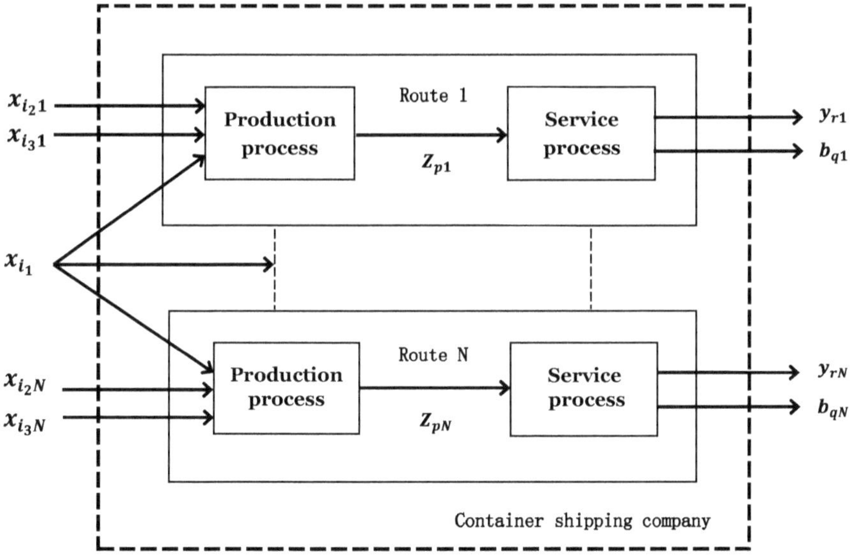

FIGURE 5.1 Operational structure.

Source: From Yu and Chen, 2016.

additional input reductions in the second phase. The proposed CNDEA model is formulated as follows:

$$\max \frac{\sum_{r=1}^{R} \theta_r}{R} + \frac{\sum_{t=1}^{T} \delta_t}{T} \tag{5.14}$$

s.t.

Common input constraints

$$\sum_{j=1}^{n} \lambda_{jk}^{1} \alpha_{i1} x_{i1j} \leq \alpha_{i1k} x_{i1}; i_1 = 1,2,\ldots,I_1; k = 1,\ldots,n$$

Specific input constraint

$$\sum_{j=1}^{n} \lambda_{jk}^{1} x_{i2j} \leq x_{i2k}; i_2 = 1,\ldots,I_2; k = 1,\ldots,n$$

Energy input constraint

$$\sum_{j=1}^{n} \lambda_{jk}^{1} x_{i3j} \leq x_{i3k}; i_3 = 1,\ldots,I_3; k = 1,\ldots,n$$

Intermediate product constraints

$$\sum_{j=1}^{n} \lambda_{jk}^{1} z_{pj} \geq z_{pk}; p = 1,\ldots,P; k = 1,\ldots,n;$$

$$\sum_{j=1}^{n} \lambda_{jk}^{2} z_{pj} \leq z_{pk}; p = 1,\ldots,P; k = 1,\ldots,n;$$

$$\sum_{j=1}^{n} \lambda_{jk}^{1} z_{pj} \geq \sum_{j=1}^{N} \lambda_{jk}^{2} z_{pj}; p = 1,\ldots,P; k = 1,\ldots,n$$

Desirable output constraints

$$\sum_{k=1}^{n}\sum_{j=1}^{n}\lambda_{jk}^{2}y_{rj} \geq (1+\theta_r)\sum_{k=1}^{n}y_{rk};r=1,\ldots,R;$$

$$\sum_{j=1}^{n}\lambda_{jk}^{2} \geq y_{rk};r=1,\ldots,R;k=1,\ldots,n$$

Undesirable output constraints

$$\sum_{k=1}^{n}\sum_{j=1}^{n}\lambda_{jk}^{2}b_{tj} = (1-\delta_t)\sum_{k=1}^{n}b_{tk};t=1,\ldots,T$$

VRS constraints

$$\sum_{j=1}^{n}\lambda_{jk}^{1} = 1;k=1,\ldots,n;$$

$$\sum_{j=1}^{n}\lambda_{jk}^{2} = 1;k=1,\ldots,n$$

Shared rate constraints

$$\tilde{\alpha}_{i1j} \leq \alpha_{i1j} \leq \hat{\alpha}_{i1j};i_1=1,\ldots,I_1;j=1,\ldots,n;$$

$$\sum_{j=1}^{n}\alpha_{i1j} = 1,i_1=1,\ldots,I_1;$$

$$\lambda_{jk}^{1},\lambda_{jk}^{2} \geq 0;j=1,..,n;k=1,\ldots,n$$

where R represents the number of desirable outputs and T represents the number of undesirable outputs. Additionally, $\lambda_{1k}^{1},\ldots,\lambda_{jk}^{1}$ represents the intensity variables for projecting route k in the production process, and $\lambda_{1k}^{2},\ldots,\lambda_{jk}^{2}$ represents the intensity variables for projecting route k in the service process. If θ_r^*, δ_q^*, and $\alpha_{i1,k}^*$ denote the optimal value from Model (5.14), the phase II model for the major adjustment policy can be solved by Model (5.15).

$$\max \sum_{i_1=1}^{I_1}\sum_{k=1}^{n}\frac{\left(s_{i1k}^- - s_{i1k}^+\right)}{\alpha_{i1k}^* x_{i1}} + \sum_{i_2=1}^{I_2}\sum_{k=1}^{n}\frac{\left(s_{i2k}^- - s_{i2k}^+\right)}{x_{i2k}} \tag{5.15}$$

s.t.

Common input constraints

$$\sum_{j=1}^{n}\lambda_{jk}^1 \alpha_{i1j}^* x_{i1} = \alpha_{i1k}^* x_{i1} - s_{i1k}^- + s_{i1k}^+; i_1 = 1,\ldots,I_1; k = 1,\ldots,n;$$

$$\sum_{k=1}^{n} s_{i1k}^- = \sum_{k=1}^{n} s_{i1k}^+; i_1 = 1,\ldots,I_1;$$

$$\sum_{j=1}^{n}\lambda_{jk}^1 x_{i2} = x_{i2} - s_{i2k}^- + s_{i2k}^+; i_2 = 1,\ldots,I_2; k = 1,\ldots,n;$$

$$\sum_{k=1}^{n} s_{i2k}^- = \sum_{k=1}^{n} s_{i2k}^+; i_2 = 1,\ldots,I_2$$

Energy input constraint

$$\sum_{j=1}^{n}\lambda_{jk}^1 x_{i3} \le \left(1-\delta_{i3q}^*\right)x_{i3k}; i_3 = 1,\ldots,I_3 k = 1,\ldots,n$$

Intermediate product constraint

$$\sum_{k=1}^{n}\lambda_{jk}^1 z_{pj} \ge \sum_{k=1}^{n}\lambda_{jk}^2 z_{pj}; p = 1,\ldots,P; k = 1,\ldots,n$$

Desirable output constraints

$$\sum_{k=1}^{n}\sum_{j=1}^{n}\lambda_{jk}^{2}y_{rj} \geq \left(1+\theta_r^*\right)\sum_{k=1}^{n}y_{rk};r=1,\ldots,R;$$

$$\sum_{j=1}^{n}\lambda_{jk}^{2}y_{rj} \geq y_{rk};r=1,\ldots,R;k=1,\ldots,n$$

Undesirable output constraint

$$\sum_{k=1}^{n}\sum_{j=1}^{n}\lambda_{jk}^{2}b_{tj} = \left(1-\delta_t^*\right)\sum_{k=1}^{n}b_{tk};t=1,\ldots,T;$$

$$\sum_{j=1}^{n}\lambda_{jk}^{2}b_{tj} = \left(1-\delta_t^*\right)b_{tk};t=1,\ldots,T;k=1,\ldots,n$$

VRS constraints

$$\sum_{j=1}^{n}\lambda_{jk}^{1} = 1;k=1,\ldots,n;$$

$$\sum_{j=1}^{n}\lambda_{jk}^{2} = 1;k=1,\ldots,n;$$

$$\lambda_{jk}^{1},\lambda_{jk}^{2} \geq 0;j=1,\ldots,n;k=1,\ldots,n;$$

$$s_{i1k}^{-},s_{i1k}^{+},s_{i2k}^{-},s_{i2k}^{+} \geq 0;i_1=1,\ldots,I_1;i_2=1,\ldots,I_2;k=1,\ldots,n$$

5.3.13 Meta-, Group-Frontiers CDEA Model

Technological heterogeneity is essential in CDEA because it significantly impacts efficiency assessment and resource allocation in diverse decision making environments. Technological heterogeneity is the variation in technology or production processes across DMUs within a centralized structure. The CDEA acknowledges that different DMUs may have unique production processes, input–output relationships, and efficiency potentials by accounting for technological heterogeneity. This recognition allows for a more accurate evaluation of efficiency levels, as it ensures that performance assessments are tailored to the specific characteristics of each DMU. Additionally, considering technological heterogeneity enables organizations to identify best practices, benchmark against peers, and implement targeted improvement strategies aligned with the unique technological profiles of individual DMUs. Ding et al. (2018) considered technological heterogeneity to propose a new CDEA model for resource allocation (Model (5.16)).

$$
\begin{aligned}
&\theta_d^{ME} = \max \frac{\sum_{o=1}^{O} u_o y_{ok}}{\sum_{i=1}^{m} v_i x_{ik}} \\
&\text{s.t.} \\
&\qquad \frac{\sum_{o=1}^{O} u_o y_{oj}}{\sum_{i=1}^{m} v_i x_{ij}} \leq 1; j = 1,\ldots,n; \\
&\qquad u_o, v_i; o = 1,\ldots,O; i = 1,\ldots,m
\end{aligned}
\tag{5.16}
$$

where θ_d^{ME} represents the meta-inefficiency score of DMU d.

$$
\begin{aligned}
&\theta_d^{GE} = \max \frac{\sum_{o=1}^{O} u_o y_{ok}}{\sum_{i=1}^{m} v_i x_{ik}} \\
&\text{s.t.} \\
&\qquad \frac{\sum_{o=1}^{O} u_o y_{oj}}{\sum_{i=1}^{m} v_i x_{ij}} \leq 1; j = 1,\ldots,N_k; \\
&\qquad u_o, v_i; o = 1,\ldots,O; i = 1,\ldots,m
\end{aligned}
\tag{5.17}
$$

where θ_d^{GE} represents the group-efficiency score of DMU d. The number of DMUs in the k-th group is N_k such that $\sum_{k=1}^{J} N_k = n$

$$\max \sum_{o=1}^{O} v_o \sum_{j=1}^{n} y_{oj}$$
$$s.t. \sum_{o=1}^{O} v_o y_{oj} - \theta_j^{GE*}\left(\sum_{i=1}^{m'} u_i x_{ij} + \hat{r}_j\right) \le 0; j = 1,\ldots,n;$$
$$\sum_{o=1}^{O} v_o y_{oj} - \theta_j^{ME*}\left(\sum_{i=1}^{m'} u_i x_{ij} + \hat{r}_j\right) \ge 0; j = 1,\ldots,n;$$
$$\sum_{i=1}^{n} u_i \sum_{j=1}^{n} x_{ij} + u_{m+1}R = 1;$$
$$\sum_{j=1}^{n} \hat{r}_j = u_{m+1}R;$$
$$u_{m+1}r_j^- \le \hat{r}_j \le u_{m+1}r_j^+; j = 1,\ldots,n;$$
$$u_i, v_r, \hat{r}_j \ge 0; i = 1,\ldots,n; r = 1,\ldots,n; j = 1,\ldots,n \tag{5.18}$$

Suppose that there is a fixed cost R that needs to be allocated. The allocated cost to DMU_j is denoted as r_j, i.e., $\sum_{j=1}^{n} r_j = R$. If the allocated cost R is treated as a new input, then the relative efficiency of DMU d can be denoted by $\frac{\sum_{o=1}^{O} v_o y_{ok}}{\sum_{i=1}^{m} u_i x_{ik} + u_{m+1} r_k}$.

5.3.14 Single-Stage CDEA Model

The necessity of a single-stage CDEA arises from its ability to streamline efficiency evaluation and resource allocation processes within centralized decision making structures. By consolidating multiple decision making stages into a single analysis, the single-stage CDEA simplifies the assessment process, reduces computational complexity, and enhances resource allocation efficiency. This approach allows decision makers to holistically evaluate the overall efficiency of the entire decision making structure rather than focusing on individual stages. Single-stage CDEA models provide a comprehensive understanding of efficiency dynamics, help organizations identify inefficiencies, optimize resource allocation, and drive performance improvement. For example, Yu and Hsiao (2018) modified the two-phase CDEA model by Yu et al. (2013) to propose a single-phase slack-based CDEA model for resource allocation (Model (5.19)).

$$\min \theta = \frac{1 - \frac{1}{m}\sum_{i=1}^{m}\left(\frac{\left|s_{ir}^{-} - s_{ir}^{+}\right|}{x_{ir}}\right)}{1 + \frac{1}{O}\sum_{r=1}^{n}\sum_{o=1}^{O}\left(\frac{t_{or}}{y_{or}}\right)}$$

$$\begin{aligned}
&s.t.\\
&\sum_{r=1}^{n}\sum_{j=1}^{n}\lambda_{jr}x_{ij} = \sum_{j=1}^{n}\left(x_{ij} + s_{ir}^{+} - s_{ir}^{-}\right); r = 1,\ldots,n; i \notin NA;\\
&\sum_{j=1}^{n}\lambda_{jr}x_{ij} = x_{ir} + s_{ir}^{+} - s_{ir}^{-}; r = 1,\ldots,n; i \notin NA;\\
&\sum_{j=1}^{n}s_{ij}^{+} \leq \sum_{j=1}^{n}s_{ij}^{-}; i \notin NA;\\
&\sum_{j=1}^{n}\lambda_{jr}x_{ij} \leq x_{ir}; r = 1,\ldots,n; i \in NA;\\
&\sum_{r=1}^{n}\sum_{j=1}^{n}\lambda_{jr}y_{oj} = \sum_{r=1}^{n}\left(y_{or} + t_{or}\right); o = 1,\ldots,O;\\
&\sum_{j=1}^{n}\lambda_{jr}y_{oj} = y_{or} + t_{or}; r = 1,\ldots,n; o = 1,\ldots,O;\\
&\sum_{j=1}^{n}\lambda_{jr} = 1; r = 1,\ldots,n;\\
&\lambda_{jr} \geq 0; j = 1,\ldots,n; r = 1,\ldots,n;\\
&s_{ir}^{+}, s_{ir}^{-}; i \notin NA; r = 1,\ldots,n;\\
&t_{or}; o = 1,\ldots,O; r = 1,\ldots,n
\end{aligned} \tag{5.19}$$

CDEA methods represent a comprehensive approach to efficiency evaluation and resource allocation within centralized decision making structures. By considering the interdependencies among decision making

DMUs and incorporating various factors, such as technological heterogeneity, nontransferable inputs, and environmental factors, the CDEA provides a holistic perspective on organizational performance. These methods enable organizations to identify inefficiencies, optimize resource allocation, and drive performance improvement initiatives more effectively. By leveraging CDEA methods, organizations can enhance efficiency, foster collaboration, and achieve sustainable growth in dynamic and interconnected settings.

5.4 OVERVIEW OF PREVIOUS STUDIES USING CDEA

This study carefully defines the criteria for selecting relevant literature focusing on inclusivity and precision. Accepted materials include journal articles, books, book chapters, and conference proceedings, ensuring a comprehensive dataset. However, trade publications, editorial reviews, and review articles were excluded from the analysis. This rigorous selection, as presented in Table 5.1, uses the Scopus database, adhering strictly to English-language sources to maintain consistency and reliability. The scope of the literature is limited to works published up to 2023, with the search concluded on 6 August 2023. With the suggested criteria, a total of 62 articles related to CDEA were included in the study.

5.4.1 Publication Trends

Figure 5.2 provides an overview of the research published using CDEA from 2001 through 2023. We observe an initial period of modest activity, followed by a significant increase in publications around the mid-2000s. This surge likely corresponds with key methodological advancements or increased recognition of the CDEA's applicability across diverse fields such as healthcare, education, and public services. A noticeable feature of the number of publications is the peak at approximately 2015–2020,

TABLE 5.1 Inclusion and Exclusion Criteria

Criterion	Inclusion	Exclusion
Literature type	Journal article, book, book chapter, conference proceeding	Trade publications, review article, editorial review
Database	Scopus	Non-Scopus
Language	English	Non-English
Timeline	<2023	>2023
Date of search	6 August 2023	>6 August 2023

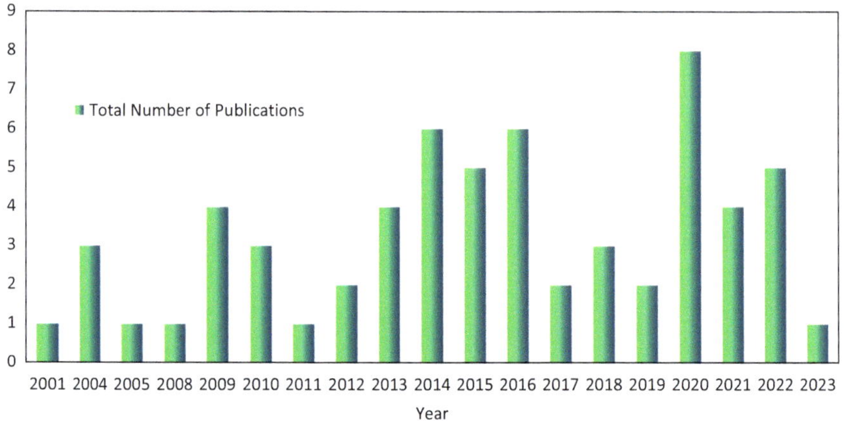

FIGURE 5.2 Total number of publications in the field.

suggesting that CDEA research may be driven by global academic and industrial interests in optimizing resource allocation and operational efficiency. The slight decline in recent years could indicate a maturation of the field or a shift toward integrating DEA with other emerging data analytics technologies.

5.4.2 Authorship

Table 5.2 highlights the authors who have contributed significantly to the literature on the CDEA model, indicating both their publication frequency and their influence through citation numbers. Notably, Lozano emerges as the leading contributor, with nearly five weighted publications and the highest citation count, suggesting his prominent role in advancing the study of the CDEA. However, Fang and Villa follow closely, both in terms of publications and citations, underscoring their substantial contributions to the field. The involvement of authors such as Hsiao and

TABLE 5.2 The Contribution to the Number of Publications of Authors

Author	Publications	Weighted Publications (%)	Citations
Lozano S	4.83 (1)	7.80	257.33 (1)
Fang L	4.00 (2)	6.45	224.50 (3)
Villa G	2.33 (3)	3.76	227.83 (2)
Hsiao B	2.03 (4)	3.28	33.63 (6)
Yu MM	2.03 (4)	3.28	46.63 (4)
Hosseinzadeh Lotfi F	1.77 (6)	2.85	39.85 (5)

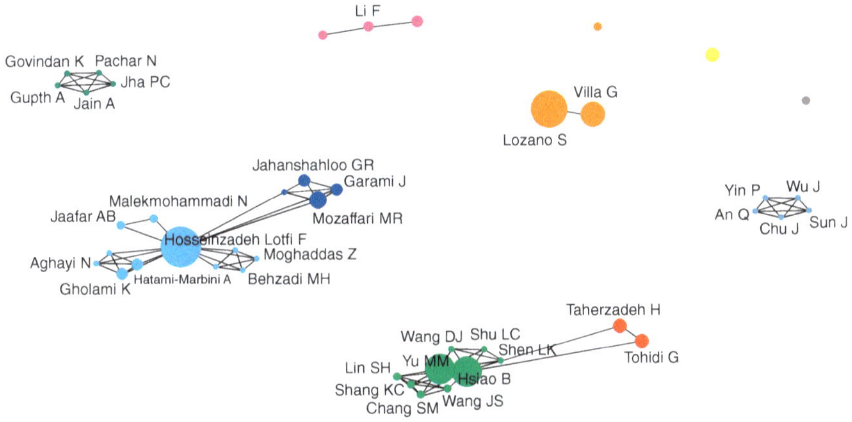

FIGURE 5.3 Author collaboration network.

Note: A total of 273 authors (nodes) are involved in the CDEA research.

Yu, who share the same number of publications but differ significantly in citation counts, invites further examination into the thematic focus of their research outputs. This could reveal the areas within the CDEA that command more academic and practical attention, thereby guiding future research directions.

Figure 5.3 depicts a detailed collaboration network among researchers specializing in the CDEA model. The key nodes in this network include Hosseinzadeh Lotfi, Yu and Hsiao, all of which have multiple linkages, indicating significant collaborative interactions. Lozano is particularly central, suggesting his role as a major contributor and influencer in the field. This central positioning could indicate that these collaborations are pivotal for advancing the methodology and applications of the CDEA model.

The network also reveals the diversity of collaboration, with nodes representing authors from different academic and geographical backgrounds. This diversity enriches the research output by integrating various perspectives and methodologies, which is crucial for the evolution of the CDEA model. The extent and strength of these collaborations can serve as indicators of the vibrancy and interdisciplinary nature of the research community in this field.

5.4.3 Institutions and Countries

Table 5.3 lists the institutions that have been most productive in terms of publications related to the CDEA model. The University of Seville stands out at the top, not only for its number of publications but also for its high

TABLE 5.3 The Ten Most Productive Authors' Institutions

Institution	Publications	Weighted Publications (%)	Citations
University of Seville	6.50 (1)	10.48 (1)	417.50 (1)
Islamic Azad University, Central Tehran Branch	4.33 (2)	6.99 (2)	21.67 (8)
Nankai University	4.00 (3)	6.45 (3)	224.50 (2)
National Taiwan Ocean University, Keelung	2.92 (4)	4.70 (4)	68.83 (3)
Islamic Azad University, Karaj Branch	2.50 (5)	4.03 (5)	15.00 (10)
Hefei University of Technology	2.00 (6)	3.23 (6)	39.50 (6)
University of Science and Technology of China	1.95 (7)	3.15 (7)	27.50 (7)
Chang Jung Christian University	1.92 (8)	3.09 (8)	15.83 (9)
Universitat Autònoma de Barcelona	1.70 (9)	2.74 (9)	43.80 (5)
Islamic Azad University, Science and Research Branch, Tehran	1.67 (10)	2.69 (10)	56.17 (4)

citation impact, indicating a leading role in the development and dissemination of DEA research. The presence of various branches of Islamic Azad University also highlights the significant contributions from Iranian academic institutions to the field, with the Central Tehran Branch and Karaj Branch being particularly notable.

The geographical diversity of these institutions—from Spain to China and Taiwan—illustrates the global interest and applicability of CDEA models in tackling efficiency and resource allocation problems across different economic and industrial landscapes. This diversity not only enriches the research by incorporating varied perspectives and methodologies but also enhances the practical relevance of the research outcomes by addressing a broader spectrum of real-world challenges.

As shown in Table 5.4, the contributions of different countries to CDEA model research are shown. China and Iran lead in terms of publication numbers, underscoring their active research communities in operations research and efficiency analysis. Spain, not far behind, tops the citation count, which might indicate seminal works that have significantly influenced the field.

The representation of countries with varying economic backgrounds, such as Taiwan, the United States, and Malaysia, highlights the universal relevance and adaptability of CDEA models to diverse economic and

TABLE 5.4 The Most Productive Authors' Countries

Country	Publications	Weighted Publications (%)	Citations
China	13.25 (1)	21.37 (1)	373.25 (2)
Iran	12.83 (2)	20.70 (2)	171.50 (4)
Spain	11.17 (3)	18.01 (3)	603.33 (1)
Taiwan	11.00 (4)	17.74 (4)	171.00 (5)
United States	3.50 (5)	5.65 (5)	103.00 (6)
Malaysia	2.00 (6)	3.23 (6)	23.00 (9)
South Korea	1.25 (7)	2.02 (7)	0.25 (10)
United Kingdom	1.17 (8)	1.88 (8)	91.33 (7)

Note: A total of 18 countries are involved in the CDEA research.

operational environments. This widespread international involvement supports the robustness of DEA methodologies and their capacity to address efficiency issues in various settings, from developed economies with complex industrial sectors to emerging markets focused on optimizing resources amidst rapid growth.

5.4.4 Publication Outlets

Table 5.5 presents the journals that are most frequently publishing research on the CDEA model. The *Journal of the Operational Research Society* and *Omega (United Kingdom)* have equal numbers of publications but differ significantly in citation numbers, with Omega receiving the most citations. This suggests that while both journals are prominent platforms for CDEA research, the articles published in Omega might be of particular

TABLE 5.5 The Most Relevant Journals

Journal	Publications	Citations
Journal of the Operational Research Society	5 (1)	198 (6)
Omega (United Kingdom)	5 (1)	335 (1)
Annals of Operations Research	4 (2)	141 (7)
Computers and Industrial Engineering	4 (2)	232 (3)
Socio-Economic Planning Sciences	3 (3)	21 (10)
European Journal of Operational Research	2 (4)	215 (4)
International Transactions in Operational Research	2 (4)	27 (9)
Journal of Computational and Applied Mathematics	2 (4)	63 (8)
Journal of Productivity Analysis	2 (4)	255 (2)

TABLE 5.6 The Ten Most Cited Articles

Article	Journal	TGC	TLC
Lozano and Villa (2004)	*Journal of Productivity Analysis*	244 (1)	49 (1)
Korhonen and Syrjänen (2004)	*Management Science*	208 (2)	27 (4)
Asmild et al. (2009)	*Omega (United Kingdom)*	140 (3)	36 (2)
Lozano et al. (2009)	*European Journal of Operational Research*	126 (4)	0 (10)
Lotfi et al. (2013)	*Computers and Industrial Engineering*	100 (5)	4 (9)
Fang (2013)	*European Journal of Operational Research*	89 (6)	22 (5)
Lozano et al. (2011)	*Computers and Industrial Engineering*	82 (7)	18 (7)
Mar-Molinero et al. (2014)	*Annals of Operations Research*	74 (8)	28 (3)
Fang (2015)	*Omega (United Kingdom)*	69 (9)	12 (8)
Lozano and Villa (2005)	*Journal of the Operational Research Society*	67 (10)	21 (6)

Note: TGC = total global citations; TLC = total local citations.

interest or relevance to the field, possibly owing to the innovative nature or practical applicability of the studies featured.

The *Annals of Operations Research* and *Computers and Industrial Engineering* also contributed significantly, highlighting the interdisciplinary nature of DEA applications, encompassing both theoretical operations research and practical industrial engineering. The variation in citation numbers across these journals could indicate the varying impact of the research on the basis of the journal's focus areas or audience.

Table 5.6 details the most cited articles within the realm of CDEA research, providing insights into the works that have shaped the field. Lozano and Villa's 2004 study in the *Journal of Productivity Analysis* holds the top position both in global and local citations, underscoring its pivotal role in advancing DEA methodology or its application. This article's prominence suggests that it might contain foundational concepts or innovative methodologies that have significantly influenced subsequent research.

5.4.5 Citation Network

Figure 5.4 illustrates the citation relationships between key publications in the field of the CDEA. This network provides a clear view of which articles have significantly influenced subsequent research. For example, if articles by Lozano S are frequently cited by newer publications, this underscores their ongoing relevance and authority in the field.

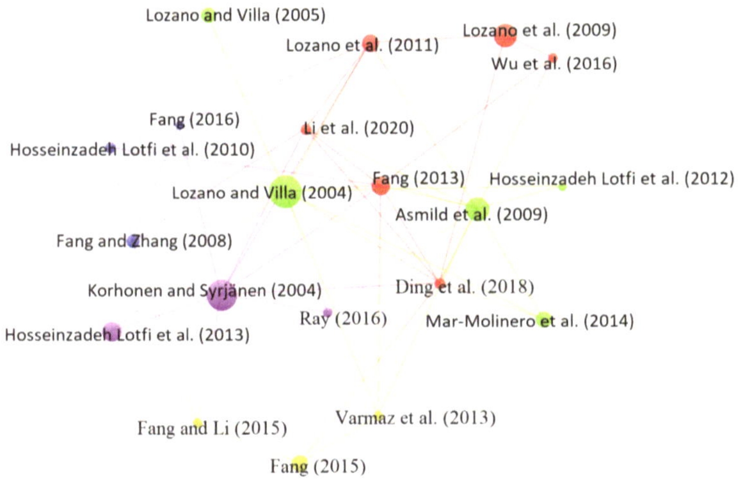

FIGURE 5.4 Citation network (25 citations—threshold).

The visualization of the citation network allows for the identification of the most influential articles and their respective journals, offering insights into where researchers might focus their literature review efforts or seek to publish their work. Observing how newer research connects to these influential articles can also provide insights into emerging trends and shifts in research focus within the CDEA community.

5.5 FUTURE DEVELOPMENTS AND CHALLENGES

CDEA stands as a formidable asset in the realm of operations management and efficiency appraisal, offering a method to gauge the effectiveness of DMUs such as firms, organizations, or departments. Several potential trends in its application signal a shift toward more comprehensive and dynamic evaluations.

First, there is a noticeable trajectory toward integrating the CDEA with the burgeoning field of big data analytics. This convergence allows for a richer analysis by assimilating diverse data sources, thus offering deeper insights into performance inefficiencies. Moreover, the evolution toward dynamic CDEA models is underway, recognizing the inherent dynamism of modern systems that are subject to constant technological advancements and market fluctuations. Such models would adeptly adapt to changes over time, ensuring a more accurate depiction of performance.

Furthermore, the landscape of the CDEA is expanding to address pressing concerns regarding its sustainability and environmental impact.

There is a growing emphasis on evaluating organizations' resource allocation in terms of environmental efficiency, encompassing aspects such as resource utilization, waste management, and the ecological footprint. This burgeoning interest may lead to the development of specialized models and metrics tailored to assess environmental efficiency using DEA methodologies. Additionally, the integration of CDEA with machine learning and AI presents an avenue for enhancing predictive capabilities and enabling more nuanced analysis of intricate systems. Hybrid models could emerge, amalgamating the strengths of CDEA with machine learning and AI techniques to refine decision making processes.

Finally, amidst the backdrop of heightened global market volatility and uncertainty, there is a discernible focus on fortifying organizational robustness and resilience. Future trends in CDEA may involve the creation of models and techniques aimed at evaluating the resilience of systems against various risks and disruptions. Such advancements would enable the identification of strategies to bolster resilience in the face of uncertainty, safeguarding organizations from potential vulnerabilities.

5.6 CONCLUSIONS

The central argument of this chapter is that in today's competitive business landscape, achieving organizational efficiency is paramount for optimizing resources and securing a competitive advantage. Centralized resource allocation is underscored as a crucial strategy to bolster efficiency across intraorganizational units and partnerships. This chapter highlights the importance of recognizing the diverse needs and interdependencies within partnerships to effectively implement centralized resource allocation. It advocates for the use of the CDEA model as a robust tool for enhancing efficiency in these contexts.

The application of the CDEA approach has attracted interest from researchers in fields such as transportation, higher education, and banking. A diverse group of 273 authors from countries such as Spain, China, Iran, Taiwan, South Korea, the United Kingdom, and the United States have published their work in prestigious journals, including the *Journal of the Operational Research Society*, *Omega*, and the *Annals of Operations Research*. Additionally, within the CDEA research community, five distinct active research groups have emerged, demonstrating the model's broad applicability and the ongoing collaborative efforts to refine and expand its use. These groups play crucial roles in advancing both the theory and practical applications of the CDEA model, enhancing its impact on organizational efficiency across various sectors.

The CDEA approach and its applications have yielded several significant findings and insights. First, by integrating centralized decision making, the CDEA facilitates optimized resource allocation across multiple business partnerships, thereby enhancing overall organizational performance through more efficient resource utilization. Additionally, the CDEA recognizes the interconnected nature of internal business partnerships and the impact of centralized decision making within complex organizational structures. This understanding leads to more accurate efficiency evaluations of centralization in resource allocation among internal business partnerships. Moreover, the CDEA addresses deficiencies in resource allocation inherent in traditional DEA methods by explicitly considering centralized resource allocation strategies.

By promoting resource allocation optimization among internal business partnerships, the CDEA aims to maximize overall organizational efficiency, taking into account potential adjustments within a centralized optimization framework. Furthermore, the CDEA approach, with its focus on centralized decision making, demonstrates proficiency in capturing the complex interdependencies and feedback loops within business partnerships. This capability enables a comprehensive understanding and effective management of these relationships. Finally, in decentralized decision making environments, inefficiencies resulting from suboptimal resource allocation based on individual unit optimization are common. Therefore, the CDEA offers a centralized resource allocation optimization framework to effectively address the challenge of enhancing the overall efficiency of internal business partnerships.

REFERENCES

Arocena, P., Cabasés, F., & Pascual, P. (2022). A centralized directional distance model for efficient and horizontally equitable grants allocation to local governments. *Socio-Economic Planning Sciences*, 81, 100947.

Asmild, M., Paradi, J. C., & Pastor, J. T. (2009). Centralized resource allocation BCC models. *Omega*, 37(1), 40–49.

Cesaroni, G. (2020). Technically and cost-efficient centralized allocations in data envelopment analysis. *Socio-Economic Planning Sciences*, 70, 100734.

Charnes, A., Cooper, W. W., & Rhodes, E. (1978). Measuring efficiency of decision making units. *European Journal of Operational Research*, 2, 429–444.

Contreras, I., & Lozano, S. (2022). Size efficiency, splits and merger gains, and centralized resource reallocation of Spanish public universities. *Socio-Economic Planning Sciences*, 81, 101190.

Davutyan, N., Demir, M., & Polat, S. (2010). Assessing the efficiency of Turkish secondary education: Heterogeneity, centralization, and scale diseconomies. *Socio-Economic Planning Sciences*, 44(1), 35–44.

Ding, T., Chen, Y., Wu, H., & Wei, Y. (2018). Centralized fixed cost and resource allocation considering technology heterogeneity: A DEA approach. *Annals of Operations Research*, 268, 497–511.

Fang, L. (2013). A generalized DEA model for centralized resource allocation. *European Journal of Operational Research*, 228(2), 405–412.

Fang, L. (2015). Centralized resource allocation based on efficiency analysis for step-by-step improvement paths. *Omega*, 51, 24–28.

Feng, C., Chu, F., Ding, J., Bi, G., & Liang, L. (2015). Carbon Emissions Abatement (CEA) allocation and compensation schemes based on DEA. Omega, 53, 78–89.

Hatami-Marbini, A., Tavana, M., Agrell, P. J., Lotfi, F. H., & Beigi, Z. G. (2015). A common-weights DEA model for centralized resource reduction and target setting. *Computers & Industrial Engineering*, 79, 195–203.

Hsiao, B., & Chen, L. H. (2022). Internal resource mobility of a railway administration: A three-phase centralized data envelopment analysis. *Research in Transportation Business & Management*, 45, 100855.

Korhonen, P., & Syrjänen, M. (2004). Resource allocation based on efficiency analysis. Management Science, 50(8), 1134–1144.

Koronakos, G., Sotiros, D., & Despotis, D. K. (2019). Reformulation of Network Data Envelopment Analysis models using a common modelling framework. *European Journal of Operational Research*, 278(2), 472–480.

López-Torres, L., & Prior, D. (2016). Centralized allocation of human resources. An application to public schools. *Computers & Operations Research*, 73, 104–114.

Lotfi, F. H., Hatami-Marbini, A., Agrell, P. J., Aghayi, N., & Gholami, K. (2013). Allocating fixed resources and setting targets using a common-weights DEA approach. *Computers & Industrial Engineering*, 64(2), 631–640.

Lotfi, F. H., Nematollahi, N., Behzadi, M. H., Mirbolouki, M., & Moghaddas, Z. (2012). Centralized resource allocation with stochastic data. *Journal of Computational and Applied Mathematics*, 236(7), 1783–1788.

Lotfi, F. H., Noora, A. A., Jahanshahloo, G. R., Gerami, J., & Mozaffari, M. R. (2010). Centralized resource allocation for enhanced Russell models. *Journal of computational and Applied Mathematics*, 235(1), 1–10.

Lozano, S., & Contreras, I. (2022). Centralised resource allocation using Lexicographic Goal Programming. Application to the Spanish public university system. *Socio-Economic Planning Sciences*, 84, 101419.

Lozano, S., & Villa, G. (2004). Centralised resource allocation using data envelopment analysis. *Journal of Productivity Analysis*, 22, 143–161.

Lozano, S., & Villa, G. (2005). Centralised DEA models with the possibility of downsizing. *Journal of the Operational Research Society*, 56, 357–364.

Lozano, S., Villa, G., & Brännlund, R. (2009). Centralised reallocation of emission permits using DEA. *European Journal of Operational Research*, 193(3), 752–760.

Lozano, S., Villa, G., & Canca, D. (2011). Application of centralised DEA approach to capital budgeting in Spanish ports. *Computers & Industrial Engineering*, 60(3), 455–465.

Mardani, A., Zavadskas, E. K., Streimikiene, D., Jusoh, A., & Khoshnoudi, M. (2017). A comprehensive review of data envelopment analysis (DEA) approach in energy efficiency. *Renewable and Sustainable Energy Reviews*, 70, 1298–1322.

Mar-Molinero, C., Prior, D., Segovia, M. M., & Portillo, F. (2014). On centralized resource utilization and its reallocation by using DEA. *Annals of Operations Research*, 221, 273–283.

Momeni, E., Lotfi, F. H., Saen, R. F., & Najafi, E. (2019). Centralized DEA-based reallocation of emission permits under cap and trade regulation. *Journal of Cleaner Production*, 234, 306–314.

Sharma, M. J., & Yu, S. J. (2010). Benchmark optimization and attribute identification for improvement of container terminals. *European Journal of Operational Research*, 201(2), 568–580.

Thomas, R. R., Barr, R. S., Cron, W. L., & Slocum Jr, J. W. (1998). A process for evaluating retail store efficiency: a restricted DEA approach. *International Journal of Research in Marketing*, 15(5), 487–503.

Tohidi, G., Taherzadeh, H., & Hajiha, S. (2014). Undesirable outputs' presence in centralized resource allocation model. *Mathematical Problems in Engineering*, 1–6.

Toloo, M., Taghizadeh-Yazdi, M., & Mohammadi-Balani, A. (2022). Multi-objective centralization-decentralization trade-off analysis for multi-source renewable electricity generation expansion planning: A case study of Iran. *Computers & Industrial Engineering*, 164, 107870.

Tone, K., & Tsutsui, M. (2010). Dynamic DEA: A slacks-based measure approach. *Omega*, 38(3–4), 145–156.

Varmaz, A., Varwig, A., & Poddig, T. (2013). Centralized resource planning and Yardstick competition. *Omega*, 41(1), 112–118.

Xiong, X., Yang, G. L., Zhou, D. Q., & Wang, Z. L. (2022). How to allocate multi-period research resources? Centralized resource allocation for public universities in China using a parallel DEA-based approach. *Socio-Economic Planning Sciences*, 82, 101317.

Yang, M., Hou, Y., & Wang, Q. (2022). Rethinking on regional CO2 allocation in China: A consideration of the carbon sink. *Environmental Impact Assessment Review*, 96, 106822.

Yu, M. M., Abdul Rashid, A., & See, K. F. (2022). Developing an innovation capital index of global airlines using a hierarchical data envelopment analysis approach. *Journal of the Operational Research Society*, 73(8), 1708–1723.

Yu, M. M., & Chen, L. H. (2016). Centralized resource allocation with emission resistance in a two-stage production system: Evidence from a Taiwan's container shipping company. *Transportation Research Part A: Policy and Practice*, 94, 650–671.

Yu, M. M., Chern, C. C., & Hsiao, B. (2013). Human resource rightsizing using centralized data envelopment analysis: Evidence from Taiwan's airports. *Omega*, 41(1), 119–130.

Yu, M. M., & Hsiao, B. (2018). Single-phase slack-based centralized DEA for resource reallocation. *International Transactions in Operational Research*, 25(2), 737–751.

Zhang, C., Bai, C., Su, W., & Balezentis, T. (2024). The data envelopment analysis model integrated with cost information and utility theory for power price setting under carbon peak strategy at the firm-level. *Energy*, 292, 130457.

Zhou, Z., Lin, L., Xiao, H., Ma, C., & Wu, S. (2017). Stochastic network DEA models for two-stage systems under the centralized control organization mechanism. *Computers & Industrial Engineering*, 110, 404–412.

Zhou, Z., Sun, L., Yang, W., Liu, W., & Ma, C. (2013). A bargaining game model for efficiency decomposition in the centralized model of two-stage systems. *Computers & Industrial Engineering*, 64(1), 103–108.

Zhu, J., & Cook, W. D. (Eds.). (2007). *Modeling data irregularities and structural complexities in data envelopment analysis*. Springer Science & Business Media.

Zhu, W., Huang, Y., & Yu, Y. (2023). DEA model for partial centralization resource allocation among independent subset of DMUs. *Computers & Industrial Engineering*, 176, 109013.

CHAPTER 6

Inverse Data Envelopment Analysis Ratio Models for Business Mergers

Mehdi Soltanifar
Department of Mathematics, Semnan Branch, Islamic Azad University, Semnan, Iran

Mojtaba Ghiyasi
Faculty of Industrial Engineering and Management Science, Shahrood University of Technology, Shahrood, Iran

Hamid Sharafi
Department of Mathematics, Science and Research Branch, Islamic Azad University, Tehran, Iran

6.1 INTRODUCTION

Data envelopment analysis (DEA) is a non-parametric method for assessing the relative efficiency of a set of homogenous decision making units (DMUs). It was initially proposed by Charnes et al. (1978) via a mathematical programming framework. Using the observed input and output data, this method measures the efficiency of DMUs by solving a set of

DOI: 10.1201/9781003345206-6

mathematical programming models. Emrouznejad and Yang (2018) provided a survey of 40 years of scholarly literature in DEA. Wei et al. (2000) looked at the DEA models from a different view and introduced a method called the inverse DEA models. This method tries to find the maximum level of outputs with a given input level (in the output orientation). Considering the input orientation, the inverse DEA model aims to find the minimum level of input that can produce a certain level of output. They used multiple objectives linear programming (MOLP) to answer these types of questions. Some extensions and modifications are introduced by different researchers. Lertworasirikul et al. (2011) dealt with the variable returns to scale (VRS) in the inverse DEA problem. Later on, Ghiyasi (2015) revised their work using the correct concept of Pareto efficiency. Amin and Emrouznejad (2007) applied the notion of inverse optimization for forecasting and provided a streamlined approach to time series analysis using inverse linear programming. Ghiyasi (2017a) considered the price of inputs and outputs and then proposed the cost-based and revenue-based inverse DEA. Ghiyasi (2017b) and Emrouznejad et al. (2019) dealt with the presence of undesirable outputs in the inverse DEA models. Soleimani-Chamkhorami et al. (2020) proposed a ranking system using the concept of inverse DEA. Sayar et al. (2021) proposed new DEA model for budgeting and planning problems. See Emrouznejad et al. (2023) for a review of the theory development and application of inverse DEA models. Shiri Daryani et al. (2021) dealt with the allocative efficiency of two-stage inverse DEA models. Ghobadi et al. (2023) dealt with the input and output estimation in the inverse DEA model over the windows analysis for periods of times. In an interesting application of the inverse DEA models, Gattoufi et al. (2014) utilized the inverse DEA for the merging problem in the banking sectors. They find the new level of the inputs and outputs for the merged bank to reach a predetermined efficiency target. This initiated another path in the inverse DEA literature for the merger analysis. Amin and Al-Muharrami (2018) proposed an inverse DEA model for merger analysis capable of dealing with negative data. Amin and Ibn Boamah (2020) dealt with the cost efficiency concept in a merger using inverse DEA analysis. In a more broad view, Amin et al. (2017) proposed inverse DEA models for modeling generalized firms' restructuring as well as anticipating the minor and major consolidation for a merger in a market. Soltanifar et al. (2024) introduced a multiplier form of inverse DEA for the merger analysis. Incorporation of prior information in the traditional DEA model helps decision makers to fit in their desires in the process of efficiency analysis. This objective is possible through weight settings in DEA models with multiplier forms, and it prevents a zero weight which is not a desired result in DEA analysis. Issues relating to the inclusion of prior information are also dealt with in the so-called pseudo-inefficiency case of DMUs; i.e., cases where the efficiency score does not reflect the real efficiency of the DMUs. In order to overcome this obstacle differently, the DEA-R model was developed by Despić et al. (2007), i.e., a combination of the DEA methodology and ratio analysis. In the traditional DEA model, inputs and outputs are given parameters, and the efficiency scores are estimated using the observed inputs and outputs data. In evaluating the performance efficiency of some organizations, certain datasets, such as financial data, may be ratio-based such as quick ratios. In such cases, traditional DEA models are not suitable models for evaluation and a suitable model should be designed and used for this type of data. This problem was solved by DEA-R models proposed by Despić et al. (2007) (Mozaffari et al., 2014a; Mozaffari et al., 2014b; Wei et al., 2011; Tohidnia and Tohidi, 2019). Mozaffari et al. (2014a) investigated DEA-R models without explicit inputs. Given the price of inputs and outputs, Mozaffari et al. (2014b) considered the cost and revenue efficiency for the DEA-R models. Tohidnia and Tohidi (2019) proposed a methodology for measuring the productivity change of production units over a period of time. Mozaffari et al. (2020) proposed an algorithm for identifying supporting hyper plans and consequently efficient hyper-plans for the DEA-R models. Gerami et al. (2022) dealt with the slack based and supper efficiency for the DEA-R models. Wanke et al. (2023) introduced a network structure for DEA-R model in the presence of stochastic ratio data. Back to the inverse modeling for the merger and partnership analysis, there exist few research on the inverse DEA-R and merger analysis for DEA-R models. Soltanifar et al. (2024) dealt with the merger analysis through DEA-R model via inverse DEA problems. They also accounted for the possibility of data negativity in the merger process of DEA-R models. Ghiyasi et al. (2022) proposed a novel inverse DEA-R model for the efficiency analysis of 130 hospitals in Iran. In this chapter, we provide a comprehensive review of the existing Inverse DEA-R models, encompassing both the models proposed for estimating inputs and outputs and the models proposed for mergers and acquisitions, considering both non-negative and negative data. This comprehensive analysis distinguishes our chapter from these previously published papers.

We have reviewed proposed DEA models, inverse DEA, and inverse DEA-R models that could be used for the merger and business partnership analysis in the remaining sections of the chapter. Section 6.2 considers and reviews both positive and negative data in DEA-R models. Section 6.3 deals with the use of DEA and inverse DEA models in mergers and acquisitions. Section 6.4 focuses on input and output estimation in mergers and business partnership analysis.

6.2 PRELIMINARIES

This section discusses ratio-based DEA models in the presence of non-negative data as well as negative data.

6.2.1 Non-Negative Data

Consider m-dimensional positive input vector of $\left(x_{1j}, x_{2j}, \ldots, x_{mj}\right) > \mathbf{0}$ for DMU$_j$, $1 \le j \le n$ that is used for producing s-dimensional positive output vector of $\left(y_{1j}, y_{2j}, \ldots, y_{sj}\right) > \mathbf{0}$. Despić et al. (2007) introduced the following models for DMU$_p$ assuming constant returns to scale (CRS), called the DEA-R max–min efficiency.

$$e_p = \max_{\sum_r \sum_i w_{ri}=1, w_{ri} \ge 0} \min_j \left(\sum_r \sum_i w_{ri} \frac{x_{ij}}{x_{ip}} \frac{y_{rp}}{y_{rj}} \right) \tag{6.1}$$

$$e_p = \max_{\sum_r \sum_i w_{ri}=1, w_{ri} \ge 0} \min_j \frac{1}{\left(\sum_r \sum_i w_{ri} \frac{y_{rj}}{y_{rp}} \frac{x_{ip}}{x_{ij}} \right)} \tag{6.2}$$

where w_{ri} are the associated weights of output–input ratio (r-th output and i-th input) for DMU$_p$. The DEA-R max–min models for DMU$_p$ assuming VRS can be presented as (6.3) and (6.4).

$$\bar{e}_p = \max_{\sum_i w_{i0} + \sum_r \sum_i w_{ri}=1, w_{ri} \ge 0} \min_j \frac{1}{\left(\sum_i w_{i0} \frac{x_{ip}}{x_{ij}} + \sum_r \sum_i w_{ri} \frac{y_{rj}}{y_{rp}} \frac{x_{ip}}{x_{ij}} \right)} \tag{6.3}$$

$$\bar{e}_p = \max_{\sum_r w_{r0} + \sum_r \sum_i w_{ri}=1, w_{ri} \ge 0} \min_j \left(\sum_r w_{r0} \frac{y_{rp}}{y_{rj}} + \sum_r \sum_i w_{ri} \frac{x_{ij}}{x_{ip}} \frac{y_{rp}}{y_{rj}} \right) \tag{6.4}$$

Models (6.1)–(6.4) were also studied and reviewed by Despić (2013). By discussing the sign of variables $w_{i0}, (\forall i)$ and $w_{r0}, (\forall r)$, Banker, Charnes, and Cooper (BCC)-Charnes Cooper and Rhodes (CCR) and CCR-BCC versions of the above models can also be presented. Models (6.3) and (6.4) can be converted into linear programming models (6.5) and (6.6), respectively.

$$
\begin{aligned}
&\bar{e}_p = \min \varphi \\
&s.t. \\
&\sum_i w_{i0} \frac{x_{ip}}{x_{ij}} + \sum_r \sum_i w_{ri} \left(\frac{\frac{y_{rj}}{x_{ij}}}{\frac{y_{rp}}{x_{ip}}} \right) \le \varphi, \ \forall j \\
&\sum_i w_{i0} + \sum_r \sum_i w_{ri} = 1 \\
&w_{ri} \ge 0, \forall r, \forall i \\
&w_{i0} \text{ is free}, \forall i.
\end{aligned}
\tag{6.5}
$$

$$
\begin{aligned}
&\bar{e}_p = \max \varphi \\
&s.t. \\
&\sum_r w_{r0} \frac{y_{rp}}{y_{rj}} + \sum_r \sum_i w_{ri} \left(\frac{\frac{x_{ij}}{y_{rj}}}{\frac{x_{ip}}{y_{rp}}} \right) \ge \varphi, \ \forall j \\
&\sum_r w_{r0} + \sum_r \sum_i w_{ri} = 1 \\
&w_{ri} \ge 0, \ \forall r, \forall i \\
&w_{r0} \text{ is free}, \forall r,
\end{aligned}
\tag{6.6}
$$

The above models are in the multiplier form. By using the rules of duality, the corresponding models, which are known as envelopment form models, can be presented as Models (6.7) and (6.8), respectively.

$$
\begin{aligned}
&\bar{e}_p = \max \theta \\
&s.t. \\
&\sum_j \lambda_j \left(\frac{\frac{y_{rj}}{x_{ij}}}{\frac{y_{rp}}{x_{ip}}} \right) \ge \theta, \ \forall r, \forall i \\
&\sum_j \lambda_j \left(\frac{x_{ip}}{x_{ij}} \right) = \theta, \ \forall i \\
&\sum_j \lambda_j = 1 \\
&\lambda_j \ge 0, \forall j \\
&\theta \text{ is free.}
\end{aligned}
\tag{6.7}
$$

$$
\begin{aligned}
&\bar{e}_p = \min \theta \\
&s.t. \\
&\sum_j \lambda_j \left(\frac{\dfrac{x_{ij}}{y_{rj}}}{\dfrac{x_{ip}}{y_{rp}}} \right) \leq \theta, \ \forall r, \forall i \\
&\sum_j \lambda_j \left(\frac{y_{rp}}{y_{rj}} \right) = \theta, \ \forall r \\
&\sum_j \lambda_j = 1 \\
&\lambda_j \geq 0, \ \forall j \\
&\theta \text{ is free.}
\end{aligned}
\tag{6.8}
$$

The λ_j is the intensity variable, and θ is the efficiency variable showing the efficiency score of the unit under evaluation.

The above models were studied by Mozaffari et al. (2020), Olesen et al. (2017), and Mozaffari et al. (2022), and more details of these models were discussed. Here, by discussing the second category of the constraints, the BCC-CCR and CCR-BCC versions can be presented. These DEA models can handle ratio data. But the models presented so far can only be used for positive data. In the next section, these models are redesigned in the presence of negative data.

6.2.2 Negative Data

In the following, the models of the previous section are rewritten in the presence of negative data. This work is done in two cases when the outputs are negative or when the inputs are negative. It should be noted that the simultaneous negative assumption of inputs and outputs is not a correct assumption due to the structure of the models.

First suppose the outputs of the problem are divided into two categories. The first category of outputs are positive $\left(\mathbf{Y}_r^P\right)$, and the second category of outputs can be negative $\left(\mathbf{Y}_r\right)$. Variables $\mathbf{Y}_r^P$ and $\mathbf{Y}_r^N$ are defined as follows (Emrouznejad et al., 2010a; Emrouznejad et al. 2010b):

$$y_{rj}^{P}=\begin{cases} y_{rj} & \text{if } y_{rj}\geq 0 \\ 0 & \text{Otherwise} \end{cases} \qquad y_{rj}^{N}=\begin{cases} -y_{rj} & \text{if } y_{rj}<0 \\ 0 & \text{Otherwise} \end{cases}, \forall j$$

In fact, $y_{rj}=y_{rj}^{P}-y_{rj}^{N}$ and $y_{rj}^{P}, y_{rj}^{N}\geq 0$. DEA-R input-oriented in the presence of negative outputs is presented in Model (6.9).

$$\begin{aligned}
&\max \theta \\
&s.t. \\
&\sum_{j}\lambda_j \frac{y_{rj}}{x_{ij}}\geq \theta \frac{y_{rp}}{x_{ip}}, \ \forall i, \forall r\in O' \\
&\sum_{j}\lambda_j \frac{y_{rj}^{P}}{x_{ij}}\geq \theta \frac{y_{rp}^{P}}{x_{ip}}, \ \forall i, \forall r\in O'' \\
&\sum_{j}\lambda_j \frac{y_{rj}^{N}}{x_{ij}}\leq \theta \frac{y_{rp}^{N}}{x_{ip}}, \ \forall i, \forall r\in O'' \\
&\sum_{j}\lambda_j \left(\frac{x_{ip}}{x_{ij}}\right)=\theta, \ \forall i \\
&\sum_{j}\lambda_j =1 \\
&\lambda_j\geq 0, \forall j \\
&\grave{e}\,\text{is free}
\end{aligned} \tag{6.9}$$

where the index set associated with the variables $\mathbf{Y}^{\mathrm{P}}$ are specified with O' and the index set associated with the variables $\mathbf{Y}^{\mathrm{P}}$ are specified with O''. By presenting this model, Soltanifar et al. (2023) showed that their model is always feasible.

In a similar way, Soltanifar et al. (2023) assumed that inputs are divided into two categories. One is a positive group, and another group can be negative. Note that the second category can be positive or negative. The following variables can be defined for second category inputs.

$$x_{ij}^{P}=\begin{cases} x_{ij} & \text{if } x_{ij}\geq 0 \\ 0 & \text{Otherwise} \end{cases} \qquad x_{ij}^{N}=\begin{cases} -x_{ij} & \text{if } x_{ij}<0 \\ 0 & \text{Otherwise} \end{cases}, \quad \forall j$$

In fact, $x_{ij} = x_{ij}^P - x_{ij}^N$ and $x_{ij}^P, x_{ij}^N \geq 0$. DEA-R output-oriented in the presence of negative inputs is presented in Model (6.10).

$$
\begin{aligned}
& \min \theta \\
& s.t. \\
& \sum_j \lambda_j \frac{x_{ij}}{y_{rj}} \leq \theta \frac{x_{ip}}{y_{rp}}, \ \forall r, \forall i \in I' \\
& \sum_j \lambda_j \frac{x_{ij}^P}{y_{rj}} \leq \theta \frac{x_{ip}^P}{y_{rp}}, \ \forall r, \forall i \in I'' \\
& \sum_j \lambda_j \frac{x_{ij}^N}{y_{rj}} \geq \theta \frac{x_{ip}^N}{y_{rp}}, \ \forall r, \forall i \in I'' \\
& \sum_j \lambda_j \left(\frac{y_{rp}}{y_{rj}} \right) = \theta, \ \forall r \\
& \sum_j \lambda_j = 1 \\
& \lambda_j \geq 0, \ \forall j \\
& \theta \text{ is free}
\end{aligned}
\tag{6.10}
$$

where the index set associated with the variables $\mathbf{I}^{\mathrm{P}}$ are specified with I' and the index set associated with the variables $\mathbf{I}^{\mathrm{P}}$ are specified with I''. By presenting this model, Soltanifar et al. (2023) proved its feasibility.

Soltanifar et al. (2023) also rewrote their DEA-R models with the modifications made by Kaffash et al. (2018) on the models presented by Emrouznejad et al. (2010a) and Emrouznejad et al. (2010b). This is presented in Models (6.11) and (6.12).

$$
\begin{aligned}
& \max \theta \\
& s.t. \\
& \sum_j \lambda_j \frac{y_{rj}}{x_{ij}} \geq (1+\theta) \frac{y_{rp}}{x_{ip}}, \ \forall i, \forall r \in O' \\
& \sum_j \lambda_j \frac{y_{rj}^P}{x_{ij}} \geq (1+\theta) \frac{y_{rp}^P}{x_{ip}}, \ \forall i, \forall r \in O'' \\
& \sum_j \lambda_j \frac{y_{rj}^N}{x_{ij}} \leq (1-\theta) \frac{y_{rp}^N}{x_{ip}}, \ \forall i, \forall r \in O'' \\
& \sum_j \lambda_j \left(\frac{x_{ip}}{x_{ij}} \right) = (1+\theta), \ \forall i \\
& \sum_j \lambda_j = 1 \\
& \lambda_j \geq 0, \forall j \\
& \theta \text{ is free}
\end{aligned}
\tag{6.11}
$$

$$\begin{aligned}
&\min \theta \\
&s.t. \\
&\sum_j \lambda_j \frac{x_{ij}}{y_{rj}} \le (1+\theta)\frac{x_{ip}}{y_{rp}}, \ \forall r, \forall i \in I' \\
&\sum_j \lambda_j \frac{x_{ij}^P}{y_{rj}} \le (1+\theta)\frac{x_{ip}^P}{y_{rp}}, \ \forall r, \forall i \in I'' \\
&\sum_j \lambda_j \frac{x_{ij}^N}{y_{rj}} \ge (1-\theta)\frac{x_{ip}^N}{y_{rp}}, \ \forall r, \forall i \in I'' \qquad (6.12)\\
&\sum_j \lambda_j \left(\frac{y_{rp}}{y_{rj}}\right) = (1+\theta), \ \forall r \\
&\sum_j \lambda_j = 1 \\
&\lambda_j \ge 0, \ \forall j \\
&\theta \text{ is free}
\end{aligned}$$

The feasibility of these models was also proved by Soltanifar et al. (2023). DEA models designed to handle ratio data are not exclusive to what was stated. For example, Emrouznejad and Amin (2009) investigated models in which some inputs or outputs are defined as ratios with specific numerator and denominator. Anyway, in this chapter, the models presented in this section are used. In the section that follows, the most important research presented for two important applications of DEA models, namely mergers and acquisitions and also inputs/outputs estimation in the presence of ratio data, is presented.

6.3 MERGERS AND ACQUISITIONS

Companies and factories with homogenous units may merge some of their units at some point in time. Economic, cultural, climatic, environmental, and demographic effects may be among the reasons for a decision to merge. DEA models are always used as decision support tools, including the decision to merge. The first question is how DEA models can be used to estimate the potential gains from mergers and acquisitions for different units and guide them on how to combine their resources to obtain maximum benefits out of the resulting synergies. To answer this question, several models have been presented which are inverse DEA-based models discussed in this chapter. But the second question is, can the presented models handle positive or negative ratio data? To answer this question, this section will present inverse DEA-based models in two subsections for non-negative and negative data.

6.3.1 Non-Negative Data

After determining the necessity of merging two specific units in a system of homogeneous units, managers usually set a target for the efficiency score of the merged unit. The pre-merger units had inputs and outputs, which after the merger based on this targeting should have new inputs and outputs in the merged unit. In other words, the outputs (inputs) of the units after the merger can be equal to the sum of the outputs (inputs) of the units before the merger, and models are provided to determine the inputs (outputs) of the units after the merger, considering the specified targeting. These two concepts are presented with input-oriented and output-oriented models.

Suppose that based on a management decision, units h and k should be merged and create a new entity called unit M. Therefore, the objective of unit M is to produce the sum of the output level of units h and k, i.e., $\mathbf{y}_{\mathrm{h}} + \mathbf{y}_{\mathrm{k}}$, with the minimum level of input $\boldsymbol{\alpha}_{\mathrm{h}} + \boldsymbol{\alpha}_{\mathrm{k}}$ and to maintain a certain level of efficiency score $\frac{1}{\bar{\theta}}$. The decision variables α_{ih} and $\alpha_{ik}, (\forall i)$ are provided to determine the inputs of units h and k, respectively. Soltanifar et al. (2024) proposed Model (6.13) to find the minimum input level of the contribution of h and k units to produce the accumulated output level with the given efficiency score of $\frac{1}{\bar{\theta}}$.

$$
\begin{aligned}
&\min \sum_{i} \left(\alpha_{ih} + \alpha_{ik}\right) \\
&s.t. \\
&\sum_{j \in F} \lambda_j \left(\frac{\dfrac{\alpha_{ih} + \alpha_{ik}}{y_{rh} + y_{rk}}}{\dfrac{x_{ij}}{y_{rj}}} \right) \geq \bar{\theta}, \ \forall r, \forall i \\
&\sum_{j \in F} \lambda_j \left(\frac{\alpha_{ih} + \alpha_{ik}}{x_{ij}} \right) = \bar{\theta}, \ \forall i \\
&\sum_{j \in F} \lambda_j = 1 \\
&0 \leq \alpha_{ih} \leq x_{ih}, \forall i \\
&0 \leq \alpha_{ik} \leq x_{ik}, \forall i \\
&\lambda_j \geq 0, \forall j \in F
\end{aligned}
\tag{6.13}
$$

where F is the set of units after merging in the system of homogeneous units. In the merger process, only the post-merger acquiring entity may continue to operate under its previous name, or both merged entities may merge into a new entity with a new name. Therefore, depending on the type of merger, F may contain one of the merged units h or k, or it may contain none and be presented in a new entity. In Model (6.13), the merged unit is shown with M regardless of the merged form and the name of the merged entity. This model actually determines the level of inherited inputs after merging. Setting targets for the efficiency score of the merged unit is very important and sensitive. Soltanifar et al. (2024) showed that this targeting depends on the efficiency of the unit resulting from the aggregation of the inputs and outputs of the units before merged.

Theorem 1

Model (6.13) is feasible if and only if $\bar{\theta} \leq \theta^*$, where $\left(\theta^*, \ddot{\mathbf{e}}^*\right)$ is the optimal solution of Model (6.14).

$$
\begin{aligned}
& \max \theta \\
& \text{s.t.} \\
& \sum_{j \in F} \lambda_j \left(\frac{\dfrac{x_{ih} + x_{ik}}{y_{rh} + y_{rk}}}{\dfrac{x_{ij}}{y_{rj}}} \right) \geq \theta, \ \forall r, \forall i \\
& \sum_{j \in F} \lambda_j \left(\frac{x_{ih} + x_{ik}}{y_{rj}} \right) = \theta, \ \forall r \\
& \sum_{j \in F} \lambda_j = 1 \\
& \lambda_j \geq 0, \ \forall j \in F \\
& \theta \text{ is free}
\end{aligned}
\tag{6.14}
$$

Proof. It is clear that Model (6.14) evaluates the efficiency of a virtual DMU with $\left(x_{ih} + x_{ik}\right)$ as the ith input $\left(\forall i\right)$ and $\left(y_{rh} + y_{rk}\right)$ as the rth output $\left(\forall r\right)$. Now, if $\bar{\theta} \leq \theta^*$, then there exists $0 < p < 1$ that $\bar{\theta} = p\theta^*$, and therefore, constraints (6.15) are established.

$$
\begin{aligned}
&\max \theta \\
&s.t. \\
&\sum_{j\in F}\lambda_j\left(\frac{\dfrac{x_{ih}+x_{ik}}{y_{rh}+y_{rk}}}{\dfrac{x_{ij}}{y_{rj}}}\right)\geq\theta,\ \forall r,\forall i \\
&\sum_{j\in F}\lambda_j\left(\frac{x_{ih}+x_{ik}}{y_{rj}}\right)=\theta,\ \forall r \\
&\sum_{j\in F}\lambda_j=1 \\
&\lambda_j\geq 0,\ \forall j\in F
\end{aligned}
\tag{6.15}
$$

Now by multiplying the first and second constraints of (6.15) by p and setting $\alpha_{ih}=px_{ih},\alpha_{ik}=px_{ik},\forall i$, the solution $\left(\boldsymbol{\lambda}^*,\boldsymbol{\alpha}_{\mathbf{h}},\boldsymbol{\alpha}_{\mathbf{k}}\right)$ can be introduced as a feasible solution of Model (6.13).

Conversely, suppose that Model (6.13) has a solution such as $\left(\boldsymbol{\lambda}^+,\boldsymbol{\alpha}_{\mathbf{h}},\boldsymbol{\alpha}_{\mathbf{k}}\right)$ and since $x_{ih}\geq\alpha_{ih},x_{ik}\geq\alpha_{ik},\forall i$, the constant numbers $p_{ir}\geq 0,\left(\forall i,\forall r\right)$ and $q_i\geq 0,\left(\forall i\right)$ are found if the constraints (6.16) hold.

$$
\begin{aligned}
&\sum_{j\in F}\lambda_r^+\left(\frac{\dfrac{\alpha_{ih}+\alpha_{ik}}{y_{rh}+y_{rk}}}{\dfrac{x_{ij}}{y_{rj}}}\right)=\sum_{j\in F}\lambda_r^+\left(\frac{\dfrac{x_{ih}+x_{ik}}{y_{rh}+y_{rk}}}{\dfrac{x_{ij}}{y_{rj}}}\right)-p_{ir}\geq\bar{\theta},\ \forall r,\forall i \\
&\sum_{j\in F}\lambda_r^+\left(\frac{\alpha_{ih}+\alpha_{ik}}{x_{ij}}\right)=\sum_{j\in F}\lambda_r^+\left(\frac{x_{ih}+x_{ik}}{x_{ij}}\right)-q_i=\bar{\theta},\ \forall i \\
&\sum_{j\in F}\lambda_r^+=1 \\
&\lambda_r^+\geq 0,\ \forall j\in F
\end{aligned}
\tag{6.16}
$$

Since $\left(\theta^*,\boldsymbol{\lambda}^*\right)$ is the optimal solution of Model (6.14), constraints (6.17) are satisfied.

$$
\theta^*=\sum_{j\in F}\lambda_r^*\left(\frac{x_{ih}+x_{ik}}{x_{ij}}\right)\geq\sum_{j\in F}\lambda_r^+\left(\frac{x_{ih}+x_{ik}}{x_{ij}}\right)=\bar{\theta}+q_i,\ \forall i \tag{6.17}
$$

This implies that $\bar{\theta} \le \theta^*$, and this ends the proof.

Model (6.13) is a nonlinear programming problem that can be converted to a linear model (6.18) by setting $\lambda_{ij}^{\dagger} = \lambda_j \left(\alpha_{ih} + \alpha_{ik}\right)\left(\forall i, \forall j \in F\right)$.

$$
\begin{aligned}
& \min \sum_{i} \left(\alpha_{ih} + \alpha_{ik}\right) \\
& s.t. \\
& \sum_{j \in F} \left(\frac{\dfrac{\lambda_{ij}^{\dagger}}{y_{rh} + y_{rk}}}{\dfrac{x_{ij}}{y_{rj}}} \right) \ge \bar{\theta}, \ \forall r, \forall i \\
& \sum_{j \in F} \left(\frac{\lambda_{ij}^{\dagger}}{x_{ij}} \right) = \bar{\theta}, \ \forall i \\
& \sum_{j \in F} \lambda_{ij}^{\dagger} = \alpha_{ih} + \alpha_{ik}, \forall i \\
& 0 \le \alpha_{ih} \le x_{ih}, \forall i \\
& 0 \le \alpha_{ik} \le x_{ik}, \forall i \\
& \lambda_{ij}^{\dagger} \ge 0, \forall i, \forall j \in F
\end{aligned}
\tag{6.18}
$$

The above linear programming can be easily solved using any standard operational solvers to find the global optimal solution. This is the minimum required input share of DMUs h and k for producing the sum of the output level of these two units, with a predefined efficiency level. In this way, by solving Model (6.18), it is possible to obtain the minimum input share required by units h and k to produce the total output level of these two units with a predefined efficiency level.

But if the inputs of the merged unit are necessarily equal to the sum of the inputs of the units before the merger and a certain level of efficiency is to be met, then the level of the outputs of the merged unit needs to be specified. In a similar way, Soltanifar et al. (2024) proposed a nonlinear model (6.19) which can be converted into a linear model (6.20) by setting $\lambda_{rj}^{\dagger} = \lambda_j \left(\beta_{rh} + \beta_{rk}\right)\left(\forall r, \forall j \in F\right)$. They also set the necessary conditions for targeting the efficiency level to achieve the feasibility conditions of Model (6.19) as $\bar{\theta} \ge \theta^*$, where $\left(\theta^*, \boldsymbol{\lambda}^*\right)$ is the optimal solution of Model (6.21).

$$\max \sum_{r} \left(\beta_{rh} + \beta_{rk}\right)$$
$$s.t.$$
$$\sum_{j \in F} \lambda_j \left(\frac{\dfrac{\beta_{rh} + \beta_{rk}}{x_{ih} + x_{ik}}}{\dfrac{y_{rj}}{x_{ij}}} \right) \le \bar{\theta}, \ \forall r, \forall i$$
$$\sum_{j \in F} \lambda_j \left(\frac{\beta_{rh} + \beta_{rk}}{y_{rj}} \right) = \bar{\theta}, \ \forall r \qquad (6.19)$$
$$\sum_{j \in F} \lambda_j = 1$$
$$y_{rh} + y_{rk} \le \beta_{rh} + \beta_{rk}, \forall r$$
$$\lambda_j \ge 0, \ \forall j \in F$$

$$\max \sum_{r} \left(\beta_{rh} + \beta_{rk}\right)$$
$$s.t.$$
$$\sum_{j \in F} \left(\frac{\dfrac{\lambda_{rj}^{\dagger}}{x_{ih} + x_{ik}}}{\dfrac{y_{rj}}{x_{ij}}} \right) \le \bar{\theta}, \ \forall r, \forall i$$
$$\sum_{j \in F} \left(\frac{\lambda_{rj}^{\dagger}}{y_{rj}} \right) = \bar{\theta}, \ \forall r \qquad (6.20)$$
$$\sum_{j \in F} \lambda_{rj}^{\dagger} = \beta_{rh} + \beta_{rk}, \forall r$$
$$y_{rh} + y_{rk} \le \beta_{rh} + \beta_{rk}, \forall r$$
$$\beta_{rh}, \beta_{rk} \ge 0, \forall r$$
$$\lambda_{rj}^{\dagger} \ge 0, \forall r, \forall j \in F$$

$$
\begin{aligned}
&\min \theta \\
&s.t. \\
&\sum_{j\in F}\lambda_j\left(\frac{\dfrac{y_{rh}+y_{rk}}{x_{ih}+x_{ik}}}{\dfrac{y_{rj}}{x_{ij}}}\right)\leq\theta,\ \forall r,\forall i \\
&\sum_{j\in F}\lambda_j\left(\frac{y_{rh}+y_{rk}}{y_{rj}}\right)=\theta,\ \forall r \\
&\sum_{j\in F}\lambda_j=1 \\
&\lambda_j\geq 0,\ \forall j\in F \\
&\theta \text{ is free}
\end{aligned}
\tag{6.21}
$$

Although the presented models were expressed for the merging of two units due to the simplicity of the presentation, these models can be easily generalized to the case where several units are supposed to merge with each other. In the following examples, the models presented by Soltanifar et al. (2024) are analyzed by a simple numerical example.

Example 1

In this section, the estimation of input and output for merging inefficient DMUs is examined. For this purpose, 12 DMUs with two inputs and one output are considered in the form of Table 6.1. In Table 6.1, the efficiency of the DMUs using Models (6.7) and (6.8) is also given.

Model (6.18) should be used to estimate the inputs of the merged unit resulting from the merging of DMU5 and DMU7 and considering the efficiency results of Model (6.7). This model gives the results presented in Table 6.2 for different targets.

Also, if DMU5 and DMU7 are to be merged based on the efficiency of Model (6.8), Model (6.21) can estimate the outputs of the merged unit. Table 6.3 shows the results of estimating outputs for different targets.

TABLE 6.1 Inputs and Outputs of DMUs in Example 1

	Input 1	Input 2	Output	Model (6.7)	Model (6.8)
DMU_1	35	120	226	1	1
DMU_2	19	229	118	0.790974	0.609672
DMU_3	17	427	112	0.682896	0.375462
DMU_4	26	125	187	0.999772	1
DMU_5	31	329	131	0.516534	0.328123
DMU_6	25	531	535	1	1
DMU_7	19	426	126	0.649399	0.366642
DMU_8	20	525	419	1	1
DMU_9	25	123	175	1	1
DMU_{10}	23	423	219	0.681075	0.505651
DMU_{11}	23	223	328	1	1
DMU_{12}	11	320	104	1	1

TABLE 6.2 Estimation Results of Inputs in Merging of DMU5 and DMU7

Targets	Input 1	Input 2
0.6	34.18152	336.45524
0.666	30.74286	302.60784
0.75	27.28817	268.60274

TABLE 6.3 Output Estimation Results in Merging of DMU5 and DMU7

Targets	Output
0.700	374.50
0.650	347.75
0.60	321.00

6.3.2 Negative Data

In a merger and acquisition process, the presence of negative data is very likely even for ratio data. The inverse DEA-R models presented in the previous section are not capable of handling negative data, so these models should be redesigned in the presence of negative data. In Section 6.2.2, DEA-R models for working with negative data were presented. These models are able to handle negative ratio data. In this section, the merger and acquisition models of the previous section will be presented in the presence of negative data.

As it was presented in Section 6.3.1, in this section, it is also assumed that after determining the necessity of merging two specific units in a system of homogeneous units, managers set a target for the efficiency score of the merged unit. The outputs (inputs) of the units after the merger are supposed to be equal to the sum of the outputs (inputs) of the units before the merger, and models are provided to determine the inputs (outputs) of the units after the merger. The merger will take place according to the specified targeting for the efficiency level of the merged unit; with the difference in this case being that the inputs (outputs) are divided into two categories. A category that only contains positive data and a category that can be positive or negative. This concept is presented with output-oriented models in the presence of negative inputs and input-oriented models in the presence of negative outputs.

Suppose that based on a management decision, units h and k should be merged to create a new entity called unit M. The purpose is to minimize the inputs of the unit M while maintaining a specific efficiency score ($\bar{\theta}$), without changing its outputs. Also, suppose that the inputs are divided into two categories; the category of positive inputs (I') and the category of inputs that can be negative (I''). Therefore, according to what was said in Section 6.2.2, ith input of the merged entity M is either $\alpha_{ih}+\alpha_{ik},\left(\forall i\in I'\right)$ or $\left(\alpha_{ih}^{P}+\alpha_{ik}^{P}\right)-\left(\alpha_{ih}^{N}+\alpha_{ik}^{N}\right),\left(\forall i\in I''\right)$ (depending on which category of inputs i belongs to). In this way, Soltanifar et al. (2023) presented inverse DEA-R output-oriented models in the presence of negative inputs based on Models (6.10) and (6.12), in Models (6.21) and (6.22) models, respectively. They also proved that these models are feasible if the merged entity M is placed inside the production possibility set (PPS) before the merger.

$$
\begin{aligned}
&\min \sum_{i\in I'}\left(\alpha_{ih}+\alpha_{ik}\right)+\sum_{i\in I''}\left[\left(\alpha_{ih}^{P}+\alpha_{ik}^{P}\right)-\left(\alpha_{ih}^{N}+\alpha_{ik}^{N}\right)\right]\\
&\text{s.t.}\\
&\sum_{j\in F}\lambda_j\frac{x_{ij}}{y_{rj}}+\lambda_M\left(\frac{\alpha_{ih}+\alpha_{ik}}{y_{rh}+y_{rk}}\right)\le\bar{\theta}\left(\frac{\alpha_{ih}+\alpha_{ik}}{y_{rh}+y_{rk}}\right),\ \forall r,\forall i\in I'\\
&\sum_{j\in F}\lambda_j\frac{x_{ij}^{P}}{y_{rj}}+\lambda_M\left(\frac{\alpha_{ih}^{P}+\alpha_{ik}^{P}}{y_{rh}+y_{rk}}\right)\le\bar{\theta}\left(\frac{\alpha_{ih}^{P}+\alpha_{ik}^{P}}{y_{rh}+y_{rk}}\right),\ \forall r,\forall i\in I''\\
&\sum_{j\in F}\lambda_j\frac{x_{ij}^{N}}{y_{rj}}+\lambda_M\left(\frac{\alpha_{ih}^{N}+\alpha_{ik}^{N}}{y_{rh}+y_{rk}}\right)\ge\bar{\theta}\left(\frac{\alpha_{ih}^{N}+\alpha_{ik}^{N}}{y_{rh}+y_{rk}}\right),\ \forall r,\forall i\in I''\\
&\sum_{j\in F}\lambda_j\left(\frac{y_{rh}+y_{rk}}{y_{rj}}\right)+\lambda_M=\bar{\theta},\ \forall r\\
&\sum_{j\in F}\lambda_j+\lambda_M=1\\
&0\le\alpha_{ih}\le x_{ih},0\le\alpha_{ik}\le x_{ik},\forall i\in I'\\
&0\le\alpha_{ih}^{P}\le x_{ih}^{P},0\le\alpha_{ik}^{P}\le x_{ik}^{P},\alpha_{ih}^{N}\ge x_{ih}^{N}\ge 0,\alpha_{ik}^{N}\ge x_{ik}^{N}\ge 0,\forall i\in I''\\
&\lambda_j\ge 0,\ \forall j\in F
\end{aligned}
\tag{6.22}
$$

$$\min \sum_{i\in I'} (\alpha_{ih} + \alpha_{ik}) + \sum_{i\in I''} \left[(\alpha_{ih}^P + \alpha_{ik}^P) - (\alpha_{ih}^N + \alpha_{ik}^N) \right]$$

$$s.t.$$

$$\sum_{j\in F} \lambda_j \frac{x_{ij}}{y_{rj}} + \lambda_M \left(\frac{\alpha_{ih} + \alpha_{ik}}{y_{rh} + y_{rk}} \right) \le (1+\bar{\theta}) \left(\frac{\alpha_{ih} + \alpha_{ik}}{y_{rh} + y_{rk}} \right), \forall r, \forall i \in I'$$

$$\sum_{j\in F} \lambda_j \frac{x_{ij}^P}{y_{rj}} + \lambda_M \left(\frac{\alpha_{ih}^P + \alpha_{ik}^P}{y_{rh} + y_{rk}} \right) \le (1+\bar{\theta}) \left(\frac{\alpha_{ih}^P + \alpha_{ik}^P}{y_{rh} + y_{rk}} \right), \forall r, \forall i \in I''$$

$$\sum_{j\in F} \lambda_j \frac{x_{ij}^N}{y_{rj}} + \lambda_M \left(\frac{\alpha_{ih}^N + \alpha_{ik}^N}{y_{rh} + y_{rk}} \right) \ge (1-\bar{\theta}) \left(\frac{\alpha_{ih}^N + \alpha_{ik}^N}{y_{rh} + y_{rk}} \right), \forall r, \forall i \in I'' \quad (6.23)$$

$$\sum_{j\in F} \lambda_j \left(\frac{y_{rh} + y_{rk}}{y_{rj}} \right) + \lambda_M \le (1+\bar{\theta}), \forall r$$

$$\sum_{j\in F} \lambda_j + \lambda_M = 1$$

$$0 \le \alpha_{ih} \le x_{ih}, 0 \le \alpha_{ik} \le x_{ik}, \forall i \in I'$$

$$0 \le \alpha_{ih}^P \le x_{ih}^P, 0 \le \alpha_{ik}^P \le x_{ik}^P, \alpha_{ih}^N \ge x_{ih}^N \ge 0, \alpha_{ik}^N \ge x_{ik}^N \ge 0, \forall i \in I''$$

$$\lambda_j \ge 0, \forall j \in F$$

Assuming that the merged unit is in the pre-merger PPS, Soltanifar et al. (2023) showed that $\lambda_M = 0$ and the nonlinear models (6.22) and (6.23) become linear models. The objective of both models is to find the minimum additional inputs that the merged unit should use in order to realize a given efficiency target $\bar{\theta}$, and F is the set of available units in the evaluation process after merging.

Similarly, if the inputs of the merged unit are necessarily equal to the sum of the inputs of the units before the merger and a certain level of efficiency is met, it is necessary to specify the level of outputs of the merged unit. Soltanifar et al. (2023) presented Models (6.24) and (6.25) based on Models (6.9) and (6.11) and assuming the merged entity M is placed in the pre-merger PPS.

$$\max \sum_{r\in O'}\left(\beta_{rh}+\beta_{rk}\right)+\left[\sum_{r\in O''}\left(\beta_{rh}^{P}+\beta_{rk}^{P}\right)-\left(\beta_{rh}^{N}+\beta_{rk}^{N}\right)\right]$$

$$
\begin{aligned}
&s.t.\\
&\sum_{j\in F}\lambda_j\frac{y_{rj}}{x_{ij}}\geq\bar{\theta}\left(\frac{\beta_{rh}+\beta_{rk}}{x_{ih}+x_{ik}}\right),\forall i,\forall r\in O'\\
&\sum_{j\in F}\lambda_j\frac{y_{rj}^{P}}{x_{ij}}\geq\bar{\theta}\left(\frac{\beta_{rh}^{P}+\beta_{rk}^{P}}{x_{ih}+x_{ik}}\right),\forall i,\forall r\in O''\\
&\sum_{j\in F}\lambda_j\frac{y_{rj}^{N}}{x_{ij}}\leq\bar{\theta}\left(\frac{\beta_{rh}^{N}+\beta_{rk}^{N}}{x_{ih}+x_{ik}}\right),\forall i,\forall r\in O''\\
&\sum_{j\in F}\lambda_j\left(\frac{x_{ih}+x_{ik}}{x_{ij}}\right)=\bar{\theta},\forall i\\
&\sum_{j\in F}\lambda_j=1\\
&\beta_{rh}+\beta_{rk}\geq y_{rh}+y_{rk},\forall r\in O'\\
&\beta_{rh}^{P}+\beta_{rk}^{P}\geq y_{rh}^{P}+y_{rk}^{P},\beta_{rh}^{N}+\beta_{rk}^{N}\leq y_{rh}^{N}+y_{rk}^{N},\forall r\in O''\\
&\lambda_j\geq 0,\forall j\in F
\end{aligned}
\tag{6.24}
$$

$$\max \sum_{r\in O'}\left(\beta_{rh}+\beta_{rk}\right)+\left[\sum_{r\in O''}\left(\beta_{rh}^{P}+\beta_{rk}^{P}\right)-\left(\beta_{rh}^{N}+\beta_{rk}^{N}\right)\right]$$

$$
\begin{aligned}
&s.t.\\
&\sum_{j\in F}\lambda_j\frac{y_{rj}}{x_{ij}}\geq\left(1+\bar{\theta}\right)\left(\frac{\beta_{rh}+\beta_{rk}}{x_{ih}+x_{ik}}\right),\forall i,\forall r\in O'\\
&\sum_{j\in F}\lambda_j\frac{y_{rj}^{P}}{x_{ij}}\geq\left(1+\bar{\theta}\right)\left(\frac{\beta_{rh}^{P}+\beta_{rk}^{P}}{x_{ih}+x_{ik}}\right),\forall i,\forall r\in O''\\
&\sum_{j\in F}\lambda_j\frac{y_{rj}^{N}}{x_{ij}}\leq\left(1-\bar{\theta}\right)\left(\frac{\beta_{rh}^{N}+\beta_{rk}^{N}}{x_{ih}+x_{ik}}\right),\forall i,\forall r\in O''\\
&\sum_{j\in F}\lambda_j\left(\frac{x_{ih}+x_{ik}}{x_{ij}}\right)\leq\left(1+\bar{\theta}\right),\forall i\\
&\sum_{j\in F}\lambda_j=1\\
&\beta_{rh}+\beta_{rk}\geq y_{rh}+y_{rk},\forall r\in O'\\
&\beta_{rh}^{P}+\beta_{rk}^{P}\geq y_{rh}^{P}+y_{rk}^{P},\beta_{rh}^{N}+\beta_{rk}^{N}\leq y_{rh}^{N}+y_{rk}^{N},\forall r\in O''\\
&\lambda_j\geq 0,\forall j\in F
\end{aligned}
\tag{6.25}
$$

The presented models were expressed for the sake of easier understanding of two units. These models can be easily extended to multi-unit merging mode. In the following, the models presented by Soltanifar et al. (2023) are analyzed with a simple numerical example.

Example 2

Consider 12 DMUs with three inputs and one output as presented in Table 6.4. The first input can have negative values. The efficiency of units using Model (6.12) is also presented in this table.

The estimation of the inputs that should be used by the merged unit resulting from the merging of DMU8 and DMU10 using Model (6.23) is presented in Table 6.5.

TABLE 6.4 Inputs and Outputs of DMUs in Example 2

	Input 1	Input 2	Input 3	Output	Model (6.12)
DMU_1	−28	128	329	74	1
DMU_2	−12	406	682	36	1
DMU_3	47	138	608	54	1
DMU_4	40	308	174	80	1
DMU_5	−29	251	523	40	1
DMU_6	−27	142	496	58	1
DMU_7	68	250	140	42	1
DMU_8	44	423	590	68	0.54438897
DMU_9	2	121	371	83	1
DMU_{10}	48	354	580	78	0.48196692
DMU_{11}	−11	221	552	75	0.59938838
DMU_{12}	4	325	702	77	0.49467951

TABLE 6.5 Estimation Results of Inputs in Merging of DMU8 and DMU10

Targets	Input 1	Input 2	Input 3
0.55	1.93	117.063855	358.931325
0.6	2.11	127.706024	391.561446
0.65	2.29	138.348193	424.191566
0.7	2.46	148.990361	456.821687

6.4 INPUTS/OUTPUTS ESTIMATION

After evaluating the performance of a system with homogeneous units, it may be necessary to change the input or output level of the units. For example, in a resource allocation problem, it is possible to change the inputs of some units. Also, according to the needs of society, increasing the output level of some units may be placed on the agenda of the managers. One of the targets that may be followed in these changes is to estimate the outputs (when the inputs are changed) or to estimate the inputs (when the outputs are changed) of the units in such a way that the efficiency score of that unit is equal to a certain value, for example, the efficiency value before the changes. Such issues have been investigated in DEA literature with inverse models. In this section, inverse DEA models that are able to handle ratio data are presented to estimate inputs or outputs.

6.4.1 Outputs Estimation

Consider a system of homogeneous units with data in ratio form evaluated using the models of Section 6.2. Assume DMUp changes its input level from x_{ip} to $\alpha_{ip} = x_{ip} + \Delta x_{ip}, \left(\Delta x_{ip} \geq 0\right)$. Considering the output-oriented CRS classical inverse models while maintaining the efficiency score, the estimation of the outputs is done with Model (6.26), where $e_{\text{CRS Op}}$ is the efficiency of the pth unit before the change Ghiyasi et al. (2022).

$$
\begin{aligned}
&\max \left\{\beta_{rp}\right\}_{\forall r} \\
&\text{s.t.} \\
&\sum_j \lambda_j \frac{y_{rj}}{x_{ij}} \geq e_{\text{CRS Op}} \frac{\beta_{rp}}{\alpha_{ip}}, \ \forall r, \forall i \\
&\sum_j \lambda_j = 1 \\
&\lambda_j \geq 0, \forall j
\end{aligned}
\tag{6.26}
$$

Ghiyasi et al. (2022) presented the following definition and theorem to prove the above claim.

Definition 1

Suppose $(\boldsymbol{\lambda}, \boldsymbol{\beta}_p)$ is a feasible solution of Model (6.26); if there is no feasible solution $(\bar{\boldsymbol{\lambda}}, \bar{\boldsymbol{\beta}}_p)$ such that $\bar{\boldsymbol{\beta}}_p > \boldsymbol{\beta}_p$, then $(\boldsymbol{\lambda}, \boldsymbol{\beta}_p)$ is called a weak efficient solution.

Theorem 2

If $\left(\bar{\boldsymbol{\lambda}},\tilde{\boldsymbol{\beta}}_p\right)$ is a weak efficient solution of Model (6.26), then the efficiency score of new unit is the same as the efficiency score of DMUp.

Proof. Let $\left(\tilde{\boldsymbol{\lambda}},\tilde{\boldsymbol{\beta}}_p\right)$ be an efficient solution of Model (6.26). Model (6.27) should be used to obtain the efficiency score of the new DMU, i.e., $\left(\boldsymbol{\alpha}_p,\tilde{\boldsymbol{\beta}}_p\right)$.

$$\begin{aligned}
&\theta^* = \max \theta \\
&s.t. \\
&\sum_j \lambda_j \frac{y_{rj}}{x_{ij}} \geq \theta \frac{\tilde{\beta}_{rp}}{\alpha_{ip}}, \forall r, \forall i \\
&\sum_j \lambda_j = 1 \\
&\lambda_j \geq 0, \forall j \\
&\theta \text{ is free}
\end{aligned} \tag{6.27}$$

It is clear that $\left(\tilde{\boldsymbol{\lambda}}, e_{\text{CRS-Op}}\right)$ is a feasible solution for Model (6.27). If $\left(\boldsymbol{\lambda}^*,\theta^*\right)$ is the optimal solution of Model (6.27), then $\theta^* \geq e_{\text{CRS Op}}$. To prove $\theta^* = e_{\text{CRS Op}}$ by contradiction, suppose that $\theta^* > e_{\text{CRS-Op}}$. Also suppose $\theta^* = ke_{\text{CRS-Op}}$, so that $k > 1$. Thus, constraints (6.28) are true.

$$\sum_j \lambda_j^* \frac{y_{rj}}{x_{ij}} \geq \theta^* \frac{\tilde{\beta}_{rp}}{\alpha_{ip}} = e_{\text{CRS-Op}} \frac{k\tilde{\beta}_{rp}}{\alpha_{ip}}, \forall r, \forall i \tag{6.28}$$

$$\sum_j \lambda_j^* = 1$$

$$\lambda_j^* \geq 0, \forall j$$

This makes $\left(\boldsymbol{\lambda}^*, k\tilde{\boldsymbol{\beta}}_p\right)$ as a feasible solution for Model (6.26), and this contradicts with weak efficiency of $\left(\tilde{\boldsymbol{\lambda}},\tilde{\boldsymbol{\beta}}_p\right)$.

In order to estimate the outputs with the VRS assumption, Ghiyasi et al. (2022) presented Model (6.29) and proved the accuracy of their model.

$$\max \left\{ \beta_{rp} \right\}_{\forall r}$$
$$s.t.$$
$$\sum_j \lambda_j \frac{y_{rj}}{x_{ij}} \geq e_{\text{VRS Op}} \frac{\beta_{rp}}{\alpha_{ip}}, \ \forall r, \forall i$$
$$\sum_j \lambda_j \frac{\alpha_{ip}}{x_{ij}} = e_{\text{VRS Op}}, \ \forall i \tag{6.29}$$
$$\sum_j \lambda_j = 1$$
$$\lambda_j \geq 0, \ \forall j$$

In the following, the presented models are analyzed with a simple numerical example.

Example 3

Consider a system of homogeneous units with 12 DMUs, each receiving two inputs and producing two outputs, as shown in Table 6.6.

Table 6.7 shows the estimated output for DMU5, DMU6, and DMU7 after increasing inputs by 16% using Model (6.26). This table also shows the efficiency of each unit in *CRS-O* mode. It should be noted that the weighting method was used to solve Model (6.26).

TABLE 6.6 Inputs and Outputs of DMUs in Example 3.

	Input 1	**Input 2**	**Output 1**	**Output 2**
DMU_1	18	156	363	136
DMU_2	8	162	193	40
DMU_3	30	158	163	44
DMU_4	16	124	354	105
DMU_5	30	157	377	159
DMU_6	16	116	208	59
DMU_7	23	141	380	121
DMU_8	27	92	154	133
DMU_9	20	119	139	162
DMU_{10}	23	144	188	76
DMU_{11}	16	123	280	157
DMU_{12}	34	115	368	127

TABLE 6.7 Estimation of Outputs Resulting From a 16% Increase in Inputs

	$e_{\text{VRS-Op}}$	α_1	α_2	β^*_1	β^*_2
DMU_5	1.1518257	34.80000	182.12000	474.88320	159.00000
DMU_6	1.6120268	18.56000	134.56000	240.89907	72.62143
DMU_7	1.0941013	26.68000	163.56000	442.83690	138.57090

6.4.2 Inputs Estimation

In a similar way to what was presented in the previous section, suppose that the outputs change in a system with homogeneous units where the data is in the form of a ratio and the estimation of the inputs is considered with the condition that the efficiency value does not change. Ghiyasi et al. (2022) presented Model (6.30) to estimate the level of inputs under the condition that the efficiency value does not change after changing the output level of the hypothetical unit p to $\beta_{rp} = y_{rp} + \Delta y_{rp}, (\Delta y_{rp} > 0)$.

$$
\begin{aligned}
&\min \{\alpha_{ip}\}_{\forall i} \\
&s.t. \\
&\sum_j \lambda_j \frac{x_{ij}}{y_{rj}} \le e_{\text{CRS-Ip}} \frac{\alpha_{ip}}{\beta_{rp}}, \ \forall r, \forall i \\
&\sum_j \lambda_j = 1 \\
&\lambda_j \ge 0, \ \forall j
\end{aligned}
\tag{6.30}
$$

Model (6.30) is an input oriented model with CRS assumption. Ghiyasi et al. (2022) showed the accuracy of this model by proving Theorem 3.

Theorem 3

If $(\tilde{\boldsymbol{\lambda}}, \tilde{\boldsymbol{\alpha}}_{\mathbf{p}})$ is a weak efficient solution of Model (6.30), then the efficiency score of new unit is the same as efficiency score of DMUp.

Proof. It is the same as proof of Theorem 2.

In order to estimate the inputs with the VRS assumption, Ghiyasi et al. (2022) presented Model (6.31) and proved the accuracy of their model.

$$\begin{aligned}
&\min \left\{ \alpha_{ip} \right\}_{\forall i} \\
&\text{s.t.} \\
&\sum_j \lambda_j \frac{x_{ij}}{y_{rj}} \le e_{\text{VRS Ip}} \frac{\alpha_{ip}}{\beta_{rp}}, \ \forall r, \forall i \\
&\sum_j \lambda_j \frac{\beta_{rp}}{y_{rj}} = e_{\text{VRS Ip}}, \ \forall r \\
&\sum_j \lambda_j = 1 \\
&\lambda_j \ge 0, \forall j
\end{aligned} \tag{6.31}$$

In the following, the presented models are analyzed with a simple numerical example.

Example 4

In this section, the system presented in Example 3 is considered with the data presented in Table 6.6. Assuming a 23% increase in outputs, the estimation of inputs based on Model (6.30) for DMU5, DMU6, DMU7, and DMU10 are presented in Table 6.8.

It should be noted that input/output estimation models designed based on inverse DEA are not limited to the presented models. Mahla et al. (2022), taking into account the model presented by Emrouznejad and Amin (2009), as well as Sohrabi et al. (2022), presented models for this purpose. In some of these models, data in a ratio form and data in a fractional form with specific numerator and denominator are considered by the authors to provide suitable inverse DEA models.

TABLE 6.8 Estimation of Inputs Resulting From a 23% Increase in Outputs

	$e_{\text{VRS-Op}}$	α^*_1	α^*_2	β_1	β_2
DMU_5	0.9071160	43.80221	184.97850	463.71000	195.57000
DMU_6	0.6240630	18.52920	143.60131	255.84000	72.57000
DMU_7	0.9369573	34.69335	165.01367	467.40000	148.83000
DMU_{10}	0.5005563	40.58111	162.17427	231.24000	93.48000

In practical case studies, confidential financial data or predefined standards often necessitate the representation of data as ratios, making the utilization of DEA-R models indispensable. The application of DEA-R models is crucial for evaluating unit performance, estimating inputs and outputs, and facilitating processes such as mergers and acquisitions in numerous real-world scenarios. The need for inverse DEA-R models arises in many cases, emphasizing the managerial significance and innovativeness of their presentation.

6.5 CONCLUSION

Partnership and merger analysis have been vital issues for businesses, and their importance has become more apparent in the last few years during the pandemic when many businesses had to close or find a suitable partner to survive in the market. We observe different structures, different types of data, different policies, etc., in different businesses. It is important to consider all types of information in the analysis of mergers and partnerships. In this study, we discussed some recent developments in the inverse DEA and merger analysis that can be used for business partnership and merger analysis in the real world. This research approach is rather new, and many extensions can be developed for real world applications. Addressing the limitations of the presented models, one notable constraint lies in the determination of inputs and outputs as ratios, which should yield a proper managerial understanding of these variables in DEA. Furthermore, an area for future research entails designing models to estimate ratio inputs and outputs in the presence of negative data, thereby providing valuable insights for further investigations.

REFERENCES

Amin, G.R., & Al-Muharrami, S. (2018). A new inverse data envelopment analysis model for mergers with negative data. *IMA Journal of Management Mathematics*, *29*(2), 137–149.

Amin, G. R., & Emrouznejad, A. (2007). Inverse forecasting: A new approach for predictive modeling. *Computers & Industrial Engineering*, *53*(3), 491–498.

Amin, G. R., Emrouznejad, A., & Gattoufi, S. (2017). Minor and major consolidations in inverse DEA: Definition and determination. *Computers & Industrial Engineering*, *103*, 193–200.

Amin, G. R., & Ibn Boamah, M. (2020). A new inverse DEA cost efficiency model for estimating potential merger gains: a case of Canadian banks. *Annals of Operations Research*, *195*(1), 21–36.

Charnes, A., Cooper, W., & Rhodes, E. (1978). Measuring the efficiency of decision making units. *European Journal of Operational Research*, *2*(6), 429–444.

Despić, O. (2013). Some properties of geometric DEA models. *Croatian Operational Research Review (CRORR)*, *4*, 2–18.

Despić, O., Despić, M., & Paradi, J. (2007). DEA-R: ratio-based comparative efficiency model, its mathematical relation to DEA and its use in applications. *Journal of Productivity Analysis*, *28*(1), 33–44.

Emrouznejad, A., & Amin, G. R. (2009). DEA models for ratio data: Convexity consideration. *Applied Mathematical Modelling*, *33*(1), 486–498.

Emrouznejad, A., Amin, G. R., Ghiyasi, M., & Michali, M. (2023). A review of inverse data envelopment analysis: origins, development, and future directions. *IMA Journal of Management Mathematics*, *34*(3), 421–440.

Emrouznejad, A., Amin, G. R., Thanassoulis, E., & Anouze, A. L. (2010a). On the boundedness of the SORM DEA models with negative data. *European Journal of Operational Research*, *206*(1), 265–268.

Emrouznejad, A., Anouze, A. L., & Thanassoulis, E. (2010b). A semi-oriented radial measure for measuring the efficiency of decision making units with negative data, using DEA. *European Journal of Operational Research*, *200*(1), 297–304.

Emrouznejad, A., Yang, G.-I., & Amin, G. R. (2019). A novel inverse DEA model with application to allocate the CO_2 emissions quota to different regions in Chinese manufacturing industries. *Journal of the Operational Research Society*, *70*(7), 1079–1090.

Emrouznejad, A., & Yang, G. L. (2018). A survey and analysis of the first 40 years of scholarly literature in DEA: 1978–2016. *Socio-Economic Planning Sciences*, *61*, 4–8.

Gattoufi, S., Amin, G. R., & Emrouznejad, A. (2014). A new inverse DEA method for merging banks. *IMA Journal of Management Mathematics*, *25*, 73–87.

Gerami, J., Mozaffari, M. R., Wanke, P., & Correa, H. (2022). A novel slacks-based model for efficiency and super-efficiency in DEA-R. *Operational Research*, 22, 3373–3410.

Ghiyasi, M. (2015). On inverse DEA model: The case of variable returns to scale. *Computers & Industrial Engineering*, *87*, 407–409.

Ghiyasi, M. (2017a). Industrial sector environmental planning and energy efficiency of Iranian provinces. *Journal of Cleaner Production*, *142*, 2328–2339.

Ghiyasi, M. (2017b). Inverse DEA based on cost and revenue efficiency. *Computers & Industrial Engineering*, *114*, 258–263.

Ghiyasi, M., Soltanifar, M., & Sharafi, H. (2022). A novel inverse DEA-R model with application in hospital efficiency. *Socio-Economic Planning Sciences*, *84*, 101427. https://doi.org/10.1016/j.seps.2022.101427

Ghobadi, S., Soleimani-Chamkhoram, K., & Zanboori, E. (2023). Simultaneous estimation of input–output levels under improving efficiency level in an assessment window. *Soft Computing*, 27, 4553–4577.

Kaffash, S., Kazemi Matin, R., & Tajik, M. (2018). A directional semi-oriented radial DEA measure: an application on financial stability and the efficiency of banks. *Annals of Operations Research*, *264*, 213–234.

Lertworasirikul, S., Charnsethikul, P., & Fang, S.-C. (2011). Inverse data envelopment analysis model to preserve relative efficiency values: The case of variable returns to scale. *Computers & Industrial Engineering*, *61*(4), 1017–1023.

Mahla, D., Agarwal, S., Amin, G. R., & Mathur, T. (2022). An inverse data envelopment analysis model to consider ratio data and preferences of decision-makers. *IMA Journal of Management Mathematics*. https://doi.org/10.1093/imaman/dpac009

Mozaffari, M., Gerami, J., & Jablonsky, J. (2014a). Relationship between DEA models without explicit inputs and DEA-R models. *Central European Journal of Operations Research*, *22*(1), 1–12.

Mozaffari, M., Kamyab, P., Jablonsky, J., & Gerami, J. (2014b). Cost and revenue efficiency in DEA-R models. *Computers & Industrial Engineering*, *78*(12), 188–194.

Mozaffari, M. R., Dadkhah, F., Jablonsky, J., & Wanke, P. F. (2020). Finding efficient surfaces in DEA-R models. *Applied mathematics and computation*, *386*, 125497.

Mozaffari, M. R., Gerami, J., Wanke, P. F., Kamyab, P., & Peyvas, M. (2022). Ratio-based data envelopment analysis: An interactive approach to identify benchmark. *Results in Control and Optimization*, *6*, 100081.

Olesen, O. B., Petersen, N. C., & Podinovski, V. V. (2017). Efficiency measures and computational approaches for data envelopment analysis models with ratio inputs and outputs. *European Journal of Operational Research*, *261*, 640–655.

Sayar, T., Ghiyasi, M., & Fathali, J. (2021). New inverse DEA models for budgeting and planning. *RAIRO-Operations Research*, *55*(3), 1933–1948.

Shiri Daryani, Z., Tohidi, G., Daneshian, B., Razavyan, S., & Hosseinzadeh Lotfi, F. (2021). Inverse DEA in two-stage systems based on allocative efficiency. *Journal of Intelligent Fuzzy Systems*, *40*(1), 591–603.

Sohrabi, A., Gerami, J., & Mozaffari, M. M. (2022). A novel inverse DEA-R model for inputs/outputs estimation. *Journal of Mathematical Extension*, *16*(8), 1–34. https://doi.org/10.30495/JME.2022.2047

Soleimani-Chamkhorami, K., Hosseinzadeh Lotfi, F., Jahanshahloo, G., & Rostamy-Malkhalifeh, M. (2020). A ranking system based on inverse data envelopment analysis. *Journal of Management Mathematics*, *31*(3), 367–385.

Soltanifar, M., Ghiyasi, M., Emrouznejad, A., & Sharafie, H. (2024). A novel model for merger analysis and target setting: A Csw-inverse DEA approach. *Expert Systems with Applications*, 249, Part A, 123326.

Soltanifar, M., Ghiyasi, M., & Sharafi, H. (2024). *Inverse DEA-R models for merger analysis of electricity distribution units.* Opsearch, https://doi.org/10.1007/s12597-024-00901-x

Soltanifar, M., Ghiyasi, M., & Sharafi, H. (2023). Inverse DEA-R models for merger analysis with negative data. *IMA Journal of Management Mathematics*, 34(3), 491–510.

Tohidnia, S., & Tohidi, G. (2019). Measuring productivity change in DEA-R: A ratio-based profit efficiency model. *Journal of the Operational Research Society*, *70*(9), 1511–1521.

Wanke, P., Ostovan, S., Mozaffari, M. R., Gerami, J., & Tan, Y. (2023). Stochastic network DEA-R models for two-stage systems. *Journal of Modelling in Management*, *18*(3), 842–875. https://doi.org/10.1108/JM2-10-2021-0256

Wei, C., Chen, L., Li, R., & Tsai, C. (2011). Using the DEA-R model in the hospital industry to study the pseudo-inefficiency problem. *Expert Systems with Applications*, *38*(3), 2172–2176.

Wei, Q., Zhang, J., & Zhang, X. (2000). An inverse DEA model for inputs/outputs estimate. *European Journal of Operational Research*, *121*, 151–163.

CHAPTER 7

New Inverse Data Envelopment Analysis Models for Optimizing Greenhouse Gas Emissions Reduction in Business Mergers

Gholam R. Amin and Mustapha Ibn Boamah
Faculty of Business, University of New Brunswick, Saint John, Canada

7.1 INTRODUCTION

Climate change caused by greenhouse gas (GHG) emissions continues to be of global concern as its effects impact on all facets of life as we know it. The evidence of climate change is apparent in increasing global warming, increasing incidence of floods and droughts, and the rising frequency of other extreme weather patterns around the globe. Governments and businesses are working independently and in various partnerships to reduce GHG emissions and adapt to the effects of climate change. Thus, climate change and its impact on our planet has been one of hot research topics in the last two decades.

 DOI: 10.1201/9781003345206-7

Research in recent years has shown the potential capacity of data envelopment analysis (DEA) and inverse DEA in solving challenging decision problems in various sectors, especially for estimating potential merger gains (Wei et al. 2000; Gattoufi et al. 2014; Amin et al. 2017; Oukil, 2022). Estimating potential merger and acquisitions gains through developing DEA and inverse DEA models has been a growing area of optimization research in the last decade. The focus in these models is on how the merged entity can minimize the inputs of merging decision making units (DMUs) and can maximize the production of additional outputs. Recently, Amin and Ibn Boamah (2023) introduced the concept of business partnerships between DMUs under the umbrella of DEA. Their study provided definition and mathematical modeling of various forms of business partnerships and strategic alliances between different DMUs by redistributing resources that improve performance and comparative efficiency. The business partnership models developed in Amin and Ibn Boamah (2023) assumed that all the input and output measures were good and desirable factors, and thus redistributing them could enhance the performance of DMUs.

However, choosing potential merging DMUs is not always based on the input minimization and/or output maximization. A merger, instead, may want to reduce an undesirable output. GHG emission, as an undesirable output, produced directly by companies is one of the key issues in environmental efficiency that must be minimized to satisfy regulatory requirements. Companies may want to merge to reduce these environmental pollutants while seeking to maximize their desirable outputs. DEA is one of the popular optimization approaches in production theory and an appropriate research tool for studying environmentally undesirable measures.

In this chapter, we develop new inverse DEA models for estimating potential merger gains with the objective of minimizing GHG emissions and advancing knowledge on business and organizational partnerships. This would address an important climate change issue that would guide business firms and organizations to create mergers and partnerships that would reduce their GHG emissions, while maximizing their productive efficiency. Thus, this chapter defines and models appropriate potential mergers to guide organizations on how redistributing their inputs and outputs can help to reduce their undesirable environmental production.

7.2 LITERATURE REVIEW

Climate change caused by GHG emissions has been significant and at the forefront of the research efforts in the past two decades. It is generally accepted that human activities are causing environmental and climate crises and governments around the world are working on a solution (Busch et al. 2022). The Paris Agreement brings all nations into a common cause to undertake ambitious efforts to combat climate change and adapt to its effects. The agreement requires nations to undertake economic and social transformation, based on the best available science, to reduce GHG emissions. We all need to take a hard look at what can realistically be produced in a reduced carbon environment and what policies would most likely succeed at mitigating climate change. This chapter contributes to the important discussion on how business organizations can transition to a cleaner environment, as it emphasizes opportunities for potential GHG emissions reduction strategies when organizations choose to merge.

As a successful application of management science, DEA was introduced in Charnes et al. (1978) and further extended in Banker et al. (1984). It is a linear programming-based approach for performance and comparative efficiency measurement of different DMUs in the presence of multiple different inputs and multiple different outputs. It allows data to identify the best practice efficiency frontier without imposing them to follow any parametric production function forms. As an empirical comparative production approach, DEA determines the efficiency of DMUs and ranks them from the least to the most relative efficient DMUs. It captures valuable insights from the data and creates a quantitative basis for determining sources of inefficiency, management, and effectiveness of polices assessment, resource allocation and sensitivity analysis, and thus creating a thorough analysis that cannot be obtained from other approaches. The number of scientific publications of DEA and its application domain has increased significantly, and continues to grow, since its introduction in 1978. These areas are in sectors such as in higher education, finance and banking, manufacturing and service, oil and gas, environmental and energy, health care and hospitals, transportation, and many more (Yan et al. 2002; Dao et al. 2014; Xiao et al. 2016; Emrouznejad & Yang, 2018; Kaffash et al. 2020; Dyckhoff and Souren, 2022).

DEA has shown great promise in recent research in the energy and environmental efficiency sector. The objective in environmental efficiency is to minimize undesirable outputs while maximizing desirable production.

Chung et al. (1997) employed a directional distance DEA model that does not require shadow prices of bad outputs as a component in a productivity index and modeled joint production of good outputs and bad outputs, crediting firms for reductions in bad outputs and increases in good outputs. Chen and Golley (2014) explored a directional distance DEA model that incorporated carbon dioxide emissions as an undesirable output directly into the production technology, crediting sectors for simultaneously reducing their emissions and increasing their output, to study total factor productivity growth of China's industrial sectors. Halkos and Tzeremes (2013) proposed conditional directional distance functions, that included the effect of regional economic growth on regions' environmental efficiency levels, to study the relationship between regions' GDP per capita and environmental inefficiency in the United Kingdom. Halkos and Polemis (2018) extended their radial and non-radial efficiency measurements in DEA to model the relationship between the calculated environmental efficiencies and economic growth in attaining sustainability of the electricity sector in the United States. Chang et al. (2013) introduced a non-radial DEA model to analyze the environmental efficiency of China's transportation sector. Xie et al. (2011) proposed a fixed-cost allocation model based on DEA and the Nash non-cooperative game theory to allocate water pollutant discharge permits in a scheme that considers environmental efficiency. Emrouznejad et al. (2019) developed a DEA approach to realize a goal of allocating carbon dioxide (CO_2) emission quota under several assumptions. Their model solves the problem of allocating CO_2 emissions reduction targets set by the Chinese government for manufacturing companies in different Chinese regions.

In particular, DEA approaches are being used in many GHG emission reduction studies, highlighting the importance of DEA as a research tool in understanding and managing the environmental impacts of GHG emissions. Lu et al. (2022) introduced a multi-stage network DEA that considered parallel production in the agricultural and industrial sectors, to assess the impact of GHG emissions on climate change and natural disaster stages in different countries. In another agricultural study, Dakpo et al. (2017) suggested a DEA model to evaluate the GHG emissions in sheep meat breeding farms in French grassland areas by finding the maximal attainable ratio of the desirable agricultural production to GHG emissions. Still in the agricultural sector, Abbas et al. (2022) introduced combined DEA and regression approach to investigate the sensitivity analysis of GHG emissions in the farm level by downgrading or upgrading the number of inputs to limit greenhouse emissions and enhance crop yield. And finally, Shen et al. (2022) developed a novel DEA approach in the European agriculture sector that includes the energy-related GHG emissions in the DEA model.

GHG emissions reduction DEA models have also been proposed in non-agricultural sectors. Wegener and Amin (2019) developed a novel DEA model for minimizing the GHG emissions. The Wegener and Amin (2019) model that was applied to the oil and gas sector for optimizing GHG emissions ascertains the minimum level of GHG emissions for a group of oil and gas companies given target production levels. Gan et al. (2022) proposed a carbon emission allocation mechanism used to apportion carbon emission quotas to public buildings to meet China's carbon emission target. Zhu et al. (2022) introduced DEA models to assess the environmental efficiency of reducing GHG emissions of each European Union member state under three different decision objectives given fixed environmental costs. Halkos and Petrou (2019), in another European study, used DEA to evaluate the environmental efficiency of reducing GHG, sulfur oxide, and nitrogen oxide emissions of 28 European Union member states.

The emphasis in the above studies is on how DMUs can reduce GHG emissions individually. This chapter introduces novel inverse DEA models to steer companies in a new direction to reduce GHG emissions by forming potential mergers. A strategic partnership, including mergers, between profit maximizing companies or DMUs, may be to work together to achieve a collective goal of minimizing cost, maximizing revenue, or reducing undesirable outputs such as GHG emissions to meet regulatory requirements.

One of the interesting applications of inverse DEA is estimating potential gains of mergers and acquisitions. This application was first developed in Gattoufi et al. (2014). To determine the optimal quantity of inherited inputs from merging DMUs to be used by the merged entity for achieving efficiency targets, Gattoufi et al. (2014) developed an input-oriented inverse DEA model. This was followed by developing an output-oriented inverse DEA model in Gattoufi et al. (2014) to identify the maximum quantity of additional outputs that the merged entity can produce by taking the entire inputs of merging DMUs. Obviously, the optimal quantity of inputs and outputs in these inverse DEA models will be obtained for given efficiency targets. The inverse DEA models in Gattoufi et al. (2014) were applied to an application in banking sector. Following the novel application of the inverse DEA in estimating merger gains, the literature has been developed in the last decade. The concept of merger and acquisitions where a set of pre-merger DMUs create a single post-merger entity was extended in Amin et al. (2017) to a new and more general structure called generalized restructuring where a set of pre-restructuring DMUs create a new set of post-restructuring DMUs. They called a generalized restructuring a consolidation if the number of pre-restructuring DMUs is strictly greater than that of the post-restructuring units. Another generalized restructuring in Amin et al. (2017) was called a split if the number of pre-restructuring DMUs is strictly less than that of the post-restructuring DMUs. Amin and Ibn Boamah (2020) suggested inverse DEA models based on cost efficiency to estimate the potential gains of mergers with an application to Canadian banks. Guijarro et al. (2020) developed a genetic algorithm for optimal implementation of the inverse DEA models for estimating potential merger gains. Their genetic algorithm was applied to case studies from banking and higher education to determine merger cases that minimize the input consumption or maximize the output production by the merged entity.

Amin and Al-Muharrami (2018) proposed inverse DEA models for estimating merger gains when there is negative data. Amin and Oukil (2019) developed a novel inverse DEA approach for generating flexible target setting in estimating potential merger gains.

Soltanifar et al. (2022) introduced inverse DEA models for estimating potential gains of merging DMUs when there is ratio and negative data. It is worth noting that when there is negative and ratio data then the base optimization DEA model(s) will be different from the standard DEA and hence Soltanifar et al. (2022) developed appropriate inverse DEA models for such a scenario.

All the above studies attempted to obtain target settings of the merged entity by focusing on input minimization or output maximization. However, this chapter introduces inverse DEA models to reduce GHG emissions as the merger gain that outperforms existing merger gains' estimation models.

7.3 ESTIMATING MERGER GAINS

Gattoufi et al. (2014) originated a novel application of inverse DEA for estimating merger gains. They developed input and output orientation inverse DEA models to identify the optimal quantity of inputs and outputs for the merged entity. In their input orientation model, the objective is to minimize the amount of the inputs from the merging DMUs that are required to produce the outputs of merging DMUs as shown below.

$$
\begin{aligned}
&\min \sum_{i=1}^{m} \left(\alpha_{ik} + \alpha_{il}\right) \\
&s.t. \\
&\qquad \sum_{j \in F} x_{ij}\lambda_j + \left(x_{ik} + x_{il}\right)\lambda_{\mathrm{M}} - \left(\alpha_{ik} + \alpha_{il}\right)\bar{\theta} \le 0 \quad i = 1,\cdots,m \\
&\qquad \sum_{j \in F} y_{rj}\lambda_j + \left(y_{rk} + y_{rl}\right)\lambda_{\mathrm{M}} \ge y_{rk} + y_{rl} \quad r = 1,\cdots,s \\
&\qquad \sum_{j \in F} \lambda_j + \lambda_M = 1 \\
&\qquad 0 \le \alpha_{ik} \le x_{ik}, 0 \le \alpha_{il} \le x_{il} \quad i = 1,\cdots,m \\
&\qquad\quad \lambda_j \ge 0, \; j \in F, \lambda_M \ge 0
\end{aligned}
\tag{7.1}
$$

where α_{ik} and α_{il} denote, respectively, the levels of the ith input from the merging DMU_k and DMU_l that will be kept by the merged entity, λ_j is the intensity variable (for each $j \in F$), and F is the set of available peers in the post-merger evaluation process. It is worth noting that the coefficient of ë$_{\mathrm{M}}$ in the original inverse DEA model in Gattoufi et al. (2014) is $\left(\alpha_{ik} + \alpha_{il}\right)$ for the ith input constraint $\left(i = 1,\cdots,m\right)$ which was modified to Model (7.1) in Amin and Ibn Boamah (2020), as all the inputs are available to the merged entity and to ensure that the merged entity is presented by itself when it cannot be written as a convex combination of the post-merger DMUs in F.

The estimation of potential merger gains in Model (7.1) is equivalent to identifying the quantity of inputs of each merging DMU for the merged entity to realize a given efficiency target, denoted by $\bar{\theta}$. Amin et al. (2017) extended the ordinary merger and acquisition scenario to a business move called generalized restructuring when the inputs of a set of pre-merger units will be redistributed to create a set of post-merger units.

Gattoufi et al. (2014) also developed an output-oriented inverse DEA model for estimating merger gains. The gains in the output orientation are to produce additional outputs by the merged entity as much as possible as shown in the model below.

$$
\begin{aligned}
& \max \sum_{r=1}^{s} \delta_r \\
& s.t. \\
& \quad \sum_{j \in F} x_{ij}\lambda_j + \left(x_{ik} + x_{il}\right)\lambda_M \le x_{ik} + x_{il} \quad i = 1,\cdots,m \\
& \quad \sum_{j \in F} y_{rj}\lambda_j + \left(y_{rk} + y_{rl} + \delta_r\right)\lambda_M \ge \bar{h}\left(y_{rk} + y_{rl} + \delta_r\right) \quad r = 1,\cdots,s \\
& \quad \sum_{j \in F} \lambda_j + \lambda_M = 1 \\
& \quad \quad \lambda_j \ge 0, \ j \in F, \lambda_M \ge 0, \ \delta_r \ge 0, \ r = 1,\cdots,s
\end{aligned}
\tag{7.2}
$$

where δ_r indicates the additional of the rth output $\left(r = 1,\cdots,s\right)$ produced by the merged entity and $\bar{h}$ is the given parameter for the output orientation efficiency target.

7.4 MINIMIZING UNDESIRABLE OUTPUTS

In the following section, we extend the inverse DEA models explained in the previous section by incorporating undesirable outputs. Let us assume there are n DMUs each uses m inputs and produces s good or desirable outputs and q bad or undesirable outputs. Specifically, let us assume DMU-j uses m inputs $x_{ij}\left(i = 1,\ldots,m\right)$ to make s desirable (good) outputs $y_{rj}^{g}\left(r = 1,\ldots,s\right)$ and q undesirable (bad) outputs $y_{pj}^{b}\left(p = 1,\ldots,q\right)$ for each $j = 1,\ldots,n$.

To obtain the efficiency score of DMU-k $\left(k = 1,\ldots,n\right)$, we use the directional distance DEA model developed in Chung et al. (1997) as shown below.

$$
\begin{aligned}
& \beta_k^* = \max \beta \\
& \quad s.t. \\
& \quad \sum_{j=1}^{n} x_{ij}\lambda_j \le x_{ik} \quad i = 1,\cdots,m \\
& \quad \sum_{j=1}^{n} y_{rj}^{g}\lambda_j \ge \left(1 + \beta\right) y_{rk}^{g} \quad r = 1,\cdots,s \\
& \quad \sum_{j=1}^{n} y_{pj}^{b}\lambda_j = \left(1 - \beta\right) y_{pk}^{b} \quad p = 1,\cdots,q \\
& \quad \sum_{j=1}^{n} \lambda_j = 1 \\
& \quad \quad \lambda_j \ge 0, \ j = 1,\ldots,n
\end{aligned}
\tag{7.3}
$$

The optimal value of Model (7.3) provides the efficiency score of the DMU under assessment as $\left(1-\beta_k^*\right)/\left(1+\beta_k^*\right)$. It is easy to verify that the DMU under evaluation is efficient if and only if the optimal value of Model (7.3), or $\beta_k^*=0$, is zero.

In what follows, we develop inverse DEA models to obtain the optimal quantity of inputs and good outputs that would minimize the quantity of undesirable outputs allowing the merged entity to reach efficiency targets. Assume without the loss of generality that two merging DMUs *k* and *l* combine their resources to create a single post-merger entity denoted by *M*. The developed models in this chapter can be easily extended to more than two merging DMUs. The merged entity has the following target setting

$$M=\left(\boldsymbol{x}_M,\boldsymbol{y}_M^g,\boldsymbol{y}_M^b\right)$$

where $\boldsymbol{x}_M=\left(\alpha_1,\ldots,\alpha_m\right)$ in which $\alpha_i\le x_{ik}+x_{il}$, $\boldsymbol{y}_M^g=\left(y_{1k}^g+y_{1l}^g+\delta_1,\ldots,y_{sk}^g+y_{sl}^g+\delta_s\right)$ and $\boldsymbol{y}_M^b=\left(y_{1k}^b+y_{1l}^b-z_1,\ldots,y_{qk}^b+y_{ql}^b-z_q\right)$. The inverse DEA model below identifies the optimal quantity of the unknowns in the target setting of the merged entity *M*.

$$\begin{aligned}
&\max\ \left(z_1,\ldots,z_q\right)\\
&s.t.\\
&\qquad\sum_{j\in F}x_{ij}\lambda_j\le\alpha_i\quad i=1,\cdots,m\\
&\qquad\sum_{j\in F}y_{rj}^g\lambda_j\ge\left(1+\hat{\beta}\right)\left(y_{rk}^g+y_{rl}^g+\delta_r\right)\quad r=1,\cdots,s\\
&\qquad\sum_{j\in F}y_{pj}^b\lambda_j=\left(1-\hat{\beta}\right)\left(y_{pk}^b+y_{pl}^b-z_p\right)\quad p=1,\cdots,q\\
&\qquad\sum_{j\in F}\lambda_j=1\\
&\qquad 0\le\alpha_i\le x_{ik}+x_{il}\quad i=1,\cdots,m\\
&\qquad 0\le z_p\le y_{pk}^b+y_{pl}^b\quad p=1,\cdots,q\\
&\qquad\quad\delta_r\ge 0\quad r=1,\cdots,s\\
&\qquad\lambda_j\ge 0,\ j\in F
\end{aligned}\tag{7.4}$$

where z_p is the amount of the pth undesirable output that can be reduced by the post-merger entity, $p=1,\cdots,q$, α_i indicates the amount of the ith input that will be used by the merged entity $(i=1,\cdots,m)$, and δ_r is the amount of the rth good output that can be additionally produced by the merged entity $(r=1,\cdots,s)$. Moreover, $\hat{\beta}$ indicates the level of inefficiency of the merged entity. This parameter needs to be given prior to solving the model due to the nature of the inverse problem.

The objective of the above inverse DEA model is to maximize the reduction of undesirable outputs as much as possible by the merged entity. When there is only one undesirable output, the above vector optimization model can be simplified to the following single objective inverse DEA model.

$$
\begin{aligned}
&\max z \\
&s.t. \\
&\qquad \sum_{j\in F} x_{ij}\lambda_j \le \alpha_i \quad i=1,\cdots,m \\
&\qquad \sum_{j\in F} y_{rj}^{g}\lambda_j \ge \left(1+\hat{\beta}\right)\left(y_{rk}^{g}+y_{rl}^{g}+\delta_r\right) \quad r=1,\cdots,s \\
&\qquad \sum_{j\in F} y_j^{b}\lambda_j = \left(1-\hat{\beta}\right)\left(y_k^{b}+y_l^{b}-z\right) \\
&\qquad \sum_{j\in F} \lambda_j = 1 \\
&\qquad 0 \le \alpha_i \le x_{ik}+x_{il} \quad i=1,\cdots,m \\
&\qquad 0 \le z \le y_k^{b}+y_l^{b} \\
&\qquad \qquad \delta_r \ge 0 \quad r=1,\cdots,s \\
&\qquad \quad \lambda_j \ge 0,\ j\in F
\end{aligned}
\tag{7.5}
$$

The application in this chapter contains oil and gas companies with only one undesirable output that is the amount of their GHG emissions. Therefore, we work with Model (7.5) hereafter. Nonetheless, similar results can be obtained for more than one undesirable output.

The following theorems reveal some properties of the proposed inverse DEA model (Model 7.5).

Theorem 1

Model (7.5) is feasible for some $\hat{\beta}$ if the following model is feasible.

$$
\begin{aligned}
&\max \ \beta \\
&s.t. \\
&\qquad \sum_{j \in F} x_{ij}\lambda_j \le x_{ik} + x_{il} \quad i = 1,\cdots,m \\
&\qquad \sum_{j \in F} y_{rj}^{g}\lambda_j \ge (1+\beta)\left(y_{rk}^{g} + y_{rl}^{g}\right) \quad r = 1,\cdots,s \\
&\qquad \sum_{j \in F} y_{j}^{b}\lambda_j = (1-\beta)\left(y_{k}^{b} + y_{l}^{b}\right) \\
&\qquad \sum_{j \in F} \lambda_j = 1 \\
&\qquad\quad \lambda_j \ge 0, \ j \in F \ \ \beta \ge 0
\end{aligned} \tag{7.6}
$$

Proof: Suppose (7.6) is feasible. It is easy to verify that the inverse DEA model (Model 7.5) is feasible for any $\hat{\beta} \le \beta^*$, where β^* is the optimal value of Model (7.6) which is sufficient for feasibility of Model (7.5). This completes the proof. ■

According to Theorem 1, the proof of the following theorem is straightforward.

Theorem 2

Assume Model (7.5) is feasible for a given $\hat{\beta}_1$. Then, it is feasible for any $\hat{\beta} \le \hat{\beta}_1$.

7.5 APPLICATION AND DATA

In this section, we apply the proposed inverse DEA models to oil and gas companies selected from North America to demonstrate the applicability of the suggested theory in this chapter. This application was first used in Wegener and Amin (2019). Table 7.1 presents the dataset consisting of 23 North American oil and gas companies.

Pre-merging DMUs often have to make decisions about minimizing inputs and maximizing outputs. However, a potential merger may also want to reduce an undesirable output such as GHG emissions. In the

TABLE 7.1 Data and Efficiency Scores of 23 Oil and Gas Firms (Year 2015)

DMUs	Employees	Cap Exp	Total Assets	Wells	Production	GHG	Efficiency
DMU-1	560	542	4,267	61	42	763,457	1.0000
DMU-2	294	376	3,948	82	31	1,284,090	1.0000
DMU-3	3,985	1,233	18,553	772	490	5,564,499	1.0000
DMU-4	1,005	1,155	12,672	700	60	2,024,889	0.4257
DMU-5	10,765	4,465	38,721	3	367	7,412,436	1.0000
DMU-6	2,891	2,232	15,644	400	248	4,474,446	0.7366
DMU-7	588	493	2,581	47	39	858,929	1.0000
DMU-8	5,552	2,162	23,779	135	239	11,260,000	1.0000
DMU-9	5,700	2,586	31,055	46	274	10,711,614	1.0000
DMU-10	449	132	3,274	7	26	1,020,934	1.0000
DMU-11	12,750	6,220	77,527	362	369	18,957,327	0.3978
DMU-12	721	350	3,028	36	20	511,213	1.0000
DMU-13	5,800	5,888	46,414	608	305	11,807,749	0.3300
DMU-14	3,860	4,748	18,842	467	195	7,100,000	0.3229
DMU-15	58,178	29,504	266,103	1,866	1,955	55,746,124	0.8439
DMU-16	15,900	10,050	97,484	498	580	26,039,254	0.5771
DMU-17	6,600	4,735	29,532	420	248	5,925,440	0.5490
DMU-18	2,760	5,013	26,975	489	209	6,723,280	0.3643
DMU-19	73,500	26,490	336,758	1,210	3,135	121,000,000	1.0000
DMU-20	2,770	4,049	34,195	196	136	5,561,176	0.4234
DMU-21	1,111	1,419	4,768	139	55	982,304	0.8864
DMU-22	2,395	2,852	24,196	219	130	2,352,253	0.7347
DMU-23	11,100	5,272	43,437	671	244	10,400,000	0.2703

context of environmental efficiency, GHG emissions are undesirable output measures that need to be minimized to meet regulatory requirements.

As discussed in Sueyoshi and Wang (2014), the oil and gas companies GHG emissions are associated with their drilling activities. Hence, the two important outputs in their production operations are the number of barrels of oil equivalent and total direct GHG emissions. As defined in Wegener and Amin (2019), there are four inputs in the dataset: (1) number of wells, (2) number of employees, (3) capital expenditures, and (4) total assets.

The number of wells represents the firm's drilling activities and ownership of the aggregate number of oil and gas producing wells drilled in a year. The number of employees and total assets both represent the size of the firm and the resources employed in the firm's operations to produce its outputs. Annual capital expenditures maintain existing property and equipment, and signal investments and development of technologies for growth and emissions reduction in the production process.

7.6 RESULT DISCUSSION

To illustrate the applicability of the proposed method in this chapter, we apply the inverse DEA model (Model 7.5) for estimating the potential merger gains taking DMUs 11 and 18 as merging units (data shown in Table 7.1).

Table 7.2 presents the optimal quantity of the inputs and outputs of the merged entity created by merging DMUs 11 and 18. As shown, Model (7.5) is infeasible for any level of inefficiency of the merged entity, $\hat{\beta} \geq 0.4$ according to Theorem 1. Different efficiencies, starting from 53.84% efficiency target of the merged entity until the merger scenario where the merged entity is fully efficient, that is, 100% efficiency, are shown in Table 7.2. The last column of the table, z, shows the amount of GHG emissions reduction denoted by the inverse DEA model (Model 7.5) for each efficiency target, if the combined barrels of oil equivalent remain unchanged. In other words, no additional barrels of oil equivalent are produced by the merged entity, $\delta_1 = 0$.

Table 7.3 presents the amount of the GHG emissions reductions obtained from Model (7.5) developed in this chapter. In the following, we

TABLE 7.2 GHG Emissions Reduction by Model (7.5)

$\hat{\beta}$	Efficiency (%)	α_1	α_2	α_3	α_4	δ_1	z
0.5	33.33			Infeasible			
0.4	42.85			Infeasible			
0.3	53.84	15,510.000	7175.971	69,386.100	851.000	0.000	3,428,093
0.2	66.67	12,738.630	5766.636	57,373.780	851.000	0.000	9,190,202
0.1	81.82	9878.310	4293.923	44,996.610	851.000	0.000	13,651,591
0	100	7240.279	2931.190	33,422.900	837.715	0.000	17,101,785

TABLE 7.3 GHG Emissions Reduction by Model (7.1)

$\hat{\beta}$	α_1	α_2	α_3	α_4	δ_1	z
0.0	6297.786	2073.309	29,139.783	786.572	0.000	16,275,531.790
0.1	7816.867	2625.238	36,093.374	796.144	0.000	12,427,678.580
0.2	9335.947	3177.168	43,046.965	805.715	0.000	7,617,862.152
0.3	10,855.027	3729.098	50,000.557	815.287	0.000	1,433,812.633

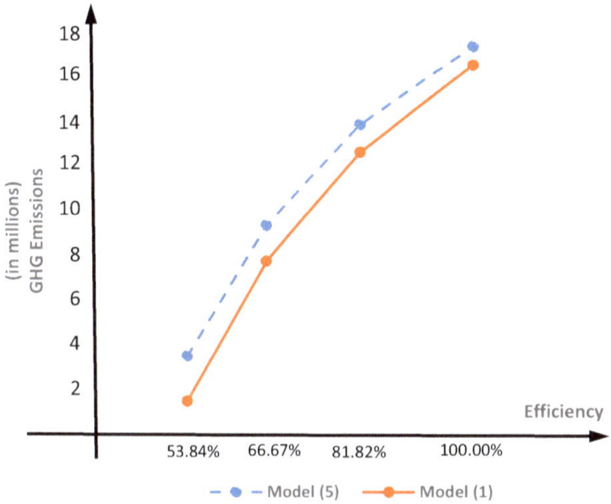

FIGURE 7.1 GHG emissions reduction: Model (7.5) versus Model (7.1).

compare the new GHG emissions reductions with the original merger gains estimation model (Model 7.1) suggested in Gattoufi et al. (2014).

The difference between the last columns in Tables 7.2 and 7.3 indicate how much GHG emissions can be additionally reduced if we employ the new inverse DEA model. For example, when $\hat{\beta} = 0.3$ then 3,428,093–1,433,812.633 = 1,994,280.367 additional GHG (metric tons CO_2 equivalent) emission reduction is achieved. Figure 7.1 presents the GHG emissions reduction, in millions of metric tons CO_2 equivalent, by the two models: the original inverse DEA model (Model 7.1) developed in Gattoufi et al. (2014) and the new model (Model 7.5) in this chapter.

As shown, Model (7.5) outperforms the original model (Model 7.1).

7.7 CONCLUSION

Greenhouse gas (GHG) emissions is a major cause of climate change and reducing GHG emissions has been the forefront of research efforts in the last decades. The economic and social transformation required to reduce GHG emissions will necessitate zero-carbon solutions across economic sectors, including energy and transportation sectors, as many more countries, regions, cities, and companies are establishing carbon neutrality targets.

Research effort in this regard has focused on how DMUs can reduce their GHG emissions individually at the business firm level. Our research in this chapter is different; it introduces an alternative that steers companies and organizations in a new direction on how additional GHG emissions reduction is possible through potential mergers.

This chapter introduced a new inverse DEA model for estimating merger gains when the objective of merging DMUs is to minimize the undesirable production. The proposed model can guide decision makers to reduce GHG emissions in a potential merger.

The results of this study are of interest to managers and decision makers in all production settings where minimizing the quantity of undesirable outputs such as GHG emissions is important. We have found that undesirable outputs such as GHG emissions can be reduced in potential mergers, and that the reduction is greater in our model compared with the original merger gains estimation model suggested by Gattoufi et al. (2014). Our proposed model provides an opportunity for DMUs to lower their undesirable outputs, including GHG emission levels, and thereby enhance the economic value of production. Our proposed model provides a tool for GHG emission regulators in different industries to recognize post-merger business and organizational entities that can improve their emission reductions strategies.

It would be useful in future research to consider and develop appropriate models that incorporate the heterogeneity in oil and gas exploration activities such as variation in drilling techniques and the differences in product quality that could be sources of inefficiencies.

REFERENCES

Abbas, A., Waseem, M., Ahmad, R., Ahmed Khan K., Zhao, C., and Zhu, J. (2022). Sensitivity analysis of greenhouse gas emissions at farm level: case study of grain and cash crops. *Environmental Science and Pollution Research*, 29, 82559–82573.

Amin, G.R., and Al-Muharrami, S. (2018). A new inverse DEA model for merger with negative data. *IMA Journal of Management Mathematics*, 29(2), 137–149.

Amin, G.R., Emrouznejad, A., and Gattoufi, S. (2017). Modelling generalized firms' restructuring using inverse DEA. *Journal of Productivity Analysis*, 48(1), 51–61.

Amin, G.R., and Ibn Boamah, M. (2020). A new inverse DEA cost efficiency model for estimating potential merger gains: a case of Canadian banks. *Annals of Operations Research*, 295(1), 21–36.

Amin, G.R., and Ibn Boamah, M. (2023). Modeling business partnerships: A data envelopment analysis approach. *European Journal of Operational Research*, 305(1), 329–337.

Amin, G.R., and Oukil, A. (2019). Flexible target setting in mergers using inverse data envelopment analysis. *International Journal of Operational Research*, 35(3), 301–317.

Banker, R.D., Charnes, A., and Cooper, W.W. (1984). Some models for estimating technical and scale inefficiencies in data envelopment analysis. *Management Science*, 30, 1078–1092.

Busch, P., Kendall, A., Murphy, C.W., and Miller, S.A. (2022). Literature review on policies to mitigate GHG emissions for cement and concrete. *Resources, Conservation & Recycling*, 182, 106278.

Chang, Y.T., Zhang, N., & Danao, D. (2013). Environmental efficiency analysis of transportation system in China: A non-radial DEA approach. *Energy Policy*, 58, 277–283.

Charnes, A., Cooper, W.W., and Rhodes, E. (1978). Measuring the efficiency of decision making units. *European Journal of Operational Research*, 2(6), 429–444.

Chen, S., and Golley, J. (2014). 'Green' productivity growth in China's industrial economy. *Energy Economics*, 44, 89–98.

Chung, Y. H., Färe, R., and Grosskopf, S. (1997). Productivity and undesirable outputs: a directional distance function approach. *Journal of Environmental Management*, 51(3), 229–240.

Dakpo, K.H., Jeanneaux, P., and Latruffe, L. (2017). Greenhouse gas emissions and efficiency in French sheep meat farming: A non-parametric framework of pollution-adjusted technologies. *European Review of Agricultural Economics*, 44(1), 33–65,

Dao, S. D., Abhary, K., and Marian, R. (2014). Optimization of partner selection and collaborative transportation scheduling in virtual enterprises using GA. *Expert Systems with Applications*, 41(15), 6701–6717.

Dyckhoff, H., and Souren, R. (2022). Integrating multiple criteria decision analysis and production theory for performance evaluation: Framework and review. *European Journal of Operational Research*, 97(3), 795–816.

Emrouznejad, A., and Yang, G. (2018). A survey and analysis of the first 40 years of scholarly literature in DEA: 1978–2016. *Socio-Economic Planning Sciences*, 61, 4–8.

Emrouznejad, A., Yang, G., and Amin, G.R. (2019). A novel inverse DEA model with application to allocate the CO_2 emissions quota to different regions in Chinese manufacturing industries. *Journal of the Operational Research Society*, 70(7), 1079–1090.

Gan, L., Ren, H., Cai, W., Wu, K., Liu, Y., and Liu, Y. (2022). Allocation of carbon emission quotas for China's provincial public buildings based on principles of equity and efficiency. *Building and Environment*, 216, 108994.

Gattoufi, S., Amin, G.R., and Emrouznejad, A. (2014). A new inverse DEA method for merging banks. *IMA Journal of Management Mathematics*, 25, 73–87.

Guijarro, F., Martínez-Gómez, M., and Visbal-Cadavid, D. (2020). A model for sector restructuring through genetic algorithm and inverse DEA. *Expert Systems with Applications*, 154(15), 113422.

Halkos, G, and Petrou, K.N. (2019). Assessing 28 EU member states' environmental efficiency in national waste generation with DEA, *Journal of Cleaner Production*, 208, 509–521.

Halkos, G., and Polemis, M. (2018). The impact of economic growth on environmental efficiency of the electricity sector: A hybrid window DEA methodology for the USA, *Journal of Environmental Management*, 211, 334–346.

Halkos, G., and Tzeremes, N. (2013). A conditional directional distance function approach for measuring regional environmental efficiency: Evidence from UK regions. *European Journal of Operational Research*, 227(1),182–189.

Kaffash, S., Azizi, R., Huang, Y., and Zhu, J. (2020). A survey of data envelopment analysis applications in the insurance industry 1993–2018. *European Journal of Operational Research*, 284(3), 801–813.

Lu, L.C., Chiu, S.Y., Chiu, Y.H., and Chang, T.H. (2022). Sustainability efficiency of climate change and global disasters based on greenhouse gas emissions from the parallel production sectors – A modified dynamic parallel three-stage network DEA model. *Journal of Environmental Management*, 317, 115401.

Oukil, A. (2022). Investigating prospective gains from mergers in the agricultural sector through Inverse DEA. *IMA Journal of Management Mathematics*. https://doi.org/10.1093/imaman/dpac004

Shen, Z., Balezentis, T., and Streimikis, J. (2022). Capacity utilization and energy-related GHG emission in the European agriculture: A data envelopment analysis approach. *Journal of Environmental Management*, 318, 115517.

Soltanifar, M., Ghiyasi, M., and Sharafi, H. (2022). Inverse DEA-R models for merger analysis with negative data. *IMA Journal of Management Mathematics*. https://doi.org/10.1093/imaman/dpac001

Sueyoshi, T., and Wang, D. (2014). Sustainability development for supply chain management in U.S. petroleum industry by DEA environmental assessment. *Energy Economics*, 46, 360–374.

Wegener, M., and Amin, G.R. (2019). Minimizing GHG emissions using inverse DEA with an application in oil and gas. *Expert Systems with Applications*, 122, 369–375.

Wei, Q., Zhang, J., and Zhang, X. (2000). An inverse DEA model for inputs/outputs estimate. *European Journal of Operational Research*, 121, 151–163.

Xiao, J., Niu, Y., Chen, P., Leung, S. C., and Xing, F. (2016). An improved gravitational search algorithm for green partner selection in virtual enterprises. *Neurocomputing*, 217, 103–109.

Xie, Q., Xu, Q., Rao, K., and Dai, Q. (2011). Water pollutant discharge permit allocation based on DEA and non-cooperative game theory. *Journal of Environmental Management*, 302, Part A, 113962, https://doi.org/10.1016/j.jenvman.2021.113962

Yan, H., Wei, Q., and Hao, G. (2002). DEA models for resource reallocation and production input/output estimation. *European Journal of Operational Research*, 136(1), 19–31.

Zhu, Y., Yang, F., Wei, F., and Wang, D. (2022). Measuring environmental efficiency of the EU based on a DEA approach with fixed cost allocation under different decision goals, *Expert Systems with Applications*, 208. https://doi.org/10.1016/j.eswa.2022.118183

CHAPTER 8

Undesirable Factors in Sustainability Improvement Using DEA

Potential of Business Partnerships

Behrouz Arabi and Sonal Choudhary
School for Business and Society, University of York, UK

8.1 INTRODUCTION

Usually, the objective of measuring efficiency is to minimize inputs and maximize outputs. On the other hand, there are instances in which it is essential to maximize some inputs and reduce outputs. These inputs or outputs are known as "undesirable" factors in the literature (Jahanshahloo et al., 2005). Numerous studies have documented a number of examples of undesirable outputs, such as the amount of unpaid debt in banking (Amirteimoori et al., 2006), delayed flights (Coli et al., 2011), poverty rates(Bruni et al., 2011), patient deaths (Yawe & Kavuma, 2008), SOx gases (Arabi et al., 2014; Zhou et al., 2007), NOx in the crop production system(Dong et al., 2018), and CO_2 in refrigeration systems (Arabi et al., 2024).

In the literature, however, there aren't many instances of undesirable inputs. Authors underline that one notable example of a recycling process' goal is to utilize as much of the recycled input as possible (Hadi Vencheh et al., 2005; Liu et al., 2009). This demonstrates how critical it is to optimize

DOI: 10.1201/9781003345206-8

those inputs that are often seen as undesirable yet are required in certain circumstances, such as recycling processes.

Undesirable factors may be added to efficiency evaluation studies using a variety of methods. Integrating environmental variables into these assessments—a process known as eco-efficiency—is one such strategy. Given the pressing environmental problems and pollution difficulties of today, eco-efficiency is an important and well-researched field. The goal is to preserve or improve desirable outputs while reducing undesirable outputs, such as pollution. Research is increasingly concentrating on creating plans that strike a balance between economic performance and environmental effect.

This dual strategy addresses the urgent need for sustainable business operations, which mitigates adverse environmental consequences while simultaneously fostering economic viability. Eco-efficiency models seek to contract environmental impacts and resource use and expand green product simultaneously so promoting a more sustainable future. To ensure that economic operations do not jeopardize the integrity and resilience of the planet, this all-encompassing approach is crucial to accomplishing the more general objectives of sustainability and environmental preservation.

8.2 ECO-EFFICIENCY

It has been shown repeatedly that more environmentally friendly products would be more efficient (Schaltegger et al., 2008). Furthermore, studies have shown that, in many situations, it is feasible to both save costs and increase environmental effects at the same time. This implies that accounting for environmental factors might potentially result in lower expenses. In Burritt et al. (2004), consequently, a new sort of efficiency—called ecological efficiency, or simply eco-efficacy—has been presented in addition to these additional characteristics and assumptions.

Various cases have shown that cleaner products are more effective (Schaltegger et al., 2008). Studies reveal that improving environmental performance is often associated with cost reductions, implying that integrating environmental considerations into corporate operations may result in lower costs. For example, Burritt et al. (2004) pointed out that businesses that use greener practices benefit financially as well as the environment by maximizing resource usage and reducing waste. This dual benefit emphasizes how crucial it is for firms to include environmental concerns into their fundamental business plan.

This has led to the emergence of a new kind of efficiency known as ecological efficiency or simply eco-efficiency. This idea incorporates environmental implications into the evaluation of company performance, going beyond typical efficiency indicators. Eco-efficiency aims to match economic activities with sustainable development objectives by producing greater value with fewer natural resources and less pollution. Businesses that embrace eco-efficiency may strike a healthy balance between environmental stewardship and economic development, assuring long-term survival and advancing the goal of sustainability.

The principle of sustainable development is the foundation of eco-efficiency. In the World Commission on Environment and Development, Brundtland (1987) said that the goal of sustainable development is "to meet the needs of the present generation without compromising the ability of future generations to meet their own needs." Bebbington et al. (1999)[1], considering the Kyoto Protocol to the United Nations Framework Convention on Climate Change, define eco-efficiency as the concept that it is feasible to be efficient, boost efficiency, and simultaneously maintain or improve a certain level of environmental performance.

Eco-efficiency can be defined in many ways, but one of the earliest definitions is provided by Schott (2007). He described it as the provision of reasonably priced products and services that enhance human well-being and quality of life while progressively reducing the environmental impacts and resource intensity of goods throughout their entire life cycles. The goal is to lower these impacts to a level where the Earth's carrying capacity is not exceeded, as estimated by the World Business Council for Sustainable Development (WBCSD).[2] Although the definition of eco-efficiency has evolved over time, certain core characteristics have remained consistent. According to Lovins (2008), these include: reducing the material or energy intensity of goods and services, decreasing the dispersion of toxic materials, improving recyclability, maximizing the use of renewable resources, enhancing the durability of products, and increasing the service intensity of goods and services. These elements collectively aim to balance economic and environmental performance, promoting sustainable development and long-term ecological health.

According to Lovins (2008), eco-efficiency is the part of the shift to sustainability that is simplest to put into practice. Schaltegger and Sturm (1990) defined efficiency as one of the first formulas in the direction of operationalization as: Eco-Efficiency = Economic Value Creation / Environmental Impact Added

TABLE 8.1 Four Basic Variants of Eco-Efficiency

	Product or Production Prime	Environmental Improvement Prime
Economy divided by environment	1 **Environmental productivity**: Production / consumption value per unit of environmental impact	3 **Environmental improvement cost**: Cost per unit of environmental improvement
Environment divided by economy	2 **Environmental intensities**: Environmental impact per unit of production/consumption value	4 **Environmental cost-effectiveness**: Environmental improvement per unit of cost

Huppes and Ishikawa (2007) provide four categories of eco-efficiency in Table 8.1 using this fractional definition.

The definitions have facilitated numerous studies on eco-efficiencies carried out by various industries. One of the most often used approaches for measuring efficiency is data envelopment analysis (DEA), which has been used in several studies on eco-efficiency in recent years. In Table 8.1, DEA applies the formula that appears in the first row.

8.3 BASIC FORMULATION

DEA is a mathematical optimization, and the objective of such a method is efficiency evaluation for each decision making unit (DMU) in a DMU group. Consider n DMUs that we want to evaluate using I similar inputs to produce J similar outputs. In fact, efficiency is the answer of a DMU to this question, how to employ multiplicative inputs for producing multiplicative outputs?

Let $\boldsymbol{x}_n = \left(x_{1n},\ldots,x_{In}\right)$ and $\boldsymbol{y}_n = \left(y_{1n},\ldots,y_{Jn}\right)$ be input and output vectors corresponding to nth DMU, respectively. Consider $\boldsymbol{x}$ as input vector of a firm and $\boldsymbol{y}$ as its output vector. The production possibility set can be defined as $P(\boldsymbol{x}) = \left\{(\boldsymbol{x},\boldsymbol{y}) \mid \boldsymbol{x} \text{ can produce} (\boldsymbol{y})\right\}$.

Also, let $\boldsymbol{b}_n = \left(b_{1n},\ldots,b_{Kn}\right)$ undesirable output vector corresponding to nth DMU. Given $\boldsymbol{b}$, as undesirable output vector of the firm, the production possibility set, is redefined as

$$P(\boldsymbol{x}) = \left\{(\boldsymbol{y},\boldsymbol{b}) : \boldsymbol{x} \text{ can produce} (\boldsymbol{y},\boldsymbol{b})\right\} \tag{8.1}$$

8.4 INCORPORATING UNDESIRABLE FACTORS INTO DEA MODELS

Various approaches have been introduced and implemented to integrate undesirable outputs using DEA. As previously discussed, undesirable outputs are those that, when increased, cause a decrease in efficiency. Similarly, undesirable inputs are those that, when increased, lead to an increase in efficiency. These variables cannot be managed as normal variables due to their inverse relationship with efficiency. In this section, we will briefly discuss a few models that have been developed to include such variables in efficiency measurement studies. These models employ different disposability considerations to effectively manage undesirable inputs and outputs, ensuring a more accurate assessment of efficiency.

Scheel (2001) categorized aforementioned models to direct and indirect. Indirect models are those that change or customize undesirable factors to include them in the model. However, direct ones treat undesirable factor with the same amount. As indirect approaches we can count, use the additive inverse of undesirable factors (Berg et al., 1992), undesirable output as an input (Tyteca, 1997), using multiplicative inverse (Knox Lovell et al., 1995), and some other approaches. However, as direct approaches that are more popular, we can name hyperbolic efficiency model (Boyd & McClelland, 1999), slacks-based measure (SBM) model (Tone, 2001), Range-adjusted measure (RAM) model (Zhou et al., 2006), and the most popular one directional distance function (DDF) model (Chung et al., 1997). This approach has found many applications among eco-efficiency measurement (Färe & Grosskopf, 2010a; Färe et al., 2007). What had been important for many is the relation between national and international environmental rules, regulation and protocols, and eco-efficiency of industries (Macpherson et al., 2010; Picazo-Tadeo et al., 2005). Yet many studies are required to address these regulations, which are so critical for policy makers.

Numerous models have been developed to incorporate undesirable elements into DEA models. Zhou et al. (2018) reviewed and compiled a comprehensive list of recent DEA models addressing both desirable and undesirable factors. In the following section, we will discuss the most popular models. However, before that, we will first address the important concept of disposability.

8.4.1 Different Types of Disposability

Disposability was first brought into the DEA literature by Färe et al. (1989). Disposability refers to how undesirable inputs (like waste materials) or undesirable outputs (like pollutants) are handled within the efficiency measurement framework. To take into consideration, the reality that not all inputs or outputs are advantageous, disposability is crucial in DEA models.

Strong or free disability: This, according to Färe et al. (1989), is the state in which undesirable outputs may be freely decreased without incurring further expenses or limitations. Stated otherwise, it is presumed that a DMU has the discretion to eliminate these undesirable outputs or inputs. This strategy could work well in situations when the reduction of undesirable inputs or outputs is not restricted by technical, regulatory, or other external considerations. Strong disposability can be formulated as:

$$P(\boldsymbol{x}) = \left\{ (\boldsymbol{x}, \boldsymbol{y}) : \boldsymbol{y} \leq \sum_{n=1}^{N} \boldsymbol{y}_n \lambda_n ; \boldsymbol{b} \leq \sum_{n=1}^{N} \boldsymbol{b}_n \lambda_n ; \boldsymbol{x} \geq \sum_{n=1}^{N} \boldsymbol{x}_n \lambda_n ; \lambda_n \geq 0; n = 1, \ldots, N \right\} \tag{8.2}$$

Weak Disposability: This acknowledges that cutting down on undesirable inputs or outputs might be expensive or demand more resources. It depicts a more grounded situation in which the mitigation of undesirable outputs is linked to technological, financial, or legal limitations. For example, cutting emissions may need spending money on new, expensive technology or procedures. Therefore, weak disposability can be written as below:

$$P(\boldsymbol{x}) = \left\{ (\boldsymbol{y}, \boldsymbol{b}) : \boldsymbol{y} \leq \sum_{n=1}^{N} \boldsymbol{y} \lambda_n ; \boldsymbol{b} = \sum_{n=1}^{N} \boldsymbol{b}_n \lambda_n ; \boldsymbol{x} \geq \sum_{n=1}^{N} \boldsymbol{x}_n \lambda_n ; \lambda_n \geq 0; n = 1, \ldots, N \right\} \tag{8.3}$$

Additionally, in weak disposability, it is assumed that both good and bad outputs are created jointly, or "null-joint," which implies that it is impossible to make good output without simultaneously creating negative output.

Later, Sueyoshi and Goto (2012) well introduced and discussed natural and managerial disposability as following.

Natural disposability: This highlights the natural relationship between reducing inputs and undesirable outputs while striving to enhance desirable outputs within the constraints of reduced production. Natural disposability refers to the idea that a DMU reduces its input vector to consequently reduce its undesirable output vector. With the decreased levels of both inputs and undesirable outputs, the firm then seeks to maximize its desirable outputs as much as possible. Then, natural disposability is formulated as follows:

$$P(\boldsymbol{x}) = \left\{ (\boldsymbol{y}, \boldsymbol{b}) : \boldsymbol{y} \le \sum_{n=1}^{N} \boldsymbol{y}_n \lambda_n ; \boldsymbol{b} \ge \sum_{n=1}^{N} \boldsymbol{b}_n \lambda_n ; \boldsymbol{x} \ge \sum_{n=1}^{N} \boldsymbol{x}_n \lambda_n ; \lambda_n \ge 0, n = 1, \ldots, N \right\} \tag{8.4}$$

Managerial disposability: This is the situation that a DMU strives to reduce the vector of undesirable outputs while concurrently increasing the vector of inputs to enhance the vector of desired outputs. This strategy emphasizes how the company actively balances the manufacturing process via management interventions and optimizations to maximize positive results and reduce adverse environmental effects. Managerial disposability is formulated as follows:

$$P(\boldsymbol{x}) = \left\{ (\boldsymbol{y}, \boldsymbol{b}) : \boldsymbol{y} \le \sum_{n=1}^{N} \boldsymbol{y}_n \lambda_n ; \boldsymbol{b} \ge \sum_{n=1}^{N} \boldsymbol{b}_n \lambda_n ; \boldsymbol{x} \le \sum_{n=1}^{N} \boldsymbol{x}_n \lambda_n ; \lambda_n \ge 0; n = 1, \ldots, N \right\} \tag{8.5}$$

Finally, to introduce the new concept of circular disposability, we refer to the notion of circular economy (CE), the null-jointness condition, and the idea of undesirable circular inputs. CE is a promising solution to confront sustainability challenges. This novel economic model involves recycling materials and reintegrating them into the supply chain, thereby decoupling economic growth from resource use (James et al., 2023).

Circular Disposability: This is a situation and in which a DMU looks for ways to simultaneously enhance desirable outputs and undesirable inputs while simultaneously decreasing undesirable outputs and desirable inputs. Assume that the data for the DEA inputs are $\boldsymbol{x}_j = \left(\boldsymbol{x}_j^D, \boldsymbol{x}_j^U\right)$ where $\boldsymbol{x}_j^D$ and $\boldsymbol{x}_j^U$ stand for desirable inputs that should be reduced and undesirable inputs that should be raised, respectively. Therefore, we can formulate the circular disposability as follows:

$$P\left(\boldsymbol{x}_j^D, \boldsymbol{x}_j^U\right) = \left\{ (\boldsymbol{y}, \boldsymbol{b}) : \boldsymbol{y} \leq \sum_{n=1}^{N} \mathrm{y}\lambda_n ; \boldsymbol{b} = \sum_{n=1}^{N} \boldsymbol{b}_n \lambda_n ; \boldsymbol{x}^D \geq \sum_{n=1}^{N} \boldsymbol{b}_n^D \lambda_n ; \boldsymbol{x}^U = \sum_{n=1}^{N} \boldsymbol{x}_n^U \lambda_n ; \lambda_n \geq 0; n = 1, \ldots, N \right\} \tag{8.6}$$

In Equation (8.6), an increase in circular inputs results in enhanced efficiency. It is worth noting that equality in $\boldsymbol{x}^U = \sum_{n=1}^{N} \boldsymbol{x}_n^U \lambda_n$, like $\boldsymbol{b} = \sum_{n=1}^{N} \boldsymbol{b}_n \lambda_n$, ensures the null-jointness property among desirable and undesirable outputs as well as undesirable inputs.

8.4.2 Hyperbolic Model

Hyperbolic efficiency emphasizes that production process performance should be evaluated against an environmentally friendly standard (Färe et al., 1989). This implies that rather of focusing on activities that create both desirable and undesirable outputs, the benchmarks for efficiency should be focused on activities that provide greater quantities of desirable while decreasing undesirable outputs. Utilizing the weak disposability assumption, the hyperbolic model can be formulated as follows:

$$\begin{aligned}
&\max \beta \\
&\text{s.t.} \\
&\sum_{n=1}^{N} \lambda_n x_{in} \leq x_{io} && i = 1, \ldots, I \\
&\sum_{n=1}^{N} \lambda_n y_{jn} \geq \beta y_{jo} && j = 1, \ldots, J \\
&\sum_{n=1}^{N} \lambda_n b_{kn} = \frac{1}{\beta} b_{ko} && k = 1, \ldots, K \\
&\lambda_n \geq 0; n = 1, \ldots, N
\end{aligned} \tag{8.7}$$

Model (8.7) is fundamentally based on the ability to simultaneously reduce undesirable outputs and increase favorable outputs proportionately. It is evident that Model (8.7) can be adapted to consider different disposability assumptions. Additionally, Model (8.7) is clearly non-linear. To address this, Färe et al. (1989) suggest approximating $\sum_{n=1}^{N} \lambda_n b_{kn} = \frac{1}{\beta} b_{ko}$ by $\sum_{n=1}^{N} \lambda_n b_{kn} = b_{ko}\left(2-\beta\right)$.

8.5 DIRECTIONAL DISTANCE FUNCTION

By using the idea of DDF introduced by Färe et al. (1994), Chung et al. (1997) developed a series of DDF DEA models. Using Shephard et al. (1970) definition and by incorporating undesirable outputs, Färe et al. (1994) rewrite the distance function as below:

$$D\left(\boldsymbol{x},\boldsymbol{y},\boldsymbol{b}\right) = \inf\left\{\varphi : \left(\left(\boldsymbol{y},\boldsymbol{b}\right)/\varphi\right) \in P\left(\boldsymbol{x}\right)\right\} \tag{8.8}$$

where φ denotes the expansion or contraction proportion of good and bad outputs, and D expands them simultaneously as much as feasible. However, Chung et al. (1997) define D as

$$D\left(\boldsymbol{x},\boldsymbol{y},\boldsymbol{b};\boldsymbol{g}\right) = \sup\left\{\theta : \left(\left(\boldsymbol{y},\boldsymbol{b}\right) + \theta\,\boldsymbol{g}\right) \in P\left(\boldsymbol{x}\right)\right\} \tag{8.9}$$

where θ plays the same role as φ in (8.8). Here if g is the vector of directions and is defined as $\boldsymbol{g} = \left(\boldsymbol{g}_y, \boldsymbol{g}_b\right)$, using this D, good outputs can be expanded while bad outputs are contracted. $\boldsymbol{g}$ can be chosen arbitrarily or based on the situation. Here, we choose to take $\boldsymbol{g} = \left(\boldsymbol{y}, -\boldsymbol{b}\right)$; thus, weak disposability implies

$$\left(\boldsymbol{y},\boldsymbol{b}\right) \in P\left(\boldsymbol{x}\right) \text{and } 0 \le \theta \le 1 \text{ imply } \left(\theta\boldsymbol{y},\theta\boldsymbol{b}\right) \in P\left(\boldsymbol{x}\right) \tag{8.10}$$

This means, to remain feasible, good outputs should be decreased with the same proportion as bad outputs. In addition, it is also assumed that good and bad outputs are produced jointly namely "null-joint"; this means it is not possible to produce good output without producing any bad output. Now according to Chung et al. (1997), $P(x)$ can be rewritten as below:

$$P(\boldsymbol{x})=\Big\{(\boldsymbol{y},\boldsymbol{b}):\sum_{n=1}^{N}\lambda_n x_{in}\le x_{io};i=1,\ldots,I;\sum_{n=1}^{N}v_n y_{jn}\ge y_{jo}+\theta y_{jo};$$
$$j=1,\ldots,J;\sum_{n=1}^{N}\lambda_n b_{kn}=b_{ko}-\theta b_{ko};k=1,\ldots,K;\lambda_n\ge 0;n=1,\ldots,N\Big\} \tag{8.11}$$

where λ_n represent intensity variables and subscript o represents the under-assessment DMU. If the direction vector is assumed to be unknown, or $\boldsymbol{g}=(\boldsymbol{g}_y,\boldsymbol{g}_b)$, we introduce the model based on Färe and Grosskopf (2010a, 2010b), as shown in Equation (8.12):

$$\begin{aligned}
&D_o(\boldsymbol{x},\boldsymbol{y},\boldsymbol{b})=\max\theta \\
&\text{s.t.} \\
&\sum_{n=1}^{N}\lambda_n x_{in}\le x_{io} && i=1,\ldots,I \\
&\sum_{n=1}^{N}\lambda_n y_{jn}\ge y_{jo}-g_{yj}.\theta && j=1,\ldots,J \\
&\sum_{n=1}^{N}\lambda_n b_{kn}=b_{ko}-g_{bk}.\theta && k=1,\ldots,K \\
&\sum_{j=1}^{J}g_{yj}+\sum_{k=1}^{K}g_{bk}=1 \\
&\lambda_n\ge 0;n=1,\ldots,N;g_{yj}\ge 0;j=1,\ldots,J;g_{bk}\ge 0; && k=1,\ldots,K
\end{aligned} \tag{8.12}$$

Model (8.12) is non-linear, requiring an appropriate method for its solution. This issue will be addressed in the following section.

8.5.1 Slacks-Based Measure of Efficiency

The SBM of efficiency calculated by DEA is one these models which was introduced by Tone (2001). The SBM and its variations were deployed to measure productivity factors in many fields which have been addressed by tone (2010). One of the models of this family which is introduced by Färe and Grosskopf (2010a, 2010b) has been presented as

$$D_o(\boldsymbol{x},\boldsymbol{y}) = \max\left(\sum_{i=1}^{I}\alpha_i + \sum_{j=1}^{J}\beta_j\right)$$
$$\text{s.t.}$$
$$\sum_{n=1}^{N}\lambda_n x_{in} \le x_{io} - \alpha_i.1 \qquad i=1,\ldots,I \tag{8.13}$$
$$\sum_{n=1}^{N}\lambda_n y_{jn} \ge y_{jo} + \beta_j.1 \qquad j=1,\ldots,J$$
$$\lambda_n \ge 0; n=1,\ldots,N; \alpha_i \ge 0; i=1,\ldots,I; \beta_j \ge 0 \qquad j=1,\ldots,J$$

Arabi et al. (2015) adopt Model (8.17) to include bad outputs as below:

$$D_o(\boldsymbol{x},\boldsymbol{y},\boldsymbol{z}) = \max\left(\sum_{j=1}^{J}\beta_j + \sum_{k=1}^{K}\gamma_k\right)$$
$$\text{s.t.}$$
$$\sum_{n=1}^{N}\lambda_n x_{in} \le x_{io} \qquad i=1,\ldots,I \tag{8.14}$$
$$\sum_{n=1}^{N}\lambda_n y_{jn} \ge y_{jo} + \beta_j.1 \qquad j=1,\ldots,J$$
$$\sum_{n=1}^{N}\lambda_n b_{kn} = b_{ko} - \gamma_k.1 \qquad k=1,\ldots,K$$
$$\lambda_n \ge 0; n=1,\ldots,N; \ge 0; \beta_j = 1,\ldots,J; \gamma_k \ge 0 \qquad k=1,\ldots,K$$

Here, $\beta_1,\ldots,\beta_I$ and $\gamma^1,\ldots,\gamma_k$ are variables. This model is also called *additive model* which was first introduced by Charnes et al. (1985) and later developed to include slack variables by Bian (2008).

Arabi et al. (2015) proved that Models (8.12) and (8.14) are equivalent. If DMU_o is located on the frontier, $\boldsymbol{g} = (\boldsymbol{g}_y, \boldsymbol{g}_b)$, the direction vector can be any direction, else if we solve (8.11) and if we take $g_{yj}.\beta = \beta_j^*$, $g_{bk}.\beta = \gamma_k^*$, we can obtain

$$\beta = \frac{\beta_1^*}{g_{y1}} = \frac{\beta_2^*}{g_{y2}} = \ldots = \frac{\beta_J^*}{g_{yJ}} = \frac{\gamma_1^*}{g_{b1}} = \frac{\gamma_2^*}{g_{b2}} = \ldots = \frac{\gamma_K^*}{g_{bk}} \tag{8.15}$$

Then we can get

$$\beta_1^*.g_{y2} = \beta_2^*.g_{y1}; \beta_2^*.g_{y3} = \beta_3^*.g_{y2},\ldots,\beta_J^*.g_{z1} = \gamma_1^*.g_{yJ},\ldots,\gamma_{K-1}^*.g_{bk} = \gamma_K^*.g_{bK-1} \tag{8.16}$$

Next, we achieve

$$
\begin{aligned}
&\beta_1^* \cdot g_{y2} - \beta_2^* \cdot g_{y2} = 0, \\
&\beta_2^* \cdot g_{y3} - \beta_3^* \cdot g_{y2} = 0, \\
&\ldots, \\
&\beta_J^* \cdot g_{b1} - \gamma_1^* \cdot g_{yJ} = 0, \\
&\ldots, \\
&\gamma_{K-1}^* \cdot g_{bk} - \gamma_K^* \cdot g_{bK-1} = 0 \\
&\sum_{j=1}^{J} g_{yj} + \sum_{k=1}^{K} g_{bk} = 1
\end{aligned} \tag{8.17}
$$

Equations (8.3–8.15) are a system of equations with first similar $J+K-1$ equations and $J+K$ unknowns. Thus, together with $\sum_{j=1}^{J} g_{yj} + \sum_{k=1}^{K} g_{bk} = 1$, we have $J+K$ equations and $J+K$ unknowns with first $J+K-1$ pairwise linearly independent equations. Furthermore, no linear combination of the first $J+K-1$ equations can generate the last equation, since first $J+K$ equations have zero in their right-hand side (RHS) but the last equation has unity in the same place. Therefore, this is a system of linear equations with a unique solution, which is $(g_{y1},\ldots,g_{yJ},g_{b1},\ldots,g_{bK})$. As a result, by solving (8.14) and (8.17), one can determine the optimal directions to the frontier, indicating the closest path for the under-assessment DMU to reach the frontier.

Model (8.18) has been utilized by Arabi et al. (2015) to calculate the Malmquist–Luenberger index without encountering the infeasibility issues discussed by Färe et al. (2004). Additionally, it is employed to calculate super-efficiency in the presence of undesirable factors when similar infeasibility problems arise (Arabi et al., 2024).

8.5.2 Range-Adjusted Model

The RAM of inefficiency was first introduced to the DEA literature by Cooper et al. (1999). RAM offers several advantages, including the ability to simultaneously adjust multiple inputs and outputs, including undesirable ones, while accommodating varying scales and units of measurement. This enhances the robustness and applicability of the analysis. By accounting for the full range of inefficiencies, RAM improves the discriminatory power of efficiency evaluations, accurately distinguishing between efficient and inefficient DMUs. Additionally, it is sensitive to the scale of operations, making it particularly useful for comparing DMUs of different sizes and capacities across diverse industries. Then, Sueyoshi and Goto (2012) introduced Model (8.18) unified efficiency under natural and managerial disposability:

$$\max \theta + \varepsilon\left(\sum_{j=1}^{J} \mathrm{R}_j^y \beta_j + \sum_{k=1}^{K} \mathrm{R}_j^y \gamma_k\right)$$

s.t.

$$\begin{aligned}
&\sum_{n=1}^{N} \lambda_n x_{in} - \alpha_i^+ + \alpha_i^- = x_{io} && i = 1,\ldots,I \\
&\sum_{n=1}^{N} \lambda_n y_{jn} - \beta_j = y_{ro} + \theta y_{ro} && j = 1,\ldots,J \qquad (8.18)\\
&\sum_{n=1}^{N} \lambda_n b_{kj} + \gamma_k = b_{ko} - \theta b_{ko} && k = 1,\ldots,K \\
&\sum_{n=1}^{N} \lambda_n = 1 \\
&\lambda_n \geq 0; n = 1,\ldots,N; i = 1,\ldots,I; \alpha_i^+ \geq 0; \alpha_i^- \geq 0; \beta_j \geq 0, && j = 1,\ldots,J, \\
&\theta : \text{Free; and } \gamma_k \geq 0; k = 1,\ldots,K
\end{aligned}$$

where the slack variables $\alpha_i^x, (i = 1,\ldots,I)$, $\beta_j^y, (j = 1,\ldots,J)$, and $\gamma_k^b, (k = 1,\ldots,K)$ are associated with inputs, desired outputs, and undesirable outputs, respectively. The inclusion of these slacks in Model (8.18) ensures that all constraints are represented as equalities, which is a crucial aspect. Model (8.18) also incorporates the small positive number $\varepsilon > 0$ into its objective function. Where a is small none-Archimedean infinitesimal. The upper and lower bounds of desirable outputs are $\bar{y}_j = \max_n \{y_{jn}\}$ and $\underline{y}_j = \min_n \{y_{jn}\}$ for all j. The range for desirable outputs becomes $R_j^y = \frac{1}{\left[(J+K)(\bar{y}_j - \underline{y}_j)\right]}$ for all j. The upper and lower bounds of undesirable outputs are mathematically expressed by $\bar{b}_k = \max_n \{b_{kn}\}$ and $\underline{b}_k = \min_n \{b_{kn}\}$ for all k. Then, $R_j^b = \frac{1}{\left[(J+K)(\bar{b}_k - \underline{b}_k)\right]}$ for all k indicate the range for undesirable outputs.

The two slacks related to the ith input are mathematically defined as $\alpha_i^+ = \left(\left|\alpha_i\right| + \frac{\alpha_i}{2}\right)$ and $\alpha_i^- = \left(\left|\alpha_i\right| - \frac{\alpha_i}{2}\right)(i = 1,\ldots,I)$. They are mutually exclusive so that a simultaneous occurrence of both $\alpha_i^+ > 0$ and $\alpha_i^- > 0(i = 1,\ldots,I)$ should be excluded from an optimal solution of Model (8.18). A computer program may provide "an unbounded solution" when the simultaneous occurrence takes place on Model (8.18) as a result of breaking the nonlinear requirements. Sueyoshi and Goto (2012) propose two computational approaches to make Model (8.18) meet the nonlinear criteria.

RAM models have significant drawbacks even though they are helpful for correcting inefficiencies with undesirable aspects. Extreme values or data outliers may have a substantial influence on efficiency scores in these models, thereby distorting the findings (Chiu et al., 2014). Furthermore, dynamic and complicated systems with ever-changing interactions between undesirable outputs, outputs, and inputs may be difficult for RAM models to handle (Cui, 2020). They could fall short of accurately representing the intricacy of real-world systems, especially when handling several interrelated variables and erratic data (Wang et al., 2021). For ambiguous data, relying only on upper and lower bounds might oversimplify the evaluation process and overlook minute system links. Moreover, it may be difficult to provide decision makers insightful information due to the inherent complexity of RAM models and the need for cautious result interpretation (Sala-Garrido et al., 2022).

8.6 MATERIAL-BALANCED PRINCIPLE (MBP) CONSIDERATIONS

The first law of thermodynamics, or the law of conservation of matter and energy, states that "what goes in any system must come out" (Baumgärtner et al., 2001). For accurate performance measurement, the production function should align with the MBP. The MBP became integral to performance measurement when environmental concerns necessitated including bad outputs or pollutants in efficiency assessments (Coelli et al., 2007). However, they demonstrated that several previous DEA models, such as the hyperbolic efficiency measure by Färe et al. (1989), the input-oriented efficiency measure by Färe et al. (1996), and three alternative efficiency measures by Färe et al. (1996), do not meet MBP requirements.

The following criterion is proposed by Coelli et al. (2007) to determine if a model is consistent with the MBP. The quantity of pollution should be expressed as follows: $b = Ax - By$, where A and B are the non-negative coefficients of the inputs x and the outputs y, representing the pollutant content of a unit of x and a unit of y.

Then, Arabi et al. (2016) introduced Model (8.19) SBM model which is MBP enabled.

$$\max\left(\sum_{l=1}^{L}\alpha l_l + \sum_{h=1}^{H}\alpha h_h + \sum_{m=1}^{M}\alpha_m + \sum_{j=1}^{J}\beta_j + \sum_{k=1}^{K}\gamma_k\right)$$

$s.t.$

$$\sum_{n=1}^{N}\lambda_n xl_{\ln} \le xl_{lo} + \alpha l_l.1 \qquad l = 1,\ldots,L$$

$$\sum_{n=1}^{N}\lambda_n xh_{hn} \le xh_{ho} - \alpha h_h.1 \qquad h = 1,\ldots,H$$

$$\sum_{n=1}^{N}\lambda_n x_n \le x_{io} - \alpha_i.1 \qquad i = 1,\ldots,I$$

$$\sum_{n=1}^{N}\lambda_n y_{jn} \ge y_{jo} + \beta_j.1 \qquad j = 1,\ldots,J$$

$$\sum_{n=1}^{N}\lambda_n b_{kn} = b_{ko} - \gamma_k.1 \qquad k = 1,\ldots,K$$

$$\sum_{l=1}^{L}\alpha l_l - \sum_{h=1}^{H}\alpha h_h = 0$$

$$\gamma_k - \sum_{j=1}^{J}b_{jk}\beta_j = \sum_{h=1}^{H}ah_{hk}\alpha h_h - \sum_{l=1}^{L}al_{lk}\alpha l_l \qquad k = 1,\ldots,K$$

$$\lambda_n \ge 0; \alpha l_l \ge 0; \alpha h_h \ge 0; \alpha_i \ge 0; \beta_j \ge 0; \gamma_k \ge 0;\ n = 1,\ldots,N$$
$$l = 1,\ldots,L; h = 1,\ldots,H; i = 1,\ldots,I; j = 1,\ldots,J;\ k = 1,\ldots,K \qquad (8.19)$$

where based on the size of their pollutant portion, $\boldsymbol{xh}$ and $\boldsymbol{xl}$ indicate high pollutants and low pollutant inputs, respectively. Capital and other non-polluting inputs are represented by the parameter $\boldsymbol{x}$. Therefore, the rates of contraction and expansion of high and low polluting inputs, respectively, are denoted by αh and αl, whereas the rate of contraction of non-polluting inputs is represented by α. Additionally, the pollutant components of the high and low pollutant inputs, respectively, are αl and αh. $\alpha h > \alpha l$ is clearly the case; if $\alpha h = \alpha l$, there is no reason to distinguish between high and low pollution levels. The total number of inputs, $H + L + M = I$, should thus be the result.

8.7 RECENT DEA MODELS FOR SUSTAINABILITY ASSESSMENT

These examples demonstrate the more inventive DEA models that have been developed recently for sustainability assessment. As research advances, more sophisticated models that include the growing complexity of achieving sustainability across several sectors are expected to emerge.

Łozowicka (2020) used DEA to evaluate the application of sustainable development policies in EU member states, with a focus on the ecological aspect. Using an intermediate DEA technique, Sueyoshi and Yuan (2018) measured energy consumption and sustainability development in Asian countries, emphasizing the need of addressing undesirable outputs in sustainability evaluations. DEA has also been used in performance enhancement and supplier selection.

Mustapa (2018) stressed resource optimization when evaluating the technical efficiency of coal-fired power plants from a CE standpoint using the BCC model in DEA. Resource productivity is a key factor that influences the efficiency of the CE, as highlighted by Lu et al. (2020), who incorporated DEA to assess the effectiveness of China's industrial CE. Arabi et al. (2024) present a new method to enhance the process of choosing sustainable refrigeration technology by fusing technique for order of preference by similarity to ideal solution (TOPSIS) and SBM methods. In order to facilitate better informed decision making in the adoption of sustainable refrigeration technology, this hybrid model offers a thorough framework for evaluating and selecting refrigeration systems based on sustainability criteria.

Xie et al. (2019) demonstrated the adaptability of DEA in sustainability assessments by using a three-stage Network DEA technique to analyze the sustainability of product packaging systems. In a supplier selection DEA model, Dobos and Vörösmarty (2023) re-examined inputs and outputs, demonstrating DEA's flexibility in handling assessment difficulties.

Assessing and enhancing the sustainability of the food manufacturing industries in the United States is the main goal of Egilmez et al. (2014), who employ the Economic Input-Output Life Cycle Assessment (EIO-LCA) model and DEA. It emphasizes how important supply networks are to overall sustainability performance and how sensitive the forest land footprint is. In order to evaluate the effectiveness of the Spanish agri-food system and suggest improvement measures to lower energy consumption and greenhouse gas emissions, Laso et al. (2018) integrated DEA and LCA.

Ma et al. (2018) used a Super-SBM DEA model to assess spatial convergence and integrated transport efficiency across China's provinces while taking undesirable outputs into account. The life cycle assessment data of French organic agricultural products is supplied by Ma et al. (2018), who also added important life cycle assessment indicators, including cumulative energy consumption and biodiversity loss to the data.

Furthermore, Chen (2022) used a super-efficiency SBM-DEA model to quantify the urban CE and show how the CE affects carbon intensity spatially. Hu et al. (2022) demonstrated the usefulness of DEA in managing uncertainties in evaluations of the CE by putting out a DEA model for calculating the effectiveness of a recycling manufacturing system using imprecise data. In order to examine how regional material flows affect the CE, Wang et al. (2023)carried out a case study in southwest China. Their findings highlight the significance of material flow analysis in evaluations of the CE.

Innovative implementations of DEA models merging SBM with network methods and multiple-criteria decision making (MCDM) methodologies have expanded sustainability evaluations across diverse industries, as shown by the recent research mentioned. An emerging trend in sustainability research is the CE performance measurement using DEA models, which reflects growing acceptance and use of holistic evaluation frameworks. These approaches not only highlight the significance of considering resource efficiency and environmental implications, but they also provide strong instruments for well-informed decision making in the direction of accomplishing global sustainable development objectives.

8.8 CE PERFORMANCE MEASUREMENT USING DEA

Sun et al. (2019)studied technology heterogeneity and efficiency of China's CE by using a game meta-frontier network DEA (NDEA) approach. In this section, we adopt the models introduced by Sun et al. (2019) to formulate a minimal two-stage CE performance measurement model. We decompose a system with waste treatment and recycled material consumption as presented in Figure 8.1.

Our decision to use DDF DEA models is made with a circular supply chain in mind. Our goal in this framework is to achieve a decrease in the production of standard input and wastes. Simultaneously, we want to increase the use of waste-derived recycled materials and the EPS's standard outputs.

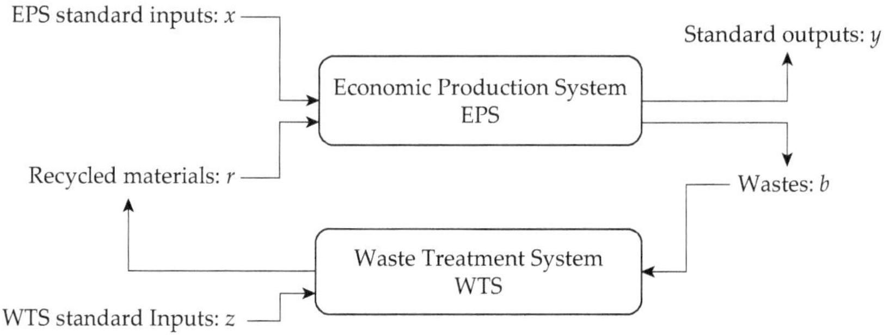

FIGURE 8.1 CE conceptual model.

Here, the basic DDF model to quantify the inefficiency of the first stage – the typical Economic Production System (EPS) – can be developed in Model (8.20) by considering weak and circular disposability properties.

$$
\begin{aligned}
&D^{EPS} = \max \beta \\
&\text{s.t.} \\
&\sum_{n=1}^{N} \lambda_n x_{in} \le x_{io} - \beta g_{xio} \qquad i = 1,\ldots I \\
&\sum_{n=1}^{N} \lambda_n r_{sn} = r_{so} + \beta g_{rso} \qquad s = 1,\ldots,S \\
&\sum_{n=1}^{N} \lambda_n y_{jn} \ge y_{jo} + \beta g_{yjo} \qquad j = 1,\ldots,J \\
&\sum_{n=1}^{N} \lambda_n b_{kn} = b_{ko} - \beta g_{bko} \qquad k = 1,\ldots,K \\
&\lambda_1,\ldots,\lambda_N \ge 0; \beta \ge 0
\end{aligned}
\tag{8.20}
$$

where r stands for recycled material, D^{EPS}, β are the intensity, EPS distance to the efficiency frontier, and inefficiency variables for $s = 1,\ldots,S$ kinds of r. The direction vector corresponding to the EPS that we selected is $g_{EPS} = (-g_x, g_r, g_y, -g_b)$, where $(g_x, g_r, g_y, g_b) = (x_o, r_o, y_o, b_o)$. We have chosen this path to make sure that our models promote the growth of desirable output while minimizing the formation of undesirable output, facilitating the use of recycled materials, and supporting the decrease of standard input consumption. In the same way, we provide the basic DDF model for the waste treatment system (WTS), as shown in Model (8.21).

$$
\begin{aligned}
&D^{WTS} = \max \theta \\
&\text{s.t.} \\
&\sum_{n=1}^{N} \mu_n z_{tn} \le z_{t0} - \theta g_{zto} && t = 1,\ldots,T \\
&\sum_{n=1}^{N} \mu_n r_{sn} \ge r_{so} + \theta g_{rso} && s = 1,\ldots,S \\
&\sum_{n=1}^{N} \mu_n b_{kn} \le b_{ko} && k = 1,\ldots,K \\
&\mu_1 \mu_2 \ldots \mu_N \ge 0; \theta \ge 0
\end{aligned}
\tag{8.21}
$$

For $t = 1\ldots T$ kinds of z, which reflect the conventional WTS inputs, the intensity, WTS distance to a perfect efficiency, and inefficiency variables are represented by μ, D^{WTS}, and θ. As in Model (8.20), we selected as the direction vector corresponding to the economic production system $g_{WTS} = (-g_z, g_r)$, where $(g_z, g_r) = (z_o, r_o)$. By using this approach, we ensure that our model supports the utilisation of recycled materials, as well as a decrease in WTS standard inputs.

Using the conceptual model shown in Figure 8.1, we develop a two-stage NDEA model. Models (8.20) and (8.21) may be combined in the manner that follows to create Model (8.22) to achieve Model (8.22) below:

$$
\begin{aligned}
&D^{Total} = \max(\beta + \theta) \\
&\text{s.t.} \\
&\sum_{n=1}^{N} \lambda_n x_{in} \le (1-\beta) x_{io} && i = 1,\ldots,I && (1) \\
&\sum_{n=1}^{N} \lambda_n r_{sn} \ge r_{so}(1+\beta) && s = 1,\ldots,S && (2) \\
&\sum_{n=1}^{N} \lambda_n y_{jn} \ge (1+\beta) y_{jo} && j = 1,\ldots,J && (3) \\
&\sum_{n=1}^{N} \lambda_n b_{kn} = (1-\beta) b_{ko} && k = 1,\ldots,K && (4) \\
&\sum_{n=1}^{N} \mu_n z_{tn} \le (1-\theta) z_{to} && t = 1,\ldots,T && (5) \\
&\sum_{n=1}^{N} \mu_n r_{sn} \ge (1+\theta) r_{so} && s = 1,\ldots,S && (6) \\
&\sum_{n=1}^{N} \mu_n b_{kn} \le b_{ko} && k = 1,\ldots,K && (7) \\
&\sum_{n=1}^{N} (\mu_n - \lambda_n) b_{kn} \le 0 && k = 1,\ldots,K && (8) \\
&\sum_{n=1}^{N} (\lambda_n - \mu_n) r_{sn} \le 0 && s = 1,\ldots,S && (9) \\
&\lambda_n \ge 0;\ \mu_n \ge 0; n = 1\ldots N;\ \beta \ge 0; \theta \ge 0
\end{aligned}
\tag{8.22}
$$

It is worth noting that in Model (8.21), Constraints (8.8) and (8.9) formulate sending wastes, *b* from EPS to WTS, and recycled materials, *r* from WTS to EPS, as illustrated in Figure 8.1.

8.9 SUSTAINABILITY REINFORCEMENT THROUGH BUSINESS PARTNERSHIPS USING DEA

To effectively leverage sustainable development, businesses need to form partnerships that allow them to pool their resources, knowledge, and networks in order to tackle intricate social and environmental issues. The sustainable development goals (SDGs) established by the United Nations, especially SDG 17, which highlights the significance of partnerships for the objectives, depend on these partnerships to be achieved. Castillo-Villar and Wright (2020) conducted research that emphasizes the critical role that cross-sector collaborations play in promoting sustainable development. These partnerships enable enterprises to combine their resources and expertise to achieve shared goals. The body of research emphasizes how crucial it is for companies to form alliances with other stakeholders, such communities, NGOs, and governments, to maximize the advantages of joint efforts toward sustainable development and to create shared value (Idemudia, 2016).

Organizations may evaluate how their activities are affecting different facets of sustainability by using frameworks and assessment criteria designed specifically for sustainable partnerships. Different criteria techniques have been developed and employed to assess how well partnerships perform in accomplishing sustainability objectives (Binda et al., 2024; Dentoni et al., 2020; Duffy & Gallagher, 2014; Loban et al., 2021; Plummer et al., 2021).

Above all, Amin and Ibn Boamah (2023) have recently introduced a fundamental technique for modeling corporate partnerships using DEA, underscoring the significance of this approach in evaluating the effectiveness and performance of such collaborations. This method paves the way for researchers to develop new DEA models that identify optimal partnerships and resource allocations, thereby enhancing both bilateral and multilateral sustainability among different DMUs.

8.10 CONCLUSION

Modern environmental, societal, and economic concerns need the integration of sustainability concepts with DEA models. This chapter has looked at how changing one's strategy is necessary when adding undesirable factors to DEA models, especially when it comes to recycling and pollution control. Conventional DEA models, which seek to minimize inputs while maximizing outputs, must be modified for systems in which maximizing certain undesirable inputs might result in improvements in sustainability. The idea of eco-efficiency, which balances economic, societal, and environmental performance, is essential to this adaptation. DEA models may improve intended outputs and reduce environmental consequences by incorporating eco-efficiency principles, which promotes sustainable business practices. These sophisticated DEA models, which emphasize waste reduction and resource efficiency, may be used in a variety of contexts, such as industrial systems, food production, power generation, and CEs. The position of DEA in assessing and maximizing sustainability via corporate relationships is a noteworthy innovation that is covered. These collaborations make use of pooled resources and knowledge, which is essential for successfully tackling challenging environmental issues. DEA models assist companies in assessing and improving cooperative endeavors, guaranteeing that financial operations are in line with sustainable objectives.

An effective foundation for striking a balance between environmental and societal impact and economic performance is provided by DEA models integrated into commercial partnerships. These models determine the best partnerships and resource allocations, promoting long-term sustainability. DEA models provide organizations the means to execute plans that advance sustainable development objectives, as companies come to understand the growing significance of eco-efficiency. Subsequent studies need to concentrate on improving DEA models by including complex environmental factors and dynamic interactions to better support sustainability objectives via corporate partnerships. By assisting companies in navigating sustainable development obstacles, DEA models are being advanced to ensure that economic activity benefits the health of the planet.

All things considered, a viable route toward a sustainable future is provided by the combination of DEA models with sustainability ideas and commercial alliances. These advanced analytical tools enhance eco-efficiency, foster effective collaborations, and ultimately contribute to environmental preservation and economic viability, ensuring economic activities support rather than compromise environmental health.

NOTES

1 STATE OF THE WORLD, 2008, Innovations for a Sustainable Economy, 25th Anniversary Edition.
2 www.wbcsd.org.

REFERENCES

Amin, G. R., & Ibn Boamah, M. (2023). Modeling business partnerships: A data envelopment analysis approach. *European Journal of Operational Research*, *305*(1), 329–337.https://doi.org/10.1016/j.ejor.2022.05.036

Amirteimoori, A., Kordrostami, S., & Sarparast, M. (2006). Modeling undesirable factors in data envelopment analysis. *Applied Mathematics and Computation*, *180*(2), 444–452. https://doi.org/10.1016/j.amc.2005.12.029

Arabi, B., Doraisamy, S. M., Emrouznejad, A., & Khoshroo, A. (2016). Eco-efficiency measurement and material balance principle: an application in power plants Malmquist Luenberger Index. *Annals of Operations Research*, *255*(1–2), 221–239. https://doi.org/10.1007/s10479-015-1970-x

Arabi, B., Munisamy, S., & Emrouznejad, A. (2015). A new slacks-based measure of Malmquist–Luenberger index in the presence of undesirable outputs. *Omega*, *51*, 29–37. https://doi.org/10.1016/j.omega.2014.08.006

Arabi, B., Munisamy, S., Emrouznejad, A., & Shadman, F. (2014). Power industry restructuringandeco-efficiencychanges:Anewslacks-basedmodelinMalmquist–Luenberger Index measurement. *Energy Policy*, *68*, 132–145. https://doi.org/10.1016/j.enpol.2014.01.016

Arabi, B., Toloo, M., Yang, Z., Zhang, P., & Xu, B. (2024). Sustainable refrigeration technology selection: An innovative DEA-TOPSIS hybrid model. *Environmental Science & Policy*, *158*, 103780. https://doi.org/10.1016/j.envsci.2024.103780

Baumgärtner, S., Dyckhoff, H., Faber, M., Proops, J., & Schiller, J. (2001). The concept of joint production and ecological economics. *Ecological Economics*, *36*(3), 365–372. https://doi.org/10.1016/s0921-8009(00)00260-3

Bebbington, J., Gray, R., & Owen, D. (1999). Seeing the wood for the trees. *Accounting, Auditing & Accountability Journal*, *12*(1), 47–52. https://doi.org/10.1108/09513579910259906

Berg, S. A., Førsund, F. R., Jansen, E. S., Berg, S. A., & Forsund, F. R. (1992). Malmquist indices of productivity growth during the deregulation of Norwegian banking, 1980–89. *The Scandinavian Journal of Economics*, *94*, 211–228. https://doi.org/10.2307/3440261

Bian, Y. (2008). Efficiency evaluation with undesirable factors based on DEA. *2008 4th International Conference on Wireless Communications, Networking and Mobile Computing*.

Binda, C. J., Adams, J., Livergant, R., Lam, S., Panchendrabose, K., Joharifard, S., Haji, F., & Joos, E. (2024). Defining a framework and evaluation metrics for sustainable global surgical partnerships: A modified delphi study. *Ann Surg*, *279*(3), 549–553. https://doi.org/10.1097/SLA.0000000000006058

Boyd, G. A., & McClelland, J. D. (1999). The impact of environmental constraints on productivity improvement in integrated paper plants. *Journal of Environmental Economics and Management*, *38*(2), 121–142. https://doi.org/10.1006/jeem.1999.1082

Brundtland, G. (1987). *Our common future: The world commission on environment and development*. Oxford University Press, Oxford.

Bruni, M., Guerriero, F., & Patitucci, V. (2011). Benchmarking sustainable development via data envelopment analysis: An Italian case study. *International Journal Environmental Research*, *5*(1), 47–56. https://doi.org/10.22059/IJER.2010.290

Burritt, R., Hahn, T., & Schaltegger, S. (2004). An integrative framework of environmental management accounting—Consolidating the different approaches of EMA into a common framework and terminology. *Environmental Management Accounting: Informational and Institutional Developments*, 21–35. https://doi.org/10.1007/0-306-48022-0_2

Castillo-Villar, R. G., & Wright, L. T. (2020). Identifying determinants of CSR implementation on SDG 17 partnerships for the goals. *Cogent Business & Management*, *7*(1), 1847989. https://doi.org/10.1080/23311975.2020.1847989

Charnes, A., Cooper, W. W., Golany, B., Seiford, L., & Stutz, J. (1985). Foundations of data envelopment analysis for Pareto-Koopmans efficient empirical production functions. *Journal of Econometrics*, *30*(1–2), 91–107. https://doi.org/10.1016/0304-4076(85)90133-2

Chen, P. (2022). The spatial impacts of the circular economy on carbon intensity – New evidence from the super-efficient SBM-DEA model. *Energy & Environment*, *35*(1), 47–63. https://doi.org/10.1177/0958305x221125125

Chiu, C. R., Chiu, Y. H., Fang, C. L., & Pang, R. Z. (2014). The performance of commercial banks based on a context-dependent range-adjusted measure model. *International Transactions in Operational Research*, *21*(5), 761–775. https://doi.org/10.1111/itor.12069

Chung, Y. H., Färe, R., & Grosskopf, S. (1997). Productivity and undesirable outputs: A directional distance function approach. *Journal of Environmental Management*, *51*(3), 229–240. https://doi.org/10.1006/jema.1997.0146

Coelli, T., Lauwers, L., & Van Huylenbroeck, G. (2007). Environmental efficiency measurement and the materials balance condition. *Journal of Productivity Analysis*, *28*(1–2), 3–12. https://doi.org/10.1007/s11123-007-0052-8

Coli, M., Nissi, E., & Rapposelli, A. (2011). Efficiency evaluation in an airline company: Some empirical results. *Journal of Applied Sciences*, *11*(4), 737–742. https://doi.org/10.3923/jas.2011.737.742

Cooper, W. W., Park, K. S., & Pastor, J. T. (1999). RAM: A range adjusted measure of inefficiency for use with additive models, and relations to other models and measures in DEA. *Journal of Productivity Analysis*, *11*(1), 5–42. https://doi.org/10.1023/a:1007701304281

Cui, Q. (2020). Airline energy efficiency measures using a network range-adjusted measure with unified natural and managerial disposability. *Energy Efficiency*, *13*(6), 1195–1211. https://doi.org/10.1007/s12053-020-09868-2

Dentoni, D., Pinkse, J., & Lubberink, R. (2020). Linking sustainable business models to socio-ecological resilience through cross-sector partnerships: A complex adaptive systems view. *Business & Society*, *60*(5), 1216–1252. https://doi.org/10.1177/0007650320935015

Dobos, I., & Vörösmarty, G. (2023). Input and output reconsidered in supplier selection DEA model. *Central European Journal of Operations Research*, *32*(1), 67–81. https://doi.org/10.1007/s10100-023-00845-5

Dong, G., Wang, Z., & Mao, X. (2018). Production efficiency and GHG emissions reduction potential evaluation in the crop production system based on emergy synthesis and nonseparable undesirable output DEA: A case study in Zhejiang Province, China. *Plos One*, *13*(11), e0206680. https://doi.org/10.1371/journal.pone.0206680

Duffy, G., & Gallagher, T. (2014). Sustaining school partnerships: The context of cross-sectoral collaboration between schools in a separate education system in Northern Ireland. *Review of Education*, *2*(2), 189–210. https://doi.org/10.1002/rev3.3034

Egilmez, G., Kucukvar, M., Tatari, O., & Bhutta, M. K. S. (2014). Supply chain sustainability assessment of the U.S. food manufacturing sectors: A life cycle-based frontier approach. *Resources, Conservation and Recycling*, *82*, 8–20. https://doi.org/10.1016/j.resconrec.2013.10.008

Färe, R., Grosskopf, S., Lovell, C. A. K., & Pasurka, C. (1989). Multilateral productivity comparisons when some outputs are undesirable: A nonparametric approach. *The Review of Economics and Statistics*, *71*(1), 90–98. https://doi.org/10.2307/1928055

Färe, R., & Grosskopf, S. (2010a). Directional distance functions and slacks-based measures of efficiency. *European Journal of Operational Research*, *200*(1), 320–322. https://doi.org/10.1016/j.ejor.2009.01.031

Färe, R., & Grosskopf, S. (2010b). Directional distance functions and slacks-based measures of efficiency: Some clarifications. *European Journal of Operational Research*, *206*(3), 702. https://doi.org/10.1016/j.ejor.2010.02.033

Färe, R., Grosskopf, S., Norris, M., & Zhang, Z. (1994). Productivity growth, technical progress, and efficiency change in industrialized countries. *The American Economic Review*, 84(1), 66–83.

Färe, R., Grosskopf, S., & Pasurka, C. (2004). Pollution abatement activities and traditional productivity: A joint production perspective. *SSRN Electronic Journal*. https://doi.org/10.2139/ssrn.593464

Färe, R., Grosskopf, S., & Pasurka, C. (2007). Environmental production functions and environmental directional distance functions. *Energy*, *32*(7), 1055–1066. https://doi.org/10.1016/j.energy.2006.09.005

Färe, R., Grosskopf, S., & Tyteca, D. (1996). An activity analysis model of the environmental performance of firms—application to fossil-fuel-fired electric utilities. *Ecological Economics*, *18*(2), 161–175. https://doi.org/10.1016/0921-8009(96)00019-5

Hadi Vencheh, A., Kazemi Matin, R., & Tavassoli Kajani, M. (2005). Undesirable factors in efficiency measurement. *Applied Mathematics and Computation*, *163*(2), 547–552. https://doi.org/10.1016/j.amc.2004.02.022

Hu, C.-F., Wang, H.-F., & Liu, T. (2022). Measuring efficiency of a recycling production system with imprecise data. *Numerical Algebra, Control & Optimization*, *12*(1), 79. https://doi.org/10.3934/naco.2021052

Huppes, G., & Ishikawa, M. (2007). An introduction to quantified eco-efficiency analysis. *Quantified Eco-Efficiency*, 1–38. https://doi.org/10.1007/1-4020-5399-1

Idemudia, U. (2016). Environmental business–NGO partnerships in Nigeria: Issues and prospects. *Business Strategy and the Environment*, *26*(2), 265–276. https://doi.org/10.1002/bse.1915

Jahanshahloo, G. R., Lotfi, F. H., Shoja, N., Tohidi, G., & Razavyan, S. (2005). Undesirable inputs and outputs in DEA models. *Applied Mathematics and Computation*, *169*(2), 917–925. https://doi.org/10.1016/j.amc.2004.09.069

James, A. T., Kumar, G., James, J., & Asjad, M. (2023). Development of a micro-level circular economy performance measurement framework for automobile maintenance garages. *Journal of Cleaner Production*, *417*, 138025. https://doi.org/10.1016/j.jclepro.2023.138025

Knox Lovell, C. A., Pastor, J. T., & Turner, J. A. (1995). Measuring macroeconomic performance in the OECD: A comparison of European and non-European countries. *European Journal of Operational Research*, *87*(3), 507–518. https://doi.org/10.1016/0377-2217(95)00226-x

Laso, J., Hoehn, D., Margallo, M., García-Herrero, I., Batlle-Bayer, L., Bala, A., Fullana-i-Palmer, P., Vázquez-Rowe, I., Irabien, A., & Aldaco, R. (2018). Assessing energy and environmental efficiency of the Spanish agri-food system using the LCA/DEA methodology. *Energies*, *11*(12), 3395. https://doi.org/10.3390/en11123395

Liu, W. B., Meng, W., Li, X. X., & Zhang, D. Q. (2009). DEA models with undesirable inputs and outputs. *Annals of Operations Research*, *173*(1), 177–194. https://doi.org/10.1007/s10479-009-0587-3

Loban, E., Scott, C., Lewis, V., & Haggerty, J. (2021). Measuring partnership synergy and functioning: Multi-stakeholder collaboration in primary health care. *Plos One*, *16*(5), e0252299. https://doi.org/10.1371/journal.pone.0252299

Lovins, L. H. (2008). Rethinking production. *State of the World 2008, ideas and opportunities for sustainable economies* (pp. 34–44). The Worldwatch Institute.

Łozowicka, A. (2020). Evaluation of the efficiency of sustainable development policy implementation in selected EU member states using DEA. The ecological dimension. *Sustainability*, *12*(1), 435. https://doi.org/10.3390/su12010435

Lu, C., Zhang, Y., Li, H., Zhang, Z., Cheng, W., Jin, S., & Liu, W. (2020). An integrated measurement of the efficiency of China's industrial circular economy and associated influencing factors. *Mathematics*, *8*(9), 1610. https://doi.org/10.3390/math8091610

Ma, F., Wang, W., Sun, Q., Liu, F., & Li, X. (2018). Integrated transport efficiency and its spatial convergence in China's provinces: A super-SBM DEA model considering undesirable outputs. *Applied Sciences*, *8*(9), 1698. https://doi.org/10.3390/app8091698

Macpherson, A. J., Principe, P. P., & Smith, E. R. (2010). A directional distance function approach to regional environmental–economic assessments. *Ecological Economics*, *69*(10), 1918–1925. https://doi.org/10.1016/j.ecolecon.2010.04.012

Mustapa, S. I. (2018). Efficiency assessment of Malaysian coal-fired power plant: A circular economy perspective. https://doi.org/10.15405/epsbs.2018.07.02.66

Picazo-Tadeo, A. J., Reig-Martínez, E., & Hernández-Sancho, F. (2005). Directional distance functions and environmental regulation. *Resource and Energy Economics*, *27*(2), 131–142. https://doi.org/10.1016/j.reseneeco.2004.07.001

Plummer, R., Witkowski, S., Smits, A., & Dale, G. (2021). Higher education institution–Community partnerships: Measuring the performance of sustainability science initiatives. *Innovative Higher Education*, *47*(1), 135–153. https://doi.org/10.1007/s10755-021-09572-8

Sala-Garrido, R., Mocholí-Arce, M., Molinos-Senante, M., & Maziotis, A. (2022). Measuring operational and quality-adjusted efficiency of Chilean water companies. *NPJ Clean Water*, *5*(1). https://doi.org/10.1038/s41545-021-00146-x

Schaltegger, S., Bennett, M., & Burritt, R. L. (2008). *Environmental management accountingforcleanerproduction*.SpringerVerlag.https://doi.org/10.1007/978-1-4020-8913-8

Schaltegger, S., & Sturm, A. (1990). Ökologische rationalität. *Die Unternehmung*, *4*(90), 273–290.

Scheel, H. (2001). Undesirable outputs in efficiency valuations. *European Journal of Operational Research*, *132*(2), 400–410. https://doi.org/10.1016/s0377-217(00)00160-0

Schott, J. (2007). Changing course: A global business perspective on development and the environment. Stephen Schmidheiney, Business Council for Sustainable Development Cambridge: MIT Press. *Business Strategy and the Environment*, *2*(1), 51–52. https://doi.org/10.1002/bse.3280020108

Shephard, R. W., Gale, D., & Kuhn, H. W. (1970). *Theory of cost and production functions*. Princeton University Press.

Sueyoshi, T., & Goto, M. (2012). Weak and strong disposability vs. natural and managerial disposability in DEA environmental assessment: Comparison between Japanese electric power industry and manufacturing industries. *Energy Economics*, *34*(3), 686–699. https://doi.org/10.1016/j.eneco.2011.10.018

Sueyoshi, T., & Yuan, Y. (2018). Measuring energy usage and sustainability development in Asian nations by DEA intermediate approach. *Journal of Economic Structures*, *7*(1). https://doi.org/10.1186/s40008-017-0100-0

Sun, J., Li, G., & Wang, Z. (2019). Technology heterogeneity and efficiency of China's circular economic systems: A game meta-frontier DEA approach. *Resources, Conservation and Recycling*, *146*, 337–347. https://doi.org/10.1016/j.resconrec.2019.03.046

Tone, K. (2001). A slacks-based measure of efficiency in data envelopment analysis. *European Journal of Operational Research*, *130*(3), 498–509. https://doi.org/10.1016/s0377-2217(99)00407-5

Tone, K. (2010). Variations on the theme of slacks-based measure of efficiency in DEA. *European Journal of Operational Research*, *200*(3), 901–907. https://doi.org/10.1016/j.ejor.2009.01.027

Tyteca, D. (1997). Linear programming models for the measurement of environmental performance of firms—Concepts and empirical results. *Journal of ProductivityAnalysis*,*8*(2),183–197.https://doi.org/10.1023/a:1013296909029

Wang, M., Chen, L., Liu, L., Zhang, R., Cheng, H., Li, X., & Su, H. (2023). The impact of regional material flows on circular economy: A case study of southwest China. *Greenhouse Gases: Science and Technology*, *13*(3), 432–449. https://doi.org/10.1002/ghg.2215

Wang, M., Chen, Y., & Zhou, Z. (2021). Assessing environmental efficiency of China's industry system using two-stage range-adjusted measure model. *Management of Environmental Quality: An International Journal*, *32*(6), 1401–1421. https://doi.org/10.1108/meq-12-2020-0282

Xie, Y., Gao, Y., Zhang, S., Bai, H., & Liu, Z. (2019). Sustainability evaluation of product packaging system with a three-stage network data envelopment analysis methodology. *Applied Sciences*, *9*(2), 246. https://doi.org/10.3390/app9020246

Yawe, B., & Kavuma, S. (2008). Technical efficiency in the presence of desirable and undesirable outputs: A case study of selected district referral hospitals in Uganda. *Health Policy and Development*, *6*(1), 37–53.

Zhou, H., Yang, Y., Chen, Y., & Zhu, J. (2018). Data envelopment analysis application in sustainability: The origins, development and future directions. *European Journal of Operational Research*, *264*(1), 1–16. https://doi.org/10.1016/j.ejor.2017.06.023

Zhou, P., Ang, B. W., & Poh, K. L. (2006). Slacks-based efficiency measures for modeling environmental performance. *Ecological Economics*, *60*(1), 111–118. https://doi.org/10.1016/j.ecolecon.2005.12.001

Zhou, P., Poh, K. L., & Ang, B. W. (2007). A non-radial DEA approach to measuring environmental performance. *European Journal of Operational Research*, *178*(1), 1–9. https://doi.org/10.1016/j.ejor.2006.04.038

CHAPTER 9

Public–Private Investment Partnership

A Case of Small and Medium-Sized Enterprises in Taiwan

Shu-Chin Huang
Department of Money and Banking, Ming Chuan University, Taoyuan, Taiwan

Chung Wei Chiu, Shao-Fang Chen, and Yu-Han Wang
New Business Department, National Innovation and Entrepreneurship Association, Taoyuan, Taiwan

9.1 INTRODUCTION

The National Development Fund of the Executive Yuan, Republic of China (NDF) is the most important public fund promoting the venture capital industry in Taiwan. It entered into partnerships with venture capital corporations (VCs) to implement equity co-investment programs since 1985. The first initiative started in 1985 and lasted until October 2007. In the programs, NDF entrusted VCs to co-invest in local small- and medium-sized enterprises (SMEs). The purpose was to promote Taiwan VC industry as well as Taiwan high-tech startups. The first initiative spent 9.27 billion Taiwanese dollars (TWD), about US$0.31 billion. Compared with Taiwan's annual GDP, the amount of money was insignificant during the period. However, the NDF cooperated with 53 VCs in total and stimulated

DOI: 10.1201/9781003345206-9

investment in more than 1,000 high-tech startups in Taiwan in the first initiative (NDF, 2007). The program won the reputation that it had guided Taiwan VC industry through its early explorations (TVCA, 2023; Chen, 2001).

The second co-investment program initiated by the NDF started in August 2007, and the program is currently ongoing. The program focusing on SMEs has been implemented by the Small and Medium Enterprise Administration of the Ministry of Economic Affairs. By the end of 2022, the program had gone through four phases of investment, marked by different policy focuses with a total investment amount of TWD 9.37 billion. Cumulatively during this period, NDF entrusted 23 VCs and co-invested in 349 cases.

The research objectives of this study are to answer the questions: What are the performances of the investment partnerships? What factors determine investment performance? And, do the NDF policy positions influence the investment outcomes?

Among the 349 invested cases in the second initiative of the NDF, a total of 190 cases (managed by 22 VCs) were exited or claimed full losses by the end of 2022. In this research, we use the 190 cases as the sample. We apply a two-stage method consisting of Data Envelopment Analysis (DEA) for computing the relative investment efficiencies of each case in the first stage followed by adopting the generalized structural equation modeling (GSEM) to detect factors determining the investment efficiencies in the second stage. The research findings can be a useful reference for VC literature. Also, the results provide information for NDF and VCs to further strengthen partnerships and cooperation to improve future investments.

9.2 LITERATURE REVIEW

Partnerships have been growing as a business practice because companies seek to flexibly respond to competition and marketing. There have been many studies in the literature emphasizing the importance of partnerships in business practices but there are limited studies for modeling business partnerships (Amin & Ibn Boamah, 2023). Often, companies are unable to acquire the necessary technologies, marketing resources, management skills, or finance by themselves to compete effectively. A strategic alliance and partnership between two or more independent decision making units (DMUs) to share their resources or to co-develop new products and technologies has become a key solution to enhance the growth of such companies.

9.2.1 Partnership Between Firms and Venture Capital

Partnership is a common practice in the VC industry. VC companies are well known for their capabilities in assisting small- and medium-sized enterprises (SMEs) in finance, management, and marketing (a review by Colombo & Grilli, 2009). For example, VCs act as scouts; they focus on specific industries and develop context-specific screening methods to judge the hidden quality of entrepreneurial talent and projects. They closely monitor the behavior of entrepreneurs to protect their investments. On the other hand, they provide the invested companies (hereafter referred to as investees) with consultancy services in strategic planning, marketing, finance, accounting, human resource management, and external networks, where the startups typically lack such capabilities. Empirical studies support that partnerships between VCs and their investees positively affect the growth of the investees. By studying 439 Italian high-tech start-ups, Colombo and Grilli (2009) show that VC investments have a dramatic positive effect on the growth of the sample firms. Using longitudinal data on 8,346 new technology-based firms from seven European countries observed from 1995 to 2008, Colombo et al. (2016) also conclude that VC-backing strongly impacts firms' participation in EU-funded R&D partnerships. Moreover, the magnitude of the impact differs considerably depending on the type of investor, with bank and government VCs exhibiting the strongest positive effects.

9.2.2 Independent Versus Corporate Venture Capital

VC investors are heterogeneous along several dimensions. Previous studies have identified two types of VCs according to their governance, namely independent and corporate VCs (Dushnitsky & Lenox, 2006; Colombo et al., 2016). Independent VCs (IVCs) are operated based on an organizational form of partnership between limited partners (LPs) and general partners (GPs). IVCs manage closed-end funds in which the fund providers act as LPs, and the managers of the funds serve as the GPs. LPs have limited responsibility for the company's debts to the amount of money they invested. On the other hand, the GPs are paid regular management fees and share some of the investment profits according to the contracts. IVCs manage closed-end funds that have a limited lifetime, usually capped at ten years. At the end of a fund's existence, the GPs cash their bonus and leave all remaining cash and securities to the LPs. Having said that, the fact that an IVC firm must continually raise new funds to assure their continuity of operations generates strong performance pressure on GPs.

They must show LPs that the IVC firm can deliver reasonable returns in order to induce LPs to inject new capital.

Corporate VCs (CVCs) refer to the VCs which are invested by established companies. They are structured as investment vehicles of their parent companies, which provide the capital to be invested by the CVC. According to Dushnitsky and Lenox (2006), CVCs apparently appeared as an investment vehicle for firms in the 1960s. CVCs have experienced rapid growth but also fluctuated up and down along with macroeconomic conditions. Although CVCs supply valuable complementary non-financial resources to the investees, to maintain pay uniformity and avoid any jealousy among departments, CVCs typically do not offer as high incentives to fund managers as IVCs. In addition, competition for limited resources within their parent companies can negatively impact CVC funds. Furthermore, if CVC funds fail to produce reasonable returns, top managers may shut them down quickly. On average, the life span of a CVC fund is shorter than that of an IVC fund.

Different types of VCs have different objectives and may adopt different investment strategies. IVC investors are more interested in having investees focus efforts and resources on commercializing promising technologies, products, and services (Colombo et al., 2016). On the other hand, CVCs often invest in newly emerging high-tech companies to monitor promising technologies. Their investments sometimes allow the parent company to learn about the technological feasibility and market potential of a new technology-based entrepreneurial idea. This option can be exercised later in the exit approach by forming an alliance with or the acquisition of the investee (Van de Vrande & Vanhaverbeke, 2013). Even though CVCs may have poor financial returns, their strategic benefits can justify the investment. Nevertheless, empirical studies have found that the performance of CVCs is as successful as that of IVCs (Dushnitsky & Lenox, 2006).

9.2.3 Public–Private Partnership

Public–private partnerships (PPPs) are cooperative institutional arrangements between public and private sector entities. This has been widely adopted in several developed and developing countries. Wang et al. (2018) conducted a literature review of 186 journal articles about PPP in the discipline of public administration, which were published before 31 December 2016 and contained in the ISI Web of Science Database. The review pointed out that PPP has been widely used in the fields of infrastructure, transportation, water and sewage, energy, environment, public health, and other public services. By using UCINET software to visualize the co-occurrence matrix of the key

PPP features written about in the 186 articles, these researchers identify measurements of PPP performance, and the influence factors emphasized in the literature. Briefly, the literature has classified PPP performance into narrow and broad measurements. The narrow measure of success is mainly about cost savings, investment returns, or the project's net present value. The broad concept of a successful PPP performance focuses on a wider range of contents and scope. It considers the project's wider benefits for partner organizations, industries, citizens, and product/service users, which have been emphasized by the governance network theory. Furthermore, they also identify factors determining a successful PPP in past literature. The most frequently appearing keywords in studies about the determinants are organizational form, contract management, management strategies, and political support.

9.2.4 Partnerships Between NDF and VCs in Taiwan

As one example of a public–private partnership, the NDF forms partnerships with VCs to co-invest in Taiwanese SMEs. The role of NDF in the partnership with VCs is akin to a LP. NDF is protected by the limits of responsibility for the investee's debts to the amount of money they invested. On the other hand, VCs function like a GP in the partnership.

By the end of 2015, all Taiwanese VCs were established as 'company limited' (Chen, 2019). Nevertheless, their functions are similar to those of Western countries' IVC and CVC. We classify Taiwanese VCs into three types. First, we have some venture capital consulting and management companies in the VC industry. They behave like GP and manage funds from various sources. By the end of 2022, these consulting and management companies had obtained most of the NDF co-investment funds. In our samples, about 93% of the co-invested cases were run by this type of VC. Hereafter, we will use IVC to refer to such a company.

The second type of VCs in Taiwan is CVC, as defined above. The parent firms use VC to diversify risks, pursue new technologies, and generate profits. They obtain funds from the parent firms and seek to be awarded co-investment funds from the NDF. Nevertheless, by the end of 2022, only a few CVCs obtained co-investment funds from the NDF. In our sample, about 7% of the NDF-VC co-invested cases are managed by CVCs.

The third type of VC in Taiwan is established by companies, individuals, and various sources of institutions. However, they entrust type one consulting and management companies to invest their funds. They act as LPs.

To the best of the authors' knowledge, there are limited empirical studies about the relationships between the NDF and Taiwanese VCs, partly due to data constraints. Huang et al. (2023) first estimated the effects of management fees paid by the NDF to the entrusted VCs on the co-investment performance. This chapter extends the literature to investigate how the partnerships between NDF and VCs directly and indirectly influence investment outcomes.

9.3 METHODOLOGY AND DATA

The NDF has accumulatively formed partnerships with VCs to co-invest in Taiwanese SMEs (Table 9.1). From August 2007 to December 2022, there were a total of 349 co-invested cases, of which 190 cases have been exited or claimed full losses. In this research, we use the 190 cases as the sample. In order to achieve our research objective, the empirical analyses are carried out in two steps. We first follow the DEA literature (Emrouznejad & Yang, 2018) to assess the investment efficiencies of these 190 cases (explained in the later paragraphs). Then, GSEM is adopted to detect factors determining the investment efficiencies. GSEM is particularly useful if we need to simultaneously estimate several co-variated equations and latent effects. It is a suitable method for our research purpose because the investment performance is likely influenced not only by management strategies but also by latent effects of NDF policy positions and individual VC's investment capability.

9.3.1 Total Factor Productivity and DEA Method

This chapter adopts DEA to measure the investment performance of individual cases. This method computes relative technical efficiency (TE)

TABLE 9.1 NDF Investment Policy Positions in Four Phases, From 2007 to 2022

Investment Phase and Starting Date	NDF Investment Policy: Strengthen Investment In	Money Invested (TWD Million)
Phase A 2007/8/1	Domestic SMEs.	3,023.1
Phase B 2010/11/1	Service industries, high-employment, early-stage, and cultural creative enterprises.	3,325.4
Phase C 2012/8/1	Early-stage enterprises.	895.7
Phase D 2016/3/1	Early-stage, social, high-employment, and developing information security enterprises; enterprises established in the Startup Terrace Linkou, and less developed regions in Taiwan.	2,127.5

Note: Policy announcements reference Ministry of Economic Affairs, R.O.C. (2007–2021).

scores for each co-invested case. We treat each investment case as a DMU. TE of a DMU is a relative term by comparing the total factor productivity (TFP) of the ith DMU (TFP_i) to the TFP_B of the most efficient DMU among all invested cases for comparison. In other words, efficiency is measured by the concept of TFP in terms of investment performance. The capital letter B denotes the best, or most efficient. In practice, the mathematic form of TE is shown in equation (9.1):

$$\mathrm{TE}_i = \frac{\mathrm{TFP}_i}{\mathrm{TFP}_B} = \frac{1}{\mathrm{TFP}_B} \times \mathrm{TFP}_i = w \times \mathrm{TFP}_i \tag{9.1}$$

DEA approach is employed to find out the weight (w) in Equation (9.1), which indeed is $\frac{1}{\mathrm{TFP}_B}$, in order to obtain one optimal solution, by using linear programming (Huang et al., 2008; Coelli, 1996).

9.3.2 DEA Models

In this chapter, we alternatively use different types of DEA models to comprehensively show co-investment efficiencies for each of the 190 cases. First, the well-known CCR model named after the developers Charnes, Cooper, and Rhodes (1978) is used to compute TEs based on the constant returns to scale (CRS) production assumption. Second, the BCC model named after Banker, Charnes, and Cooper (1984) is employed. The model developers argued that imperfect competition, constraints on finance, etc., may cause a DMU to not operate at optimal CRS. Thus, they proposed an extension of the CRS CCR model to account for variable returns to scale (VRS) situations. They claimed that the difference between the CRS technical efficiency (hereafter briefly denoted as TEcrs) and VRS technical efficiency (TEvrs, known as pure technical efficiency) is a result of scale efficiency (SE). These DEA models have been widely applied by many studies in different fields (Emrouznejad & Yang, 2018).

Next, the meta-frontier concept (O'Donnell et al., 2008) is applied to compare TEs of individual cases in different NDF investment policy regimes (A, B, C, and D in Table 9.1). Please note that the investment phase starting date was not the end of the co-investment application date. A VC could apply for a co-investment fund from the NDF based on policy positions declared in different phases at any time until the fund was exhausted. Thus, a case in phase A was not necessarily invested before the cases in phases B, C, or D.

Also note that a dynamic DEA model or the Malmquist total factor productivity change measurement (Nedaei et al., 2020; Färe et al., 1994; Ray & Desli, 1997) is not feasible in our case, due to limitations of available data. Although the NDF and VCs monitored the performance of investees from time to time, an investment outcome was only generated after the case was exited or claimed a full loss. Moreover, once NDF partially or fully sold out the equities of holding investees, the income went back to the NDF. The NDF did not reuse the returns to finance another round of investment during the entire program period. In other words, every dollar budgeted was used to invest in an SME only once.

Furthermore, unlike Amin and Ibn Boamah (2023) swapping inputs and/or outputs in a DEA program to improve efficiencies, we adopt a two-stage approach. In the second stage, the study proposes the GSEM to detect factors determining investment TEs. We use GSEM to control the co-variate and latent effects of NDF's four investment-policy regimes and VC investment capabilities on co-investment decisions. At the same time, GSEM is used to simultaneously control co-variates among different investment strategies that directly affect investment efficiencies. In this case, the pre-assumption that partners can switch inputs and/or output resources is not visible because there were pre-determined rules for the implementation of NDF funds. Alternatively, the two-stage approach used allows policymakers to understand the determinants of investment efficiencies. Partners can further modify investment strategies to improve investment efficiencies based on this understanding.

9.3.2.1 Output-Oriented DEA Models

The DEA method has two forms, the output-oriented and the input-oriented DEA methods. We adopt the output-oriented model for this study because the research question is: 'How much can output quantities, such as investment returns and cases invested in line with NDF policy positions, be proportionally expanded without altering the input quantities used?' To highlight the model, let us assume that there are n DMUs $\left(k=1,2,\ldots,n\right)$, and each of them uses m inputs $\left(i=1,2,\ldots,m\right)$ to produce s outputs $\left(r=1,2,\ldots,s\right)$. The mathematical form of the output-oriented CRS CCR model is shown in the following linear program (2):

$$\max_{\varphi,\delta} \varphi$$

Subject to $\delta_1 Y_{r1} + \delta_2 Y_{r2} + \ldots + \delta_n Y_{rn} \geq \varphi Y_{rk} \; (r = 1, 2, \ldots, s)$

$$\delta_1 X_{i1} + \delta_2 X_{i2} + \ldots + \delta_n X_{in} \leq X_{ik} \; (i = 1, 2, \ldots, m)$$

$$\delta_1, \delta_2, \ldots, \delta_n \geq 0 \tag{9.2}$$

where Y_{rk} and X_{ik}, respectively, are the rth output and the ith input of the kth DMU. δ is $n \times 1$ vector of weights from δ_1 to δ_n for both input and output items of DMUs. CRS CCR model uses linear program (2) to find an optimal solution for both δ and φ. Let φ^* be the optimal solution of φ. Each DMU has its own optimal solution of φ^*. The output-oriented TE (TE_k) score of DMU$_k$ is $\frac{1}{\varphi_k^*}$. We can use $(1 - TE_k)$ to show the proportion of outputs that DMU$_k$ can expand to catch up the frontier DMU performance.

The CRS linear programming problem can be modified to account for VRS by adding the convexity constraint (3) into the program (2) to obtain technical efficiency scores (*TEvrs*) for each DMU for comparison:

$$\delta_1 + \delta_2 + \ldots + \delta_n = 1 \; (k = 1, 2, \ldots, n) \tag{9.3}$$

Moreover, scale efficiency scores (SE_k) can be calculated by

$$\mathrm{SE}_k = \frac{\mathrm{TE}_{crs_k}}{\mathrm{TE}_{vrs_k}} \tag{9.4}$$

9.3.3 Meta-Frontier

Investment TEs among different NDF investment-policy regimes are compared. The NDF-VC co-investment program went through four phases of investment, each with different policy focuses, as they are described in Table 9.1. The meta-frontier concept developed by O'Donnell et al. (2008) is used to compare. The distinction between meta-frontier and group frontiers is illustrated in Figure 9.1. VCs in different NDF policy regimes (A, B, C, and D four phases) face different policy constraints and investment opportunities. Technically, they make choices from different sets of feasible or predicted investment input–output combinations. These so-called investment technology sets differ because of differences in available NDF co-investment amounts, NDF policy focuses, NDF practices,

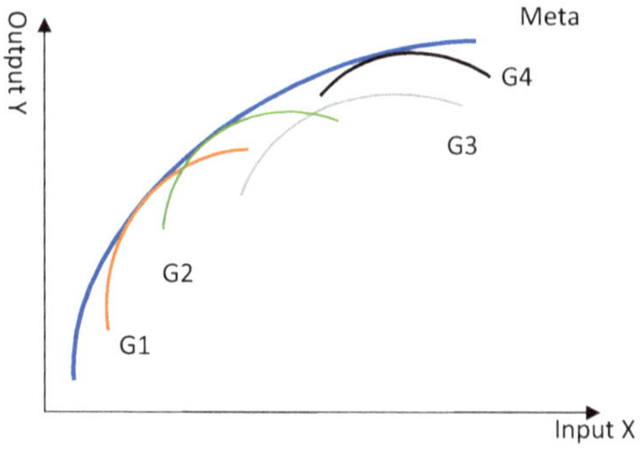

FIGURE 9.1 Meta-frontier envelops group frontiers.

VC investment capabilities, investment strategies, and macro-economic conditions. Thus, investment frontiers are separated for different groups of invested cases according to the four phases of NDF investment-policy regime.

Figure 9.1 shows that the meta-frontier (denoted as meta) envelops the four groups of investment technological frontiers (denoted as G1 to G4). The most efficient investment outcomes are DMUs (invested cases) standing on the meta-frontier, which is drawn by pooling all 190 invested cases. Individual groups include only DMUs belonging to the respective NDF policy regimes. Because the investment TE score is a relative measurement, the most investment efficient DMU in a group ($TE_{k\text{-best}}$) is not necessarily also the most efficient DMU standing on the pooled meta-frontier ($TE_{meta\text{-best}}$). Moreover, a DMU's TE_k group score is higher or equal to its TE_{meta} meta score because the subgroups' investment input–output sets are part of the meta input–output sets.

Mathematically, the output set $\left(P(x)\right)$ is defined for any input vector x, $(x \geq 0)$ that can produce investment output y, $(y \geq 0)$, given the meta investment technology (T) as

$$P(x) = \{y : (x, y) \in T\} \tag{9.5}$$

satisfies the standard regularity properties listed in O'Donnell et al. (2008). The investment technology using the meta-distance function is defined as

$$D(x, y) = \inf_\theta \{\theta > 0 : (y / \theta) \in P(x)\} \tag{9.6}$$

where θ is the output efficiency. This function gives the maximum amount by which an invested case (DMU) can radially expand its output vector, given an input vector. An observation (x, y) is considered technically efficient with respect to the meta-frontier if $D(x, y) = 1$.

Similarly, the G sub-group specific technologies can also be represented by the group-specific investment output sets and output distance functions:

$$P^g(x) = \{y : (x, y) \in T^g\} \qquad g = 1, \ldots, G \tag{9.7}$$

$$D^g(x, y) = \inf_\theta \{\theta > 0 : (y / \theta) \in P^g(x)\} \tag{9.8}$$

We refer to the boundaries of the group-specific output set as group frontiers.

9.3.3.1 Input–Output Selections

In terms of investment outputs, the objective of the NDF investment for SMEs is to promote Taiwan's VC industry and, at the same time, through VCs pick winners to enhance the growth of SMEs to lead Taiwan's industrial development. Thus, investment revenues are not the only criterion to assess the investment performance. In line with previous studies, we define investment revenues (denoted as *orev*) as a narrow measure of investment success (Vecchi et al., 2010). At the same time, we incorporate the broad concept of public–private partnership performance to include a wider range of contents and scopes (Wang et al., 2018). Policy measures, such as whether or not the co-invested case was in line with NFD policy positions as the broad measure of investment benefit to the partner of NFD (denoted as *opolicy*), are included. In practice, *opolicy* is constructed if the co-invested case satisfied the NDF investment policy positions = 2, otherwise = 1 (Table 9.2). Because all invested cases are SMEs, they obtain at least one credit. Also, we consider investment benefits to the investees and society. If the co-invested case is publicly offered after being financed by the NDF-VC, then the case receives two points, otherwise one point (denoted as *oipo*). In the study, publicly offered includes stock-market listed, *TPEx* Mainboard listed, Emerging-stock-market listed, and Go Incubation Board (GIB).

From the NDF point of view, a comprehensive technical investment efficiency score using the DEA approach is, therefore, a result of the relative investment input–output ratios of each invested case. In input-measure selection, we consider the investment costs, including the investment amount (denoted as *iinvcost*) and the opportunity cost of foregoing interest

TABLE 9.2 Descriptive Statistics

Variable	Mean	Std. Dev.	Min	Max
invcost (TWD millions)	25.70	17.50	2.25	90.00
interest (TWD millions)	3.08	2.94	0.01	18.10
itotal (TWD millions)	28.80	20.00	2.39	106.00
orev (TWD millions)	32.80	59.50	0.00	461.00
opolicy	1.84	0.37	1	2
orevratio	1.21	1.75	0	11.45
oprofratio	0.27	1.80	−1	10.45
oipo	1.32	0.47	1	2
oiratio	1.09	1.57	0	10.61
industry	4.73	2.84	1	11
coinvratio	1.84	2.82	0.2	35.52
priceratio	2.79	2.37	0.5	15
invscale	2.17	0.86	1	5
overseas	0.52	0.50	0	1
stage	2.31	1.09	1	4
yearinv	5.54	2.58	0.04	13.44
region	1.95	0.39	1	3
Exitype	3.81	1.38	1	5
ownership	2.28	1.13	1	4
cvc	0.07	0.26	0	1
vccapital	0.41	0.49	0	1
vcage	24.08	7.22	11.7	42.5
firmage	7.04	8.24	0.08	42.73
abcd	2.03	1.02	1	4
vcid	11.33	6.29	1	22

revenues if the amount of money had been deposited or used to buy risk-free assets. Thus, Taiwan's ten-year government bond yield rates are used to compute the interest-revenue foregone and denote it as iinterest. By adding the interest-revenues foregone input cost in the assessment of investment efficiency, the cost of the time value of the money invested in each case is indeed included. Finally, to remove the problem due to differences in data units and scales, *z*-score method is used to standardize all input and output data, and then, 2.5 was added to each input and output standardized values to make all I-O data positive.

9.3.4 Generalized Structural Equation Modeling

In the second stage of the analysis, we apply GSEM to investigate how the partnerships between NDF and VCs directly and indirectly influence investment outcomes. The two main features of GSEM are the ability to simultaneously perform multiple and interrelated regression estimates and that it is an estimate that removes estimation errors. The latter feature refers to the fact that GSEM estimates using the estimated value after removing estimation errors. So, GSEM is generally more efficient than other estimation methods (Hair et al., 2018; Stata, 2017). Furthermore, this study adopts GSEM instead of traditional SEM because the estimation of GSEM does not require assuming that the data distribution conforms to normality, while the SEM does. DEA investment TE scores are generally between 0 and 100, which might not exactly follow the form of normal distribution.

The novelty of our GSEM model is to construct two layers of latent variables that influence investment decisions. As shown in Figure 9.2, the top layer of the latent variable represents the four phases of NDF investment policy positioning and practices (denoted as *Lndf[abcd]*). The second layer of the latent variable is the investment ability of individual VCs (denoted as *Lvc[abcd>vcid]*), which operate under the NDF policy regimes. In addition,

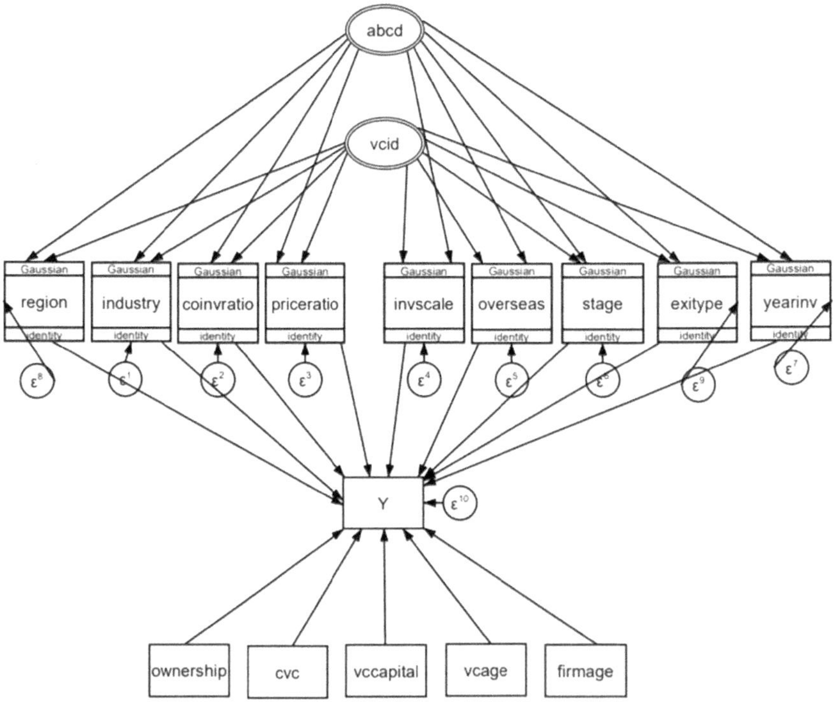

FIGURE 9.2 The case-strategy-VC-NDF four-level GSEM model.

these two latent variables are correlated in GSEM. This design intends to capture VC investment decisions that are influenced by the policies and practices set by the NDF. Moreover, NDF also modified its practices due to learning from interactions with VCs over time.

Nevertheless, it is the investment strategies chosen by VCs and approved by NDF that directly affect investment outcomes. In practice, investment strategies include selecting which industry to invest in (noted as *industry*), making co-investment ratio decisions (*coinvratio*), purchasing price decisions (*priceratio*), investment scale decisions (*invscale*), considering investee's overseas Chinese ownership (*overseas*, 1 if Chinese 0 otherwise), investee's development stage (*stage*), years invested (*yearinv*), investee's location (*region*), and exiting type (*exitype*). Table 9.2 presents the descriptive statistics.

Moreover, a number of measures commonly used in the analysis of the financial performance of firms are included in the GSEM. In terms of VC-specific characteristics, VC type by ownership, VC type by governance (IVC, CVC), VC capital, VC age, etc., have been discussed in the literature (Dushnitsky & Lenox, 2006; Berger & Ofek, 1995; Hu et al., 2022). For example, some studies argue that CVCs may be better positioned to provide complementary capabilities and resources to investees. In addition, higher age and capital relate to the VCs positioned to reap lucrative financial returns because they may be skilled in selecting valuable ventures and often have expert personnel with in-depth knowledge of markets and technology. Also, larger firms may experience economies of scale and scope in managerial effort that allow them to perform superior due diligence at lower costs.

In practice, VCs are classified into four types according to their ownership, including industrial company, financial holding company, public or semi-public organization, and others (denoted as *ownership*) in this study. VC types are also distinguished by governance (*cvc*) and included in the GSEM. Other VC specific characteristics such as VC size (measured by the company's capital, denoted as *vccpaital*) and VC age (*vcage*) are also included in the model. To control investee characteristics, investee's age (*firmage*), industry (*industry*), development stage (*stage*), location (*region*), and overseas ownership (*overseas*) are included in the GSEM. Note that the study controls for Industry type. Differences in industry profitability may have to do with any number of factors including industry structure, technological innovation, and government regulation.

The latent variables and covariate structures are important to answer the research questions because a single equation might not capture the fact that investment outcomes are determined by not only investment strategies but also policy regimes and VC investment abilities. In addition, the collinearity of predictors is no longer a nuisance in the GSEM model. Instead, it provides that an appropriate measuring model is built, which is usually lost in generalized linear models (GLMs) to reduce variance inflation (Lombardi et al., 2017). The general structural equations are modeled as the case-strategy-VC-NDF four-level model shown in Figure 9.2.

Mathematically, the GSEM simultaneous regression functions are presented as follows:

$$\text{industry strategy} = \alpha_1 + \gamma_1 Lndf\left[abcd\right] + \delta_1 Lvc[abcd > vcid] + \varepsilon_1$$

$$\text{coinvratio strategy} = \alpha_2 + \gamma_2 Lndf\left[abcd\right] + \delta_2 Lvc[abcd > vcid] + \varepsilon_2$$

$$\text{priceratio strategy} = \alpha_3 + \gamma_3 Lndf\left[abcd\right] + \delta_3 Lvc[abcd > vcid] + \varepsilon_3$$

$$\text{invscale strategy} = \alpha_4 + \gamma_4 Lndf\left[abcd\right] + \delta_4 Lvc[abcd > vcid] + \varepsilon_4$$

$$\text{investee overseas ownership} = \alpha_5 + \gamma_5 Lndf\left[abcd\right] + \delta_5 Lvc[abcd > vcid] + \varepsilon_5$$

$$\text{investee stage strategy} = \alpha_6 + \gamma_6 Lndf\left[abcd\right] + \delta_6 Lvc[abcd > vcid] + \varepsilon_6$$

$$\text{yearinv strategy} = \alpha_7 + \gamma_7 Lndf\left[abcd\right] + \delta_7 Lvc[abcd > vcid] + \varepsilon_7$$

$$\text{investee region strategy} = \alpha_8 + \gamma_8 Lndf\left[abcd\right] + \delta_8 Lvc[abcd > vcid] + \varepsilon_8$$

$$\text{exitype strategy} = \alpha_9 + \gamma_9 Lndf\left[abcd\right] + \delta_9 Lvc[abcd > vcid] + \varepsilon_9$$

$$\begin{aligned} Y = {} & \beta_0 + \beta_1 \text{industry} + \beta_2 \text{coinvratio} + \beta_3 \text{priceratio} + \beta_4 \text{invscale} \\ & + \beta_5 \text{overseas} + \beta_6 \text{stage} + \beta_7 \text{yearinv} + \beta_8 \text{region} + \beta_9 \text{exitype} \\ & + \tau_1 \text{ownership} + \tau_2 \text{cvc} + \tau_3 \text{vccapital} + \tau_4 \text{vcage} + \tau_5 \text{firmage} + \varepsilon_{10} \end{aligned} \tag{9.9}$$

where Y is investment performance indicator. In GSEM, we alternatively use narrow and broad measures of investment performance indicators to represent Y. For example, *Y1* denotes profit–investment ratio (*oprofratio*), *Y2* represents revenue–cost ratio (*oiratio*), *Y3* is simple super-efficiency scores based on CRS, and *Y4* is CCR TE scores computed by using input investment amount plus interest revenues foregone opportunity cost (*itotal*) and output investment revenues (*orev*). *Y5* is CCR TE scores computed using the same inputs, but outputs augment NDF policy objective and benefits to investees and society into the measurement (*orev, opolicy & oipo*). The estimated results are reported in Table 9.8.

Note that the investment length (*yearinv*) of each case is used as an explanatory variable in GSEM to control the time effect on the investment outcomes. The design may mitigate the fact that a dynamic or Malmquist DEA approach could not be used in the first stage of the DEA analysis.

9.3.5 Data Analyses

Variables used in our empirical analyses are summarized in Table 9.2. Among the 190 co-invested cases, the maximum amount of investment (iinvcost) was TWD 90 million, and the minimum was TWD 2.25 million. On average, the investment scale was TWD 25.7 million per case. The length of investment per case (yearinv) averaged 5.5 years, minimum about four months and maximum 13.4 years. Including interest revenues foregone opportunity costs (iinterest), the total investment cost per case by the NDF (itotal) averaged 28.8 million, about a 12% mark up from the original investment amount.

Investment outcomes measured by revenues averaged TWD 32.8 million per case, maximum TWD 461 million and minimum zero dollars (total loss). A total of 57 cases claimed full losses, accounting for 30% of the 190 sample cases, which were exited or claimed full losses by the end of 2022. Without considering the interest revenues' foregone opportunity cost, among the 190 cases, 108 cases (56.8%) suffered losses, 49 cases (25.8%) made profits of less than twice the investment amount, 31 cases (16.3%) generated revenues between two to ten times of investment amount, and two cases (1.1%) created revenues more than 10 times of investment amount. In general, the NDF investment revenues are akin to previous VCs' performances in the United States, except that NDF produced a lower percentage of winning cases with revenues more than ten times of the investment amount. According to Kupor (2019), a study of the United States' VC industry, about 50% of VC investments end up losing money, only 20%–30% of investments may produce profits of less than twice the investment amount, and 10%–20% of investments have the opportunity to make a tenfold profit or more.

TABLE 9.3 Samples Spread Throughout A, B, C, and D Phases

	Phases of NDF Investment				
	A	B	C	D	Total
Exited cases with some revenues	66	14	46	7	133
(%)	78	54	68	64	70
Cases with full losses	19	12	22	4	57
(%)	22	46	32	36	30
Total cases	85	26	68	11	190
(%)	100	100	100	100	100

Pearson Chi-squared test for independence of rows and columns:
Chi2(3) = 5.99 Prob. = 0.11

Table 9.3 shows that the 190 sample cases were spread throughout the A, B, C, and D investment phases, with 85, 68, 26, and 11 cases, respectively. They account for 54.4% of the total 349 co-invested cases in the four phases. The shares of sample cases in each phase are 85%, 57.6%, 53.1%, and 13.4%, respectively. Those open cases remain unstudied because their final investment outcomes are still uncertain. Based on the 190 sample cases, Phase B had about 46% of cases ending up with full losses, which was relatively high compared with Phase A at 22%, Phase C at 32%, and Phase D at 36%. We apply the Pearson Chi-squared test for independence of phase and full loss. The resulting probability of 11% approaches the 10% significance level.

Pearson's correlations among inputs and outputs are reported in Table 9.4. They measure linear correlations between two variables without control of other variables. Results show that the narrow measure of output, such as investment revenues (*orev*), is positively and significantly correlated with input factors, including NDF investment amount (*iinvcost*) and interest revenues foregone during the invested period for individual cases (*iinterest*). A broad measure of output (benefit to investee and society) such as *oipo* that indicates the investee was publicly offered after being financed by the NDF-VC investment, is also significantly and positively correlated with these input factors. These positive correlations meet the requirement for using the DEA approach, which assumes more inputs lead to more outputs.

Notably, another broad output measure, *opolicy* represents benefits to NDF and society if the invested case was in line with NDF policy positions. The policy output is insignificantly correlated with inputs. The insignificance indeed illustrates the fact that a case in line with policy focus does not necessarily demand a large or a small amount of

TABLE 9.4 Pearson's Correlation: Inputs Versus Outputs

	iinvcost	iinterest	orev	opolicy	oipo	oprofit	oiratio	itotal
iinvcost	1							
iinterest	0.82 (0.00)***	1						
orev	0.42 (0.00)***	0.42 (0.00)***	1					
opolicy	−0.02 (0.79)	0.005 (0.95)	−0.09 (0.24)	1				
oipo	0.12 (0.09)*	0.17 (0.02)**	0.43 (0.00)***	−0.14 (0.05)**	1			
oprofit	0.14 (0.05)**	0.20 (0.01)***	0.96 (0.00)***	−0.09 (0.20)	0.43 (0.00)***	1		
oiratio	0.04 (0.59)	0.06 (0.43)	0.75 (0.00)***	−0.05 (0.48)	0.44 (0.00)***	0.80 (0.00)***	1	
itotal	1.00 (0.00)***	0.87 (0.00)***	0.43 (0.00)***	−0.016 (0.83)	0.13 (0.07)*	0.15 (0.03)**	0.06 (0.42)	1

Note:

* Denotes $p(.)<0.1$

** $p(.)<0.05$, and

*** $p(.)<0.001$

investment. Nevertheless, the policy output '*opolicy*' does significantly and negatively correlate with the output '*oipo*'. In other words, a case in line with policy focus was less likely to become publicly offered, although the size of coefficient −0.14 is not high.

9.4 EMPIRICAL RESULTS

9.4.1 Investment Possibilities Frontier (Meta Concept)

If we consider that every single dollar invested is equally important to NDF, then CRS of production scale should be applied in the measurement of investment TE, which would be CCR DEA method. Production Possibilities Frontiers (PPFs) based on CRS are illustrated in Figures 9.3–9.5. In Figure 9.3, the horizontal axle measures input total investment costs (*itotal*), which includes investment amount and interest revenues foregone opportunity cost. The vertical axle shows the return ratio (*oiratio*) of investment revenues to investment total costs. As Pearson's correlation indicates, Table 9.4 Column 8, the return ratio does not significantly correlate with input total investment costs. Nevertheless, the plots in Figure 9.3 illustrate that DMU111 is the most efficient invested case. It was a case invested in Phase B with a small investment scale of TWD 3 million. The

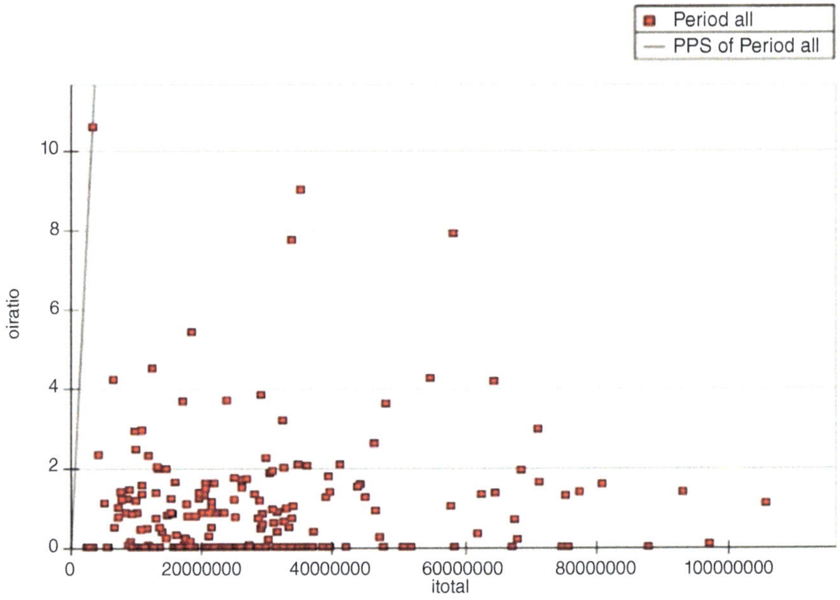

FIGURE 9.3 Meta investment possibilities frontier: Revenues to total cost ratio (oiratio) versus total costs (itotal).

case turned out with a return ratio of 10.6 times of invested costs. It was invested during its startup stage, which was in line with NDF policy positions. The performance of DMU111 is far ahead of others. If we exclude it from the comparison, then the revenue–cost ratio plots appear a pattern close to normal distribution but skewed to the left. Decreasing returns is apparent especially when the invested scale is above TWD 60 million (Figure 9.3).

Instead of using revenue–cost ratio, the vertical axle measures investment revenues (*orev*) and plot the I-O PPF in Figure 9.4. DMU111 is still the most efficient case standing on the frontier. If we exclude it, then both Figures 9.3 and 9.4 show similar pictures that decreasing returns to scale seems happening when the invested scale is beyond TWD 60 million.

Figure 9.5 (a) plots inputs versus policy output (*opolicy*). Most cases met the NDF investment policy positions. Only 16% of the sample cases do not follow the policy positions. Nevertheless, all of them are SMEs. Figure 9.5 (b) draws inputs versus the output benefits to investees and society (*oipo*). A total of 31.6% of the sample cases were finally publicly offered after they were invested by NDF-VC.

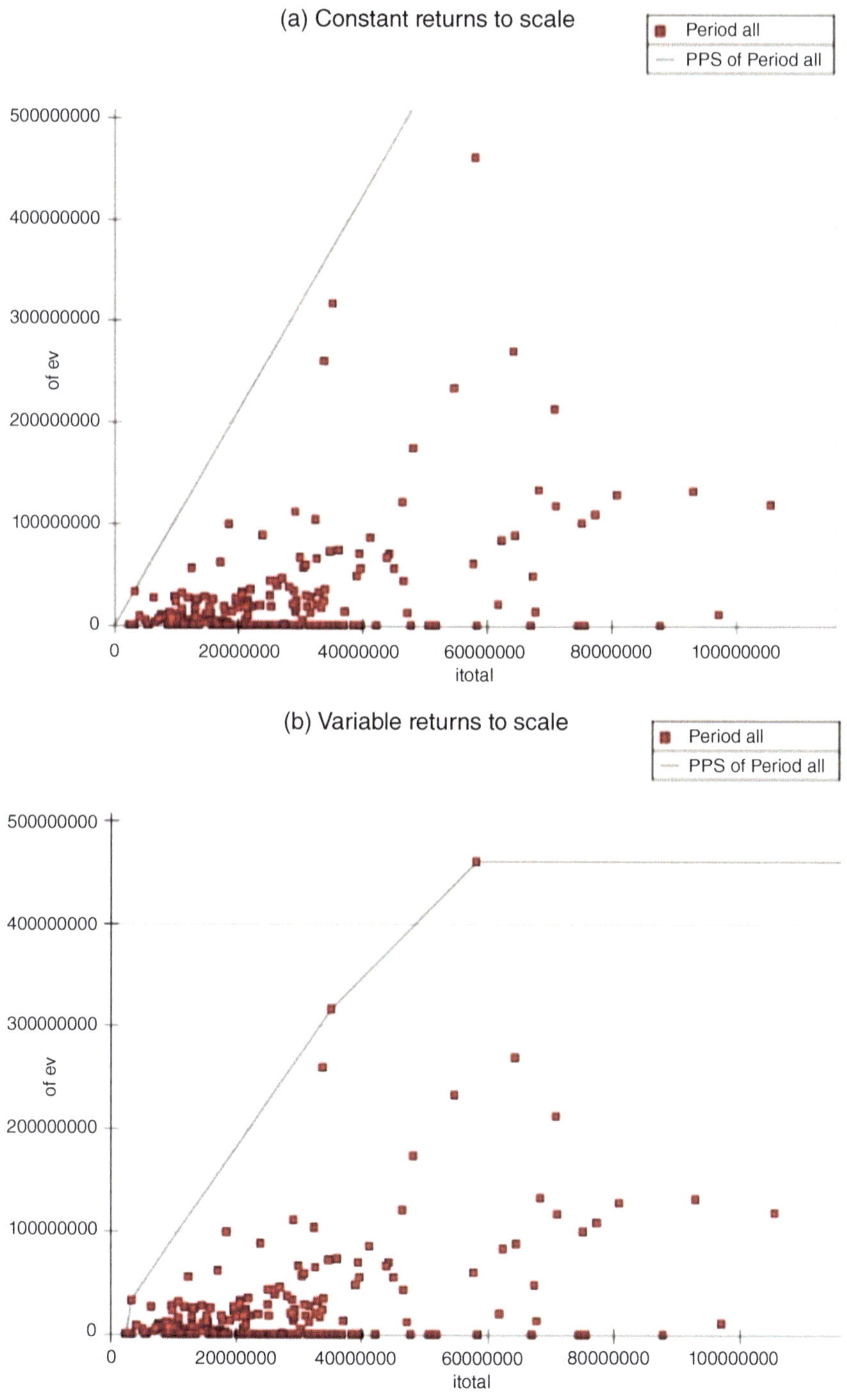

FIGURE 9.4 Meta investment possibilities frontier: Revenues (*orev*) versus total costs (*itotal*). (a) CRS, (b) VRS.

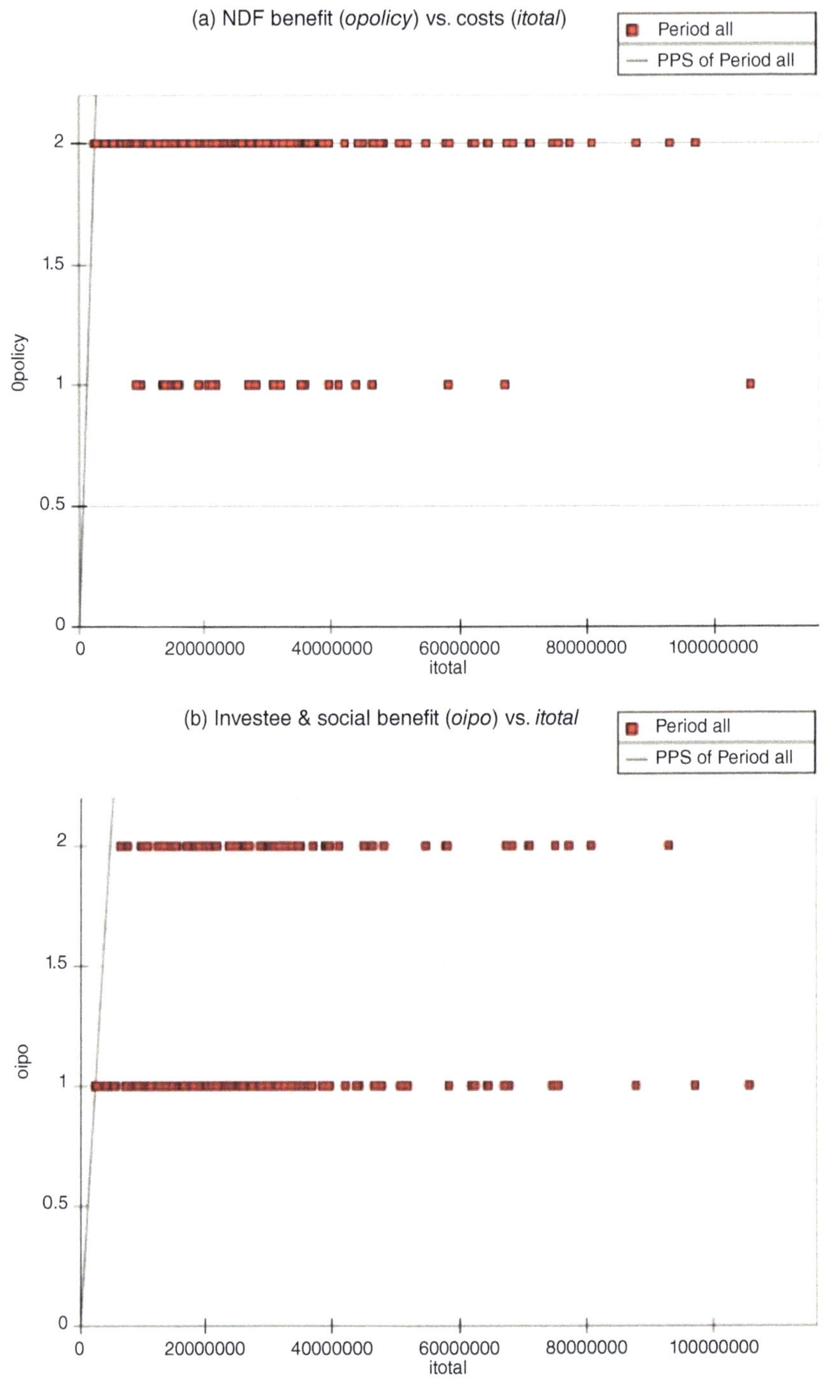

FIGURE 9.5 Inputs versus broad measures of outputs. (a) NDF benefit (*opolicy*) versus costs (*itotal*), (b) Investee & social benefit (*oipo*) versus *itotal*.

9.5 INVESTMENT EFFICIENCIES

Due to differences in data scales, all inputs and outputs standardized and 2.5 is added to each to make all data values positive. All input and output factors have equal weighting. In addition, output *opolicy* (meet NDF investment position or not) is treated as a non-discretionary variable because the policy output is fixed at the beginning when NDF and VC decide to invest in the case (Banker & Morey, 1986). Similarly, output *oipo* (publicly offered or not after investment) is also treated as a non-discretionary variable because the value is fixed at 1 or 2, which does not vary with input amounts. Table 9.6 reports investment TEs by using CCR constant returns to scale (CCRcrs), which is a result of pure investment TE (BCCvrs, using BCC VRS) times scale efficiency (SE). The list of inputs and outputs used in Table 9.6 are Inputs={X_1=iivcost, X_2=itotal} and Outputs={y_1=orev, y_2=opolicy, y_3=oipo}.

9.5.1 Super-Efficient Cases

The simple super-efficiency concept developed by Andersen and Petersen (1993) is applied to detect super-efficient invested cases (DMUs). The super-efficiency of a DMU is computed by assessing it relative to all DMUs, excluding itself. This means that some DMUs may have efficiency scores above 100 and are super-efficient cases. By assuming CRS, the results indicate that DMU6 scoring 106.4 is a super-efficient case if only conventional input investment amount and output revenues (Table 9.6 Column 2, Model 1) are considered. In Model 2, interest revenues foregone are added as another opportunity cost (*iinterest*) and allow for VRS. The results show that in addition to DMU6, DMU69 and DMU111 are also super-efficient cases (Table 9.6 Column 3). The main characteristics of the three super-efficient cases are listed in Table 9.5. Two of them were invested when the firms were at the expansion stage. Nevertheless, it was the one invested at the startup stage that finally brought the highest return ratio, 10.6 times the investment amount.

9.5.2 Investment Technical and Scale Efficiencies

To give the rest of DMUs a more realistic set of attainment targets, the three super-efficient cases are dropped from the sample for the CCR and BCC analyses. In addition, we drop DMU175 as an outlier because it cannot be solved in the super-efficiency VRS linear program. Results show that the average investment meta CCR technical score is 45.7 (Table 9.6

TABLE 9.5 Characteristics of Super-Efficient Cases

Super-Eff. Case	Inv. Phase	Firm Stage	Years Inv.	iinvcost (TWD Million)	Industry	Location	Exit Type	orev/itotal	Meet Policy	IPO After Inv.
DMU6	A	Expan.	8.2	30	life sci. & pha.	North	TPEx Main-board	9.0	yes	yes
DMU69	B	Expan.	7.9	50	inf. & tel.	North	TWSE	7.9	no	yes
DMU111	B	startup	3.9	3	software	North	not IPO	10.6	yes	no

TABLE 9.6 Investment Efficiencies: Meta Mean Scores

Invest. Phase	(1) Sup. eff. crs	(2) Sup. eff. vrs	(3a) Scale eff.	(3b) BCC vrs	(3c) CCR crs	(4a) Scale eff.	(4b) BCC vrs	(4c) CCR crs	(5a) Scale eff.	(5b) BCC vrs	(5c) CCR crs	(6a) Scale eff.	(6b) BCC vrs	(6c) CCR crs
A	39.3	43.3	87.4	51	44.3	86.9	51.3	44.1	91.9	51.3	47.2	91.8	51.6	47.4
B	40.8	46.9	86.4	53.2	45.4	85.9	53.6	45.4	90.8	53.6	48.8	92.1	53.8	49.8
C	44.3	49.8	90.4	57.3	50.1	89.4	57.7	49.9	99.8	57.7	57.6	99.8	57.7	57.6
D	42.9	50.6	87.4	58.3	49.3	86.7	59.4	49.8	96	59.4	57.2	96.9	59.8	58.2
Total	40.7	45.9	87.4	53.1	45.7	86.9	53.4	45.7	92.8	53.4	49.8	92.3	53.7	50.3
Inputs	X_1	X_2	X_1	X_1	X_1	X_2	X_2	X_2	X_2	X_2	X_2	X_2	X_2	X_2
Outputs	Y_1	Y_1	Y_1	Y_1	Y_1	Y_1	Y_1	Y_1	Y_1, Y_2	Y_1, Y_2	Y_1 Y_2	Y_1 Y_2, Y_3	Y_1, Y_2 Y_3	Y_1 Y_2 Y_3
Average orev target required					143%			143%			132%			131%

Notes: Model (1) assumes CRS, and the super-efficient DMU is DMU6, with a score of 106.4. Model (2) assumes VRS and exclude the observation of DMU 175. The super-efficient DMUs are DMU 6, 69, and 111, with scores of 109, 133, and 121, respectively. Models (3a) to (6c) exclude the observations of DMU 6, 69, 111, and 175.

Model 3c), given the input investment amount (*iinvcost*) and output investment revenues (*orev*). The score is a relative TFP score and the comparative peer is DMU12, which was invested in Phase A with an investment amount of about TWD 30 million. DMU12 was invested when it was a startup in the pharmaceutical industry, and finally brought a return to the NDF at 8.7 times of the investment amount. Interest revenues foregone of the investment amount during the investment period is also included as an additional opportunity cost. Given the new total input (*itotal* = *iinvcost* + *iinterest*) and output *orev*, the average meta CCRcrs is the same 45.7 in which SE is 86.9 and BCCvrs is 53.4 (Table 9.6, models 4a, 4b, and 4c).

Next, the outputs are enlarged to include the NDF policy goal (*opolicy*) and treat the variable as a non-discretionary variable. Given the same inputs *itotal* and new outputs *orev* and *opolicy*, the average meta CCRcrs score increases by 4.1 points to 49.8 (Table 9.6, Model 5c). Finally, another non-discretionary output variable, *oipo*, expressing the broad benefit to the investees and society, is included in the assessment. The average meta CCRcrs score becomes 50.3 in which BCCvrs score is 53.7 and SE score is 92.3 (Table 9.6, models 6a, 6b, and 6c). Peers of the comparison are DMU12, DMU132, and DMU181. As noted in the previous paragraph, DMU12 achieved a good return. In addition, it met the NDF policy position and was listed on TPEx Mainboard after being financed by NDF-VC. DMU132 was invested in Phase D with an investment amount of TWD 6 million. The investment scale was relatively small, while it brought a return of 4.6 times of the investment amount. The case also satisfied NDF policy positions and traded in Taiwan Emerging Market Stock market after being financed by NDF-VC investment. It was invested when the firm was a startup in the semiconductor industry. Surprisingly, DMU181, a case of full loss, becomes one of the peers when the broad measurements *opolicy* and *oipo* are included in the performance assessment. Investment in DMU181 occurred in Phase C with a small amount of investment, about TWD 2.4 million. It met NDF policy position that was invested in the seed stage in the innovative culture industry. Given the small amount of investment, NDF made its commitment to foster a seed company in the government-focused industry, even though it did not achieve any monetary return.

Overall, if the public–private partnership of co-investment in SMEs is measured by the narrow output of money returns, then the output target of *orev* has to be further increased by an average of 143% in order to achieve the investment TE (akin to the performance of DMU12). However, if the broad output measurements of *opolicy* and *oipo* is included in the assessment, then the output revenue target of *orev* needs only an average of 131% improvement to reach an efficiency level. Clearly, there is considerable room for the investment team to raise investment efficiency.

9.6 GSEM FINDINGS – DETERMINANTS OF INVESTMENT EFFICIENCIES

Tables 9.7 and 9.8 report the GSEM regression results. The estimated coefficients reported are unstandardized coefficients. The estimated errors represent the variance of the indicated dependent variable (*strategy*) that is not measured by the two latent variables (*Lndf[abcd]* and *Lvc[abcd>vcid]*) shown in Table 9.7. Also, note that we use different indicators to measure the investment performance of each invested case (the dependent variable

TABLE 9.7 Part I: GSEM Results – NDF and VC Latent Effects on Investment Strategies Fitting Fixed-Effects Model; Family: Gaussian; Link: Identity

Dep. Var.	**Indep. Var.**	**GSEM 1 Coef.**
industry (7.46)	*Lndf[abcd] (0.31)*	1.00 (cons.)
	Lvc[abcd>vcid] (0.38)	1.00 (cons.)
coinvratio (6.56)	*Lndf[abcd]*	−0.52
	Lvc[abcd>vcid]	−1.82**
priceratio (5.15)	*Lndf[abcd]*	−0.6
	Lvc[abcd>vcid]	0.96*
invscale (0.69)	*Lndf[abcd]*	−0.52*
	Lvc[abcd>vcid]	−0.05
overseas (0.24)	*Lndf[abcd]*	0.02
	Lvc[abcd>vcid]	0.14
stage (1.03)	*Lndf[abcd]*	−0.83**
	Lvc[abcd>vcid]	−0.19
yearinv (6.35)	*Lndf[abcd]*	−0.84
	Lvc[abcd>vcid]	−0.51
region (0.15)	*Lndf[abcd]*	0.11
	Lvc[abcd>vcid]	−0.03
exitype (1.7)	*Lndf[abcd]*	0.84*
	Lvc[abcd>vcid]	−0.26

Notes: *(**, ***) show a significant level at 10% (5%, 1%). Estimated error variance values are in brackets of the first column. The variance values of latent variables *Lndf[abcd] and Lvc[abcd>vcid] are 0.31 and 0.38, respectively*. The latent effects for GSEM 2 to GSEM 5 are the same as GSEM 1, and thus, we do not repeat the results in the table.

TABLE 9.8 GSEM Results – Strategic Effects on Investment Efficiency

	GSEM 1	GSEM 2	GSEM 3	GSEM 4	GSEM 5
	Y1	Y2	Y3	Y4	Y5
Industry:	Compare to Optical industry				
Semiconductor	0.85*	0.78**	8.62***	6.47**	8.04**
Innovative culture	−0.18	−0.13	1.24	3.22	6.59
Mechanical	1.56***	1.05**	7	4.2	3.8
Life sci., pha.	0.49	0.41	3.96	2.76	2.65
Green energy	0.29	0.21	0.02	−0.23	−2.1
Internet	−0.31	−0.3	2.21	−2.67	−3.05
Information	0.92*	0.69	7.05*	3.22	4.49
Software	1.37**	1.26**	5.66	1.93	2.8
Communications	0.02	0.03	3.67	6.41	11.81**
Others	0.16	0.06	1.35	0.89	2.6
coinvratio	0.01	0.01	0.03	−0.12	−0.27
priceratio	0.06	0.06	0.75*	0.44	0.49
invscale:	Compare to under 10 million				
10–30 million (inc.)	−0.44	−0.35	−7.95***	−15.66***	−27.61***
30–50 million (inc.)	−0.48	−0.43	−13.09***	−25.10***	−39.70***
50–80 million (inc.)	−0.98**	−0.87**	−20.36***	−31.83***	−46.29***
above 80 million	−0.95	−0.82	−30.85***	−38.81***	−52.83***
overseas	0.26	0.27	−0.03	−2.13	−3.07
stage:	Compare to Startup				
Late	−1.65***	−1.48***	−11.79***	−7.34**	−7.77*
Expansion	−0.27	−0.28	−1.38	−2.37	−2.44
Seed	−0.2	−0.18	2.52	1.57	3.18
yearinv	0.05	0.02	−1.62***	−0.23	−0.35
region:	Compare to Central Taiwan				
North	0.18	0.06	−2.43	−1.97	−2.96
South	0.02	−0.07	−2.52	−4.09	−5.34
exitype:	Compare to Stock market list				
TPEx Mainboard list	−0.26	−0.12	−0.11	1.91	2.07
EMS	−1.04**	−0.78*	−5.42	−3.13	−2.9
GIB	−2.00***	−1.62***	−11.89***	−10.19***	−11.83***
Not IPO	−2.13***	−1.77***	−12.54***	−10.16***	−11.25***
ownership:	Compare to Others				
Government sup.	−0.99	−0.88	−6.61	−3.28	−4.5
Industrial co.	0.64	0.58*	−0.2	−0.79	−1.35
Financial co.	−0.06	−0.02	−2.43	−0.89	−1.11

(Continued)

TABLE 9.8 (Continued)

	GSEM 1	GSEM 2	GSEM 3	GSEM 4	GSEM 5
vccapital	0.94*	0.82*	3.82	1.58	2.69
Log-likelihood	−3190	−3163	−3572	−3515	−3550
AIC	6523	6470	7289	7175	7245
BIC	6757	6704	7523	7408	7478

Notes: *(**, ***) show a significant level at 10% (5%, 1%). Each GSEM sub-equation includes a constant variable and control variables cvc, vcage, and firmage. These three control variables are insignificant to explain Ys. Their coefficients are not reported in the table due to space limitations.

for function Y). A total of five GSEM models are estimated. However, the regression results of the latent effects of NDF and VC on the investment strategies are the same across these GSEM models.

In addition, values of log-likelihood, AIC, and BIC are reported at the bottom of Table 9.8. These statistics are used to compare the fitness of alternative GSEMs with different *Y* measurements from *Y1* to *Y5*. In principle, a larger log-likelihood value indicates that the model offers a better fit to the data. In contrast, a smaller value of AIC or BIC shows that the model is a better fit for the data. Nevertheless, adding more explanatory variables to the model will almost always increase the log-likelihood value, even if the additional variables are statistically insignificant. Thus, the model has to be used with caution. Overall, the three statistical values are stable across the 5 alternative GSEMs.

9.6.1 NDF and VC Partnership to Co-Determine Investment Decisions

Results in Table 9.7 show the role of NDF and VCs in the partnership to influence strategic investment decisions. Although the NDF is somewhat similar to an LP in the partnership, its policy positions and practices, including game design, investment scales, and the power of final decision to grant the co-investment funds, could influence VCs' investment decisions. In the coefficient column of the table, *p*-values indicate that the latent NDF variable (*Lndf[abcd]*) influences the choice of investment-scale strategy and the selection of exit-type strategy, each at a 10% significance level. It also explains the selection of the investee's development stage strategy at a 5% significance level.

On the other hand, the 22 VC companies make investment decisions within the four phases of NDF policy regimes. The coefficients of the VC latent variable (*Lvc[abcd>vcid]*) in the nine types of strategic decision functions show that VC capability explains the co-investment ratio at a 5% significance level and purchase-price to face-value ratio at a 10% significance level. Notably, the negative signs of the coefficients illustrate that VC with higher investment capability can lower the purchase price, and their investment amounts share in the co-investment funds are higher. Overall, the estimated variance of the top layer of latent variable (*Lndf[abcd]*) expressing unobserved values of four phases of NDF policy positions and practices is 0.31. The variance of the second layer of latent variable (*Lvc[abcd>vcid]*) representing unobserved values of 22 VC corporations' investment capabilities is 0.38.

9.6.2 Effects of Investment Strategies

Table 9.8 reports the GSEM estimates of function *Y*. As mentioned in the methodology, we alternatively use different investment performance indicators for the dependent variable Y (from *Y1* to *Y5*). The estimates are consistent across the five alternative functions. Among the nine types of investment strategies, we found that selecting industries, investment scales, investee's development stage, and exit type do make the investment performance significantly different.

9.6.2.1 Industrial Strategy

Given the sample and with the controls on VC characteristics and the investee's age, the results show that choosing the semiconductor industry for the investment generates a significantly higher profit ratio, revenue–cost ratio, and all types of TEs studied in our GSEMs. The results echo Taiwan's industrial performance in recent decades wherein the semiconductor industry has been leading the growth of the economy, even during the Covid-19 pandemic period. Next, when policy objective (*opolicy*) and investment broad benefits to investees and society (*oipo*) are also included in the measurement (Table 9.8, GSEM 5, *Y5*), the communications industry appears to be an efficient investment champion as well. Only five cases are in this industry; most of them have small investment amounts and meet the NDF policy positions. One of them was listed on the GIB after the investment. The experience of investing in the communications industry provides the message that the investment amount does not necessarily have to be large to achieve the objective of NDF for promoting SMEs to develop high-tech industries.

9.6.2.2 Investment Scale Strategy

The coefficients of investment scales show that the investment scale under TWD 10 million produces the highest investment TEs. The finding is consistent across *Y4* and *Y5* functions. Moreover, the results show that the higher the scale of investment, the lower the efficiency. This result applies to all efficiency measures estimated. The finding bursts a myth that a high investment scale yields a high investment efficiency. In contrast, limited by public finance, many small amounts of funding allocated to a larger number of SMEs are proved to yield more investment efficiency, not only in terms of revenue outcomes but also policy objectives and social benefits.

9.6.2.3 Investee's Development Stage Strategy

The investment stage is important in determining investment results. There is no sign to show that investment in the expansion or late stage can produce a better profit ratio or investment TE than investment in the startup and the seed stages. In contrast, the negative coefficients of the late stage and the expansion stage show that investment in the startup stage, on average, generates a higher profit ratio, revenue–cost ratio, and investment TE scores, and it is significantly higher than investment in the late stage. If policy objectives and broad investment benefits to investees and society are augmented, then results show that investment in the seed stage increases investment TEs even no less than investment in the startup stage (functions Y3 to Y5). These findings are different from conventional thinking that investment in the investee's expansion stage or late stage may make it easier to obtain higher returns.

9.6.2.4 Exit Type Strategy

Choosing an exit type is important for achieving investment profit and policy goals. The coefficients show that exit type by listing in the Taiwan Stock Market or TPEx Mainboard generates higher investment performance, followed by listing in the Emerging Market Stock (EMS). Joining GIB or not becoming publicly offered (Not IPO) produces the lowest investment outcomes. The findings are constant across functions *Y1* to *Y5* results. The implication for fund managers is to help investees list in stock markets or start trading in Mainboard or EMS to realize investment goals.

9.7 CONCLUSION

This chapter studies the public–private partnership between the National Development Fund of R.O.C. (Taiwan) and venture capital corporations to invest in SMEs during the 2007–2022 time period. The investment performance is assessed not only by conventional investment profit ratio but also by applying Data Envelopment Analysis, including CRS CCR approach, VRS BCC approach, scale efficiency approach, and super efficiency approach. In the DEA approach, the innovation is to include the investment amount and interest revenues foregone during the investment period as input factors. On the other hand, the investment outputs include the narrow measure of investment revenues, as well as the broad measures of meeting policy objectives and producing benefits to investees and society.

A two-stage method involves computing investment efficiencies using DEA in the first stage and then applying GSEM to detect factors determining the success of investment efficiencies in the second stage. The study uses GSEM to evaluate the effects of nine types of investment strategies on the investment efficiency scores obtained from the first stage. The strategies include selecting an industry, co-investment ratio, purchase price-face value ratio, investment scale, investee's overseas ownership, investee's development stage, the length of investment, investee's location, and exit type. The results indicate that the four phases of NDF policy positions and practices do play a latent role in influencing investment decisions, particularly significant in determining investment scale, choosing the investee's development stage, and the exit type. At the same time, VCs make investment decisions under the NDF policy regimes and play the second layer of latent role to explain the co-investment ratio and purchase price significantly. Furthermore, although the power of making decisions on investment length and choosing investees' locations and overseas ownership is not significantly explained by either the NDF or VC latent variables, the relative sizes of coefficients and P-values do show that the NDF has a higher influence on the former two choices and VCs have more influence on overseas ownership selection. The overall empirical evidence illustrates that the NDF and VCs indeed co-determine the investment strategies.

Finally, investment strategies directly affect investment performance. The GSEM estimates confirm previous studies that investment outcomes are not random. Instead, different investment strategies can produce significantly different results. With controls on VC characteristics, and investee age, we find that investing in the semiconductor industry and the communications industry significantly produces higher investment efficiency than investment in the other industries. Unlike the conventional impression, the results show that investment in the life science, pharmaceutical, and medical materials industries does not generate significantly higher efficiency than investment in the optical industry (base for the comparison). This might be because investment in the life industry is usually on a large scale. When we consider every dollar invested, each dollar's output should be treated equally (CRS), then the investment efficiency of the life industry does not outperform the semiconductor and communications industries.

Important determinants of investment performance also include investment scale, selecting the investee's development stage, and exit type. Results show that the NDF investment scale under TWD 10 million produces the highest investment TE, followed by an investment scale between 10 and 30 million, and efficiency declines as the investment scale increases. The finding is true no matter the outputs, including investment revenues alone or also policy objectives and benefits to investees and society. No empirical study has been published, to the best of the authors' knowledge, that shows the relationship between investment scale and investment performance in the VC industry. However, observed from venture capital investment data of OECD member states and the Russian Federation in 2007–2016, venture capital invested in small businesses with less than 100 employees, accounts for approximately 85% of total investment in enterprises (OECD, 2017). Our finding echoes the rational practice in the OECD VC industries.

Regarding the selection of the investee's development stage for the investment, the GSEM results depict that late-stage investment generates a significantly lower profit ratio and revenues-cost ratio. On the other hand, investments in the seed and startup stages, on average, produce higher investment efficiencies once we include the NDF policy objective and benefits to investees and society in the assessment. This research result differs from conventional thinking that early-stage investment has a higher loss rate. In contrast, the negative coefficients of expansion-stage and late-stage investments consistently indicate that they do not outperform the startup-stage and seed-stage investments across *Y1* to *Y5* functions.

Next, investment efficiency is found to be strongly linked with exit type. Exit type by becoming listed in the Taiwan Stock Exchange market or TPEx Mainboard generates higher profit ratio and investment TEs. Exit by the GIB or not becoming an IPO produces the least investment efficiencies. Trading in the Emerging Market Stock generally produces better investment efficiency than exit by GIB or not becoming an IPO.

The research findings have policy implications in that NDF, and VCs may focus on important investment strategies to increase investment efficiency further. This paper adds to the literature on public–private partnerships, the venture capital industry, and investment efficiency. Future studies may investigate individual policy practices and their effects on VC investment behaviors to provide more specific policy suggestions that go beyond the scope of this paper.

ACKNOWLEDGMENTS

This research was conducted by the authors through an industry–academia collaboration project between Ming Chuan University and the National Innovation and Entrepreneurship Association, Republic of China (Taiwan), and sponsored by the National Innovation and Entrepreneurship Association (NIEA). The authors of this article sincerely thank the late Professor Susila Munisamy for her guidance on meta-frontier theory. We also appreciate the editors of this book, Professor Gholam R. Amin and Professor Mustapha Ibn Boamah, for the suggestions to improve this paper. The views expressed are those of the authors, who are also responsible for any errors or omissions.

REFERENCES

Amin, G. R. & Ibn Boamah, M. (2023). Modeling business partnerships: A data envelopment analysis approach. *European Journal of Operational Research*, 305, 329–337.

Andersen, P. & Petersen, N. C. (1993). A procedure for ranking efficient units in data envelopment analysis. *Management Science*, 39 (10), 1261–1264

Banker, R. D., Charnes, A. & Cooper, W. W. (1984). Some models for estimating technical and scale inefficiencies in data envelopment analysis. *Management Science*, 30(9), 1078–1092.

Banker, R. D. & Morey, R. C. (1986). Efficiency analysis for exogenously fixed inputs and outputs. *Operations Research*, 34 (4), 513–521.

Berger, P. & Ofek, E. (1995). Diversification effect on firm value. *Journal of Financial Economics*, 37(1), 39–65.

Charnes, A., Cooper, W. W. & Rhodes, E. (1978). Measuring the efficiency of decision making units. *European Journal of Operational Research*, 2, 429–444.

Chen, C-D. (2019). Law of limited partnership and the organization of venture capital. *Judicial Weekly*, No. 1973 (Part I), 2–3 and No. 1974 (Part II), 2–3.

Chen, Q-W. (2001). Year 2000 news review—Venture capital. *iThome News*, https://www.ithome.com.tw/news/11147
Coelli, T. (1996). A guide to DEAP version 2.1: A data envelopment analysis (computer) program. *CEPA Working Paper* 96/08.
Colomboa, M. G., D'Adda, D. & Pirelli, L. H. (2016). The participation of new technology-based firms in EU-funded R&D partnerships: The role of venture capital. *Research Policy*, 45(2), 361–375.
Colomboa, M. G. & Grilli, L. (2009). A capital partnership: how human and venture capital affect the growth of high-tech start-ups. *Strategic Change*, 18, 231–239.
Dushnitsky, G. & Lenox, M.J. (2006). When does corporate venture capital investment create firm value? *Journal of Business Venturing*, 21(6), 753–772.
Emrouznejad, A. & Yang, G-L. (2018). A survey and analysis of the first 40 years of scholarly literature in DEA: 1978–2016. *Socio-Economic Planning Sciences*, 61, 4–8.
Färe, R., Grosskopf, S. & Lovell, C. A. K. (1994). *Production frontiers*. Cambridge: Cambridge University Press.
Hair, J. F., Black, W. C., Babin, B. J. & Anderson, R. E. (2018). *Multivariate Data Analysis*, 8th ed. Singapore: Cengage.
Hu, J., Hu, L., Hu, M. & He, Q. (2022). Machine learning-based investigation on the impact of Chinese venture capital institutions' performance: evaluation factors of venture enterprises to venture capital institutions. *Systems*, 10(4), 92, pp. 1–23.
Huang, C. J., Fu, T. T. & Huang, M. Y. (2008). *Performance evaluation: Efficiency and productivity*. Taipei: Shinlou Corporation, Ltd.
Huang, S.-C., Chiu, C. W., Chen, S.-F. & Wang, Y.-H. (2023, Sep). Investment performance of Taiwan venture capital management consulting corporations and managing fees. *International Conference on Data Envelopment Analysis*, Guildford, UK.
Kort, M., Verweij, S. & Klijn, E. H. (2016). In search for effective public-private partnerships: An assessment of the impact of organizational form and managerial strategies in urban regeneration partnership using FsQCA. *Environment and Planning C: Government and Policy*, 34 (5), 777–794.
Kupor, S. (2019). *Secret of sand hill road: Venture capital and how to get it* (Chinese and English bilingual version), Taipei: Master 60'.
Lombardi, S., Santini, G., Marchetti, G. M. & Focardi, S. (2017). Generalized structural equations improve sexual-selection analyses. *PLoS ONE* 12(8), e0181305. https://doi.org/10.1371/journal.pone.0181305
Ministry of Economic Affairs, R.O.C. (2007–2021). *Ministry of economic affairs—Small and medium enterprise administration (MOEA-SMEA) directions of enhancing capital investment in SMEs*. Taipei: Ministry of Economic Affairs, R.O.C.
National Development Fund, Executive Yuan, R.O.C. (NDF). (2007, November 27). *National Development Fund press release — Reports on the plan of investment in venture capital enterprises*. Taipei: NDF.
Nedaei, H., Naini, S. G. J. & Makui, A. (2020). A dynamic DEA model to measure the learning rates of efficient frontier and DMUs: An application to oil and gas wells drilling. *Computers and Industrial Engineering*, 144, 1–14.

O'Donnell, C. J., Rao, D. S. P. & Battese, G. E. (2008). Metafrontier frameworks for the study of firm-level efficiencies and technology ratios. *Empirical Economics*, 34, 231–255.

Organization of Economic Cooperation and Development (OECD) (2017). *Entrepreneurship at a Glance 2017*. Paris: OECD Publishing.

Ray, S. C. & Desli, E. (1997). Productivity growth, technical progress, and efficiency change in industrialized countries: Comment. *The American Economic Review*, 87(5), 1033–1039.

Stata (2017). *Stata 15 manual*. USA: StataCorp LLC.

Taiwan Venture Capital Association (TVCA) (2023). *Venture capital information*. Taipei: TVCA. https://www.tvca.org.tw/invest-01.html (2023.12.12).

Van de Vrande, V. & Vanhaverbeke, W. (2013). How prior corporate venture capital investments shape technological alliances: A real options approach. *Entrepreneurship Theory and Practice*, 37, 1019–1043.

Vecchi, V., Hellowell, M. & Longo, F. (2010). Are Italian healthcare organizations paying too much for their public–private partnerships? *Public Money & Management*, 30 (2), 125–132.

Wang, H., Xiong, W., Wu, G. & Zhu, D. (2018). Public–private partnership in Public Administration discipline: A literature review. *Public Management Review*, 20(2), 293–316.

CHAPTER 10

Economic Efficiency of Mergers

Subhash C. Ray
University of Connecticut, USA

10.1 INTRODUCTION

While high value mergers between firms like America Online and Time Warner worth $360 billion in 2000, Dow Chemical and DuPont worth $130 billion in 2017, or Exxon and Mobil in 1999 worth $81 billion capture headlines, domestic or global mergers between firms are quite common in the business world. Such mergers may be between firms in the same industry (called horizontal mergers) or firms across industries. In the second category when the firms involved in a merger are parties in a supply chain, these are typically known as vertical mergers (line Netflix, which originally was a DVD rental company moving into online streaming). Even outside of such supply chain mergers are those where companies merge to diversify their output bundles. Banks acquiring insurance companies (like Citibank and Travelers) are examples of such diversification.[1]

Regulating authorities are primarily concerned about the increased market power of the acquiring firm and the resulting effect of any merger upon consumers' welfare. In this chapter, desirability or otherwise of a merger is judged by whether the output producible from the combined inputs of the participants as a single firm after the merger is higher than the total output that can be produced by the firms separately. Conversely,

DOI: 10.1201/9781003345206-10

one can look at whether their combined output can be produced with less input thereby lowering cost if they merge.

Baumol (1977), Baumol and Fischer (1978), and Baumol, Panzar, and Willig (1982) have made seminal contributions to the Industrial Organization literature in economics focusing on subadditivity of the cost function. Maindiratta (1990) introduced the concept of size efficiency (as distinct from scale efficiency) to determine whether a firm was 'too large' in the sense that its observed output can be produced with less input by two or more (identical) smaller firms. Building on Baumol and Fischer (1978) and Maindiratta (1990), Ray and Hu (1997) constructed nonparametric models to determine the 'technically optimal' organization of an industry using data envelopment analysis (DEA). Bogetoft and Wang (2006) examined sources of potential gains from merger and identified a 'harmony' effect (due to convexity) and a 'size or scale effect' as the two contributing factors. Ray and Sethia (2022) look at potential cost saving from a merger both in the short run where the firms entering a merger already have varying quantities of a fixed input and in the long run when the merged firm can choose all inputs to minimize cost.

In this chapter, we provide an overview of the literature on potential economic gains from mergers and formulate appropriate nonparametric linear and mixed integer programming problems from DEA and free disposal hull (FDH) analysis. The rest of the chapter unfolds as flows. Section 10.2 and its sub-sections provide the theoretical background starting with the basic assumptions about the technology, defining the production possibility set, input and output sets, input- and output-oriented technical efficiencies and explain how merger efficiencies can be measured with diagrams. Section 10.3 provides the relevant DEA models and also the FDH models when convexity assumption is relaxed.[2] A potential problem of infeasibility of the combined post-merger output in DEA with variable returns to scale (VRS) is resolved in an FDH model assuming additivity. Section 10.4 looks into potential cost saving through merger identifying convexity, ray-subadditivity, and a capital expansion effect (in the short run) as the sources of cost savings. Relevant DEA models are also included. Section 10.5 measures potential cost savings from mergers without the convexity assumption but resorting to additivity when the merged output is infeasible for a simple FDH technology. Section 10.6 concludes.

10.2 THE PRODUCTION TECHNOLOGY

Consider an industry producing vectors of m outputs from n inputs. Let $x \in R_+^n$ be an input vector and $y \in R_+^m$ be an output vector, then the input–output combination (x, y) is a feasible production plan if output y can be produced from input x. The production possibility set T consists of all feasible production plans and can be defined as

$$T = \{(x, y) : x \in R_+^n \text{can produce } y \in R_+^m\} \quad (10.1)$$

In a single output case, the production possibility set is typically defined as the hypograph of a production function $y = f(x)$ 10.as

$$T = \{(x, y) : y \leq f(x); x \in R_+^n, y \in R_+\} \quad (10.2)$$

In the multiple output case, one can use a transformation function $F(x, y) = \alpha$ to define the production possibility set as

$$T = \{(x, y) : F(x, y) \leq 0; x \in R_+^n, y \in R_+^m\} \quad (10.3)$$

In parametric models using stochastic frontier analysis (SFA), the production or the transformation function is explicitly specified for empirical estimation to analyze input substitution, returns to scale, and other related characteristics of the technology. By contrast, in nonparametric approaches like DEA, one typically makes a number of fairly general assumptions about the technology but does not specify any explicit functional form of the production or transformation function.

The four common assumptions are:

(A1) Inclusion: every observed input–output bundle is feasible.

(A2) Input Disposability: If $(x^0, y^0) \in T$ and $x^1 \geq x^0$, then $(x^1, y^0) \in T$.

(A3) Output Disposability: If $(x^0, y^0) \in T$ and $y^1 \leq y^0$, then $(x^0, y^1) \in T$.

(A3) Convexity: If $(x^0, y^0), (x^1, y^1) \in T$ then $(\lambda x^0 + (1-\lambda)x^1, \lambda y^0 + (1-\lambda) y^1) \in T$ for all $\lambda \in [0,1]$

When the production or the transformational function is continuous, (A2) implies $\frac{\partial f}{\partial x_i} > 0 (i = 1, 2, ..m)$ in the single output case, while (A2) and (A3) together imply $\frac{\partial F}{\partial x_i} \leq 0, \frac{\partial F}{\partial y_r} \leq 0, (i = 1, 2, \ldots, n; r = 1, 2, \ldots, m)$ in the case of multiple outputs.

An additional assumption made in specific cases is

(A5) Constant Returns to Scale (CRS): If $(x, y) \in T$, then for any $k \geq 0$, $(kx, ky) \in T$.

Next suppose that the data set $D = \{(x^j, y^j); j = 1, 2, \ldots, N\}$ consists of observed input–output bundles for N firms in an industry.

An empirical approximation of the production possibility set using the data and assumptions (A1–A4) is

$$S = \left\{ (x, y) : x \geq \sum_{j=1}^{N} \lambda_j x^j, y \leq \sum_{j=1}^{N} \lambda_j y^j, \sum_{j=1}^{N} \lambda_j = 1,; \lambda_j \geq 0; j = 1, 2, \ldots, N \right\} \quad (10.4)$$

The set S is known as the free disposal convex hull of the observed data points. If additionally one assumes CRS, the restriction $\sum_{j=1}^{N} \lambda_j = 1$ is deleted.

10.2.1 Input and Output Sets

First consider the *producible output set*

$$Y = \{ y : \exists x \in R_+^n \in \text{such that} (x, y) \in T \}. \quad (10.5)$$

For any output vector $y \in Y$, the input set with reference to the production possibility set S defined in (10.4) is

$$V(y^0) = \left\{ x : x \geq \sum_{j=1}^{N} \lambda_j x^j, y^0 \leq \sum_{j=1}^{N} \lambda_j y^j, \sum_{j=1}^{N} \lambda_j = 1,; \lambda_j \geq 0; j = 1, 2, \ldots, N \right\}$$

(10.6)

The input-oriented (radial) technical efficiency of an input vector $x^0 \in V(y^0)$ 10.is

$$\tau_x(x^0, y^0) = \min \theta : \theta x^0 \in V(y^0). \quad (10.7)$$

In an analogous way, the output set of an input bundle x^0 consists of all output bundles that can be produced from x^0 and can be constructed as

$$P(x^0) = \left\{ y : x^0 \geq \sum_{j=1}^{N} \lambda_j x^j, y \leq \sum_{j=1}^{N} \lambda_j y^j, \sum_{j=1}^{N} \lambda_j = 1,; \lambda_j \geq 0; j = 1, 2, \ldots, N \right\}$$

(10.8)

The output-oriented (radial) technical efficiency is

$$\tau_y\left(x^0, y^0\right) = \frac{1}{\varphi^*}$$

where

$$\varphi^* = \max \varphi : \varphi y^0 \in P\left(x^0\right). \tag{10.9}$$

10.2.2 Returns to Scale, Additivity, and Convexity

Consider two firms A and B with input–output bundles $\left(x_A . y_A\right)$ and $\left(x_B . y_B\right)$ from a single-output single-input industry with the production function $y = f\left(x\right)$. Furthermore, both firms are assumed to be efficient. Thus, $y_A = f\left(x_A\right)$ and $y_B = f\left(x_B\right)$. Now suppose that the two firms merge and their combined input–output post-merger is $\left(x_T, y_T\right)$ where $y_T = f\left(x_T\right) = f\left(x_A + x_B\right)$. Clearly, merger is technically beneficial if

$$y_T = f\left(x_A + x_B\right) > y_A + y_B = f\left(x_A\right) + f\left(x_B\right). \tag{10.10}$$

A function $y = f\left(x\right)$ is *super-additive* if $f\left(\sum_{j=1}^{k} x_j\right) > \sum_{j=1}^{k} f\left(x_j\right)$. Similarly, it is *subadditive* if $f\left(\sum_{j=1}^{k} x_j\right) < \sum_{j=1}^{k} f\left(x_j\right)$. In the special case when each $x_j = x$, super additivity implies $f\left(kx\right) > kf\left(x\right)$. Thus, increasing returns to scale implies global super-additivity. Similarly, global sub-additivity implies diminishing returns to scale globally.

We now consider a variable returns to scale (VRS) production function which exhibits locally increasing, constant, or diminishing returns to scale along the frontier. We return to the two firm examples and define $\bar{x} = \frac{1}{2}\left(x_A + x_B\right)$. Thus, $x_T = x_A + x_B = 2\bar{x}$. Furthermore, assume arbitrarily, $x_A = \bar{x} - \delta$ and $x_B = \bar{x} + \delta$. Thus, $f\left(x_A + x_B\right) = f\left(x_T\right) = f\left(2\bar{x}\right), f\left(x_A\right) = f\left(\bar{x} - \delta\right)$, and $f\left(x_B\right) = f\left(\bar{x} + \delta\right)$. Now, taking a second-order Taylor's series approximation,

$$f\left(x_A\right) \approx f\left(\bar{x} - \delta\right) = f\left(\bar{x}\right) - \delta f'\left(\bar{x}\right) + \frac{1}{2}\delta^2 f''\left(\bar{x}\right) \tag{10.11a}$$

and

$$f(x_B) \approx f(\bar{x}+\delta) = f(\bar{x}) + \delta f'(\bar{x}) + \frac{1}{2}\delta^2 f''(\bar{x}) \quad (10.11b)$$

Hence,

$$f(x_A) + f(x_B) \approx 2f(\bar{x}) + \delta^2 f''(\bar{x}) \quad (10.11c)$$

Thus,

$$f(x_A + x_B) - [f(x_A) + f(x_B)] \approx [f(2\bar{x}) - 2f(\bar{x})] - \delta^2 f''(\bar{x}) \quad (10.12)$$

The two terms on the right identify two distinct components of the measured gain/loss from merger of the two firms. If the production function is concave and, therefore, the production possibility set is convex, $f''(\bar{x}) < 0$, and the second term on the right is positive. The first term, on the other hand, is related to returns to scale at the average input bundle. If the first term is also positive, then gain from merger will be decidedly positive. But even when it is negative, but relatively smalls, merger will lead to superadditivity if the production function is concave.

Several points should be emphasized. First, globally increasing returns to scale or concavity of the production function are sufficient but not necessary for the above results to hold. All that is needed is that the function should be concave at the average input level. Also, in this example, what is needed is that output from the total input bundle should be greater than twice the output from the average input.[3]

Secondly, the reasoning generalizes to multiple inputs and merger of multiple firms. Consider the merger of k firms with input bundles $x^j \in R^n_+, (j = 1,2,\ldots,k)$. The corresponding outputs are $y_j = f(x^j)$. Now the average input vector is $\bar{x} = \frac{1}{k}\sum_{j=1}^{k} x^j$, and the deviation vectors are $\delta^j = x^j - \bar{x}$. Note that $\sum_{j=1}^{k} \delta^j = 0$.

Then, $f(x^j)$ can be approximated as

$$f(x^j) \approx f(\bar{x}) + \delta^{j\prime} \left.\frac{\partial f}{\partial x}\right|_{\bar{x}} + \frac{1}{2}\delta^{j\prime} \left.\frac{\partial^2 f}{\partial x \partial x'}\right|_{\bar{x}} \delta^j. \quad (10.13a)$$

Hence,

$$\sum_{j=1}^{k} f\left(x^{j}\right) \approx kf\left(\bar{x}\right) + \frac{1}{2}\sum_{j=1}^{k} \delta^{j\prime} \frac{\partial^2 x}{\partial x \partial x'}\Big|_{\bar{x}}\, \delta^{j} \tag{10.13b}$$

Thus,

$$f\left(\sum_{j=1}^{k} x^{j}\right) - \sum_{j=1}^{k} f\left(x^{j}\right) \approx \left[f\left(k\bar{x}\right) - kf\left(\bar{x}\right)\right] - \frac{1}{2}\sum_{j=1}^{k} \delta^{j\prime} \frac{\partial^2 x}{\partial x \partial x'}\Big|_{\bar{x}}\, \delta^{j} \tag{10.13c}$$

As in the single-input case, if $y = f\left(x\right)$ is concave with a negative definite Hessian at the mean input bundle, the second term on the right is positive. Global concavity is not necessary. Similarly, if $f\left(k\bar{x}\right) > kf\left(\bar{x}\right)$, the first term is also positive. These are point specific requirements for positive gains from mergers.

10.2.3 Measurement and Decomposition of Gains From Merger

10.2.3.1 Output-Oriented Merger Efficiency

Bogetoft and Wang (2006) provide a measure of merger efficiency and its multiplicative decomposition into a harmony effect and a return to scale effect.[4]

Consider the single-output, two-input technology defined by the production function $y = f\left(x_1, x_2\right)$. Firms 1 and 2 use the two different input bundles[5] $x^{j} = \left(x_{1j}, x_{2j}\right)\left(j = 1,2\right)$. Assume, initially, that both firms produce the same level of output y_0, and both are technically efficient. Thus, both bundles lie on the boundary of the input requirement set $V\left(y_0\right) = \left\{x : f\left(x\right) \geq y_0\right\}$. The points A $\left(x^1\right)$ and B $\left(x^2\right)$ in Figure 10.1 show the two input bundles.

The point *D* represents the sum of the two input bundles x^1 and x^2. Merger of the two firms will be efficient if the bundle *D* produces greater output than $2y_0$. First, define the point *C* representing the average input bundle $\bar{x} = \frac{1}{2}\left(x^1 + x^2\right)$. Bogetoft and Wang (2006) conceptualize the merger of the two firms as a two-step process. In the first step, they consider a firm that uses the average input bundle $\bar{x}$. In the second step, this average firm is doubled in scale to become the merged firm shown by the point *D*. Note that if the production function $y = f\left(x\right)$ is strictly concave,

$$f\left(\lambda x^1 + \left(1-\lambda\right)x^2\right) > \lambda f\left(x^1\right) + \left(1-\lambda\right) f\left(x^2\right); 0 < \lambda < 1.$$

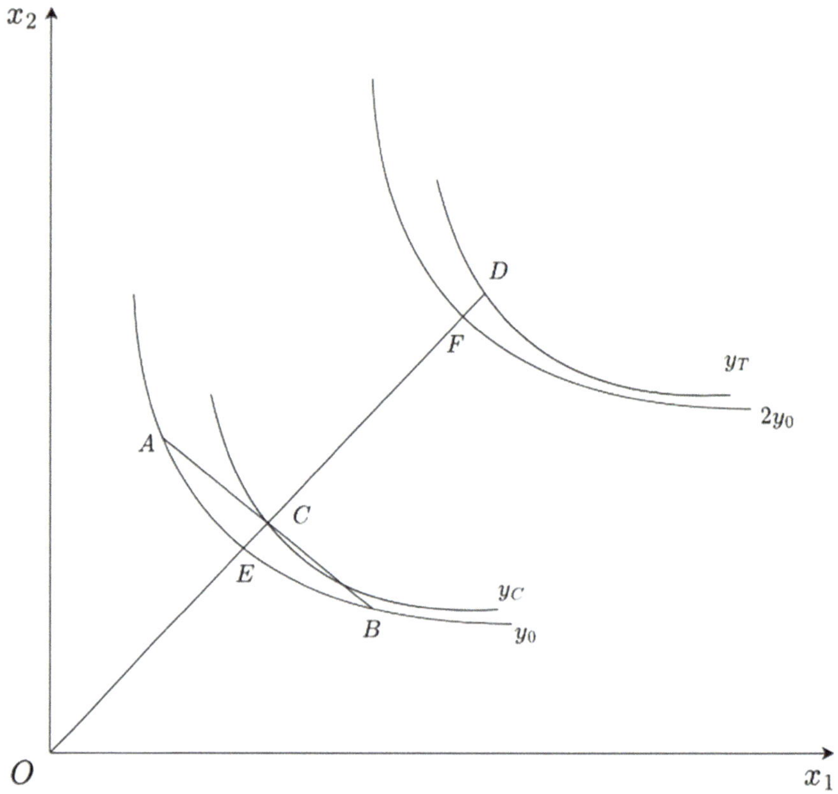

FIGURE 10.1 Merger of two firms with equal outputs.

Setting $\lambda = \frac{1}{2}$, one can trivially get $f(\bar{x}) > y_0$. That is, both firms can increase their output by selecting the average input bundle instead of their different bundles. Thus, the output of each firm from this average input bundle $y_C = f(\bar{x})$ exceeds y_0. Bogetoft and Wang call this increase in output the *harmony effect* because if the firms shared the combined input equally and used the identical (average) bundle, each would produce this higher level of output. Thus, their combined output of *two identical average firms* would be $2y_C$. Next, consider the output y_T produced by a single firm from the combined input bundle $\left(x^T = x^1 + x^2 = 2\bar{x}\right)$ shown by the point *D*. The output-oriented efficiency of merger can be measured as

$$ME_y\left(x^1, x^2\right) = \frac{y_T}{2y_0}. \tag{10.14}$$

When ME_y exceeds unity, potential gains from merger of the two firms would be positive. We may further decompose the merger efficiency as

$$ME_y\left(x^1, x^2\right) = \left(\frac{y_C}{y_0}\right) \cdot \left(\frac{y_T}{2y_C}\right). \tag{10.15}$$

The output-oriented harmony effect is

$$H = \frac{y_C}{y_0} \tag{10.16}$$

and the scale effect is

$$S = \frac{y_T}{2y_D}. \tag{10.17}$$

As noted above, the harmony effect is greater than unity whenever isoquants are strictly convex. But the scale effect may be greater than, equal to, or less than unity. If the scale effect is much lower than 1, ME_y can be less than 1 implying that the post-merger output from the combined input bundles would be lower.[6]

Note that the results hold true even when the two firms 1 and 2 do not produce the same quantity of output. Suppose that $f\left(x^1\right) = y_1$ and $f\left(x^2\right) = y_2$. Strict concavity of the production function would still imply that

$$f\left(\frac{1}{2}x^1 + \frac{1}{2}x^2\right) = f\left(\bar{x}\right) > \frac{1}{2}f\left(x^1\right) + \frac{1}{2}f\left(x^2\right) = \frac{1}{2}\left(y_1 + y_2\right) = \bar{y}.$$

Thus, the harmony effect will be positive.

This is shown in Figure 10.2 where the firms produce different quantities of output. As in Figure 10.1, points A and B show the input bundles x^1 and x^2 located on the isoquants for outputs y_1 and y_2, respectively. Point C is the average input bundle $\tilde{x}$ which produces the output y_C. Due to strict concavity of the production function, $y_C > \bar{y} = \frac{1}{2}\left(y_1 + y_2\right)$. This is shown by the lower isoquant for $\bar{y}$. The point D shows the input bundle $x_T = x^1 + x^2 = 2\bar{x}$ which produces the output y_T. By contrast, the isoquant for output $2\bar{y} = y_1 + y_2$ is the one through point F. In Figure 10.2, if $OD > 2OC, y_T = f\left(2\bar{x}\right) > 2f\left(\bar{x}\right) = 2y_C$.

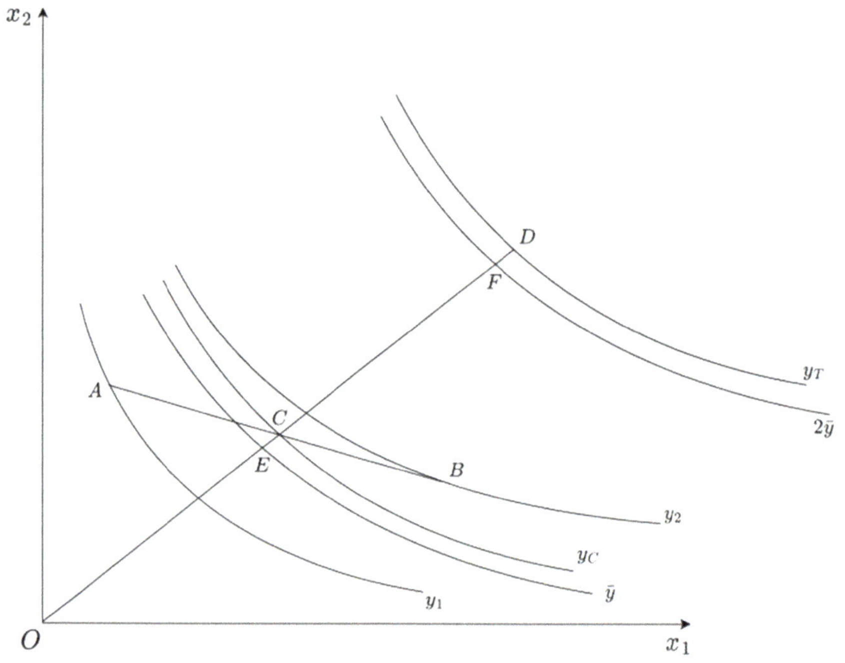

FIGURE 10.2 Merger of firms with unequal outputs.

10.2.3.2 Input-Oriented Merger Efficiency

The above is an output-oriented measure of merger efficiency where we examine the gain or loss of aggregate output due to merger with no increase in the combined input of the merging firms. Alternatively, we could measure the input-oriented efficiency of a merger and focus primarily on the change in the required input (saving or increase) without reducing the combined output of the firms.

In Figure 10.1, where both firms produce the same output quantity y_0. The total output is $2y_0$. This can be produced by the merged firm using the input bundle $x^M = \theta_M x^T = 2\theta_M \bar{x}$. shown by the point F. Producing independently, they are using the total input bundle $x^T = \left(x^1 + x^2\right) = 2\bar{x}$. Thus, an input-oriented measure of merger efficiency is

$$ME_x = \frac{OF}{OD} = \theta_M. \tag{10.18}$$

If each firm switch to the common input $\bar{x} = \frac{1}{2}\left(x^1 + x^2\right)$ but produced independently, each would produce a higher output y_C and could scale down their input bundles to the bundle $x^H = \theta_H \bar{x}$ shown by point E. Thus, the input-oriented harmony effect is

$$\theta_H = \frac{OE}{OC}. \tag{10.19}$$

The input-oriented merger efficiency can be decomposed as

$$ME_x = \frac{OF}{OD} = \left(\frac{OF}{OE}\right)\left(\frac{OE}{OC}\right)\left(\frac{OC}{OD}\right) \tag{10.20}$$

Recall, at this point, that

$$\begin{aligned} OD &= x_T = 2\bar{x}; \\ OF &= \theta_M x_T = 2\theta_M \bar{x}; \\ OC &= \bar{x}, \\ OE &= \theta_H \bar{x}. \end{aligned}$$

Thus,

$$ME_x = \theta_M = \left(\frac{OF}{2OE}\right)\theta_H = \theta_S.\theta_H. \tag{10.21}$$

In this decomposition, θ_H is the harmony effect and will be less than 1 for convex isoquants. Point F on the $2y_0$ isoquant shows the minimum radial expansion of the average input bundle $\bar{x}$ that is required to produce the output $2y_0$. If it is less than $2\bar{x}$, the scale effect is positive. On the other hand, if $OF > 2\bar{x}$, the scale effect is negative and increases θ_M.

It should be noted that a value of θ_M less than unity implies gains from the merger in the form of input saving. Also, unlike the usual input-oriented technical efficiency, the input-oriented merger efficiency θ_M can be greater than unity which indicates that more input would be required to produce the total output produced by the firms operating independently after the merger. In that case, merger would be technically inefficient.

10.3 DEA MODELS FOR MEASURING MERGER EFFICIENCY

We first consider the DEA model for measuring the output-oriented merger efficiency and its components in the multiple-output, multiple-input case. Suppose that we are considering the potential gains from the merger of k firms out of N firms in the industry. The input–output bundles of the individual firms in the data set are $\left(x^j, y^j\right)\left(j = 1,2,..,N\right)$. The total input bundle of the merging firms is $x_T = \sum_{j=1}^{k} x^j$, and their total pre-merger output is $y_T^k = \sum_{j=1}^{k} y^j$. Their post-merger output can be found by solving the DEA LP problem

$$
\begin{aligned}
& ME_y = \max \varphi^T \\
& s.t \sum_{i=1}^{N} \lambda_j y^j \geq \phi^T y_T; \\
& \sum_{i=1}^{N} \lambda_j x^j \leq x^T; \\
& \sum_{i=1}^{N} \lambda_j = 1; \\
& \lambda_j \geq 0; (j = 1, 2, \ldots, N)
\end{aligned}
\tag{10.22}
$$

If $\varphi_T > 1$, post-merger output is higher and gains from the merger are positive. If, on the other hand, $\varphi_T < 1$, merger will be economically inefficient.

To determine the harmony effect, one needs to first define the average input and output bundles of the merging firms $\overline{x} = \frac{1}{k}\sum_{j=1}^{k} x^j$ and $\overline{y} = \frac{1}{k}\sum_{j=1}^{k} y^j$ and solve the DEA problem

$$
\begin{aligned}
& \max \varphi^H \\
& s.t \sum_{i=1}^{N} \lambda_j y^j \geq \varphi^H \overline{y}; \\
& \sum_{i=1}^{N} \lambda_j x^j \leq \overline{x}; \\
& \sum_{i=1}^{N} \lambda_j = 1; \\
& \lambda_j \geq 0; (j = 1, 2, \ldots, N)
\end{aligned}
\tag{10.23}
$$

Finally, the size or scale effect is obtainable as

$$
\varphi^S = \frac{\varphi^T}{\varphi^H}. \tag{10.24}
$$

In an analogous way, the input-oriented merger efficiency can be measured as

$$
\begin{aligned}
& ME_x = \min \theta^M \\
& s.t \sum_{i=1}^{N} \lambda_j y^j \geq y_T; \\
& \sum_{i=1}^{N} \lambda_j x^j \leq \theta^M x^T; \\
& \sum_{i=1}^{N} \lambda_j = 1; \\
& \lambda_j \geq 0; \left(j = 1, 2, \ldots, N \right)
\end{aligned}
\tag{10.25}
$$

The input-oriented harmony effect is

$$
\begin{aligned}
& ME_x = \min \theta^H \\
& s.t \sum_{i=1}^{N} \lambda_j y^j \geq \bar{y}; \\
& \sum_{i=1}^{N} \lambda_j x^j \leq \theta^M \bar{x}; \\
& \sum_{i=1}^{N} \lambda_j = 1; \\
& \lambda_j \geq 0; \left(j = 1, 2, \ldots, N \right)
\end{aligned}
\tag{10.26}
$$

Finally, the input-oriented scale or size effect is

$$
S_x = \frac{\theta^M}{\theta^H}. \tag{10.27}
$$

10.3.1 Additivity and Replication Without Convexity: A Digression

The nonparametric method of FDH analysis introduced by DePrins, Simar, and Tulkens (1984) retains the assumptions of inclusion and free disposability of inputs and output but exclude convexity. Any observed input–output bundle is demonstrably feasible and clearly is an element of the production possibility set. Furthermore, free disposability of inputs and outputs amounts to assuming the possibility of keeping inputs idle or inefficient utilization of resources. Both assumptions are quite realistic. By contrast, convex combinations of observed input–output bundles are usually not observed and cannot be regarded as necessarily feasible.[7]

Based only upon assumptions A1–A3, the non-convex FDH production possibility set constructed from the data is

$$T_{\text{FDH}} = \left\{ (x, y) : \sum_{j=1}^{N} \lambda_j x^j ; y \le \sum_{j=1}^{N} \lambda_j y^j ; \sum_{j=1}^{N} \lambda_j = 1; \lambda_j \in \{0,1\}; j = 1,2,\ldots,N. \right\} \tag{10.28}$$

Consider the FDH problem for the input-oriented merger efficiency

$$\begin{aligned} & \min \theta^M \\ & s.t \\ & \sum_{j=1}^{N} \lambda_j x^j \le \theta^M x_T ; \\ & \sum_{j=1}^{N} \lambda_j y^j \ge y_T ; \\ & \sum_{j=1}^{N} \lambda_j = 1; \lambda_j \in \{0,1\}; j = 1,2,\ldots,N \end{aligned} \tag{10.29}$$

The optimization problem in (10.29) is similar to the corresponding DEA problem in (10.25) except that the constraint $\lambda_j \ge 0$ in (10.25) is replaced in (29) by $\lambda_j \in \{0,1\}$. As a result, the LP problem in (10.25) becomes a mixed integer programming problem in (10.29).

A potential problem often encountered in DEA with the VRS technology approximated by the free disposal convex hull of the observed input–output vectors is that the aggregated output vector of the merging firms may lie outside the empirically constructed production possibility set. This infeasibility problem arises whenever the output is larger than the largest observed output in the single output case, or any element of the output vector is greater than the largest observed quantity of the corresponding element in the multiple output case. Consider the one-output one-input example shown in Figure 10.3. The observed input–output pairs

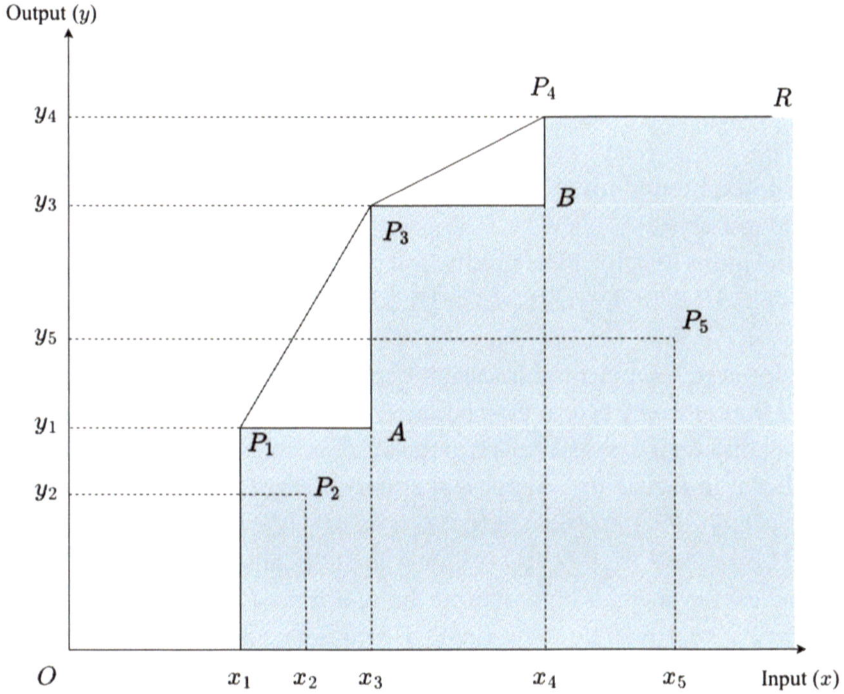

FIGURE 10.3 DEA and FDH frontiers.

are $(x_j, y_j)(j=1,2,3,4,5)$. The post-merger output of any two firms in this example will lie above the VRS DEA frontier whenever the combined output is greater than y_4. No matter how much the input is increased, the attainable output will not exceed y_4. Suppose that we wanted to measure the input-oriented efficiency of merger between firms 2 and 4. The relevant DEA problem would be

$$\begin{aligned}
&\min \theta^M \\
&s.t. \\
&\sum_{j=1}^{5} \lambda_j y_j \geq (y_2 + y_4); \\
&\sum_{j=1}^{5} \lambda_j x_j \leq \theta^M (x_2 + x_4); \\
&\sum_{j=1}^{5} \lambda_j = 1; \lambda_j \geq 0; (j=1,2,\ldots,5)
\end{aligned} \tag{10.30}$$

Note that even when θ^M is increased indefinitely, there will be no way to satisfy the output constraint simultaneously with the constraint $\sum_{j=1}^{5} \lambda_j = 1; \lambda_j \geq 0; (j = 1,2,\ldots,5)$. Of course, under CRS, the constraint that the λ-weights add up to 1 is removed, and this infeasibility problem does no longer arise.

In Figure 10.3, the FDH production possibility frontier is the step function $x_1P_1AP_3BP_4R$. By contrast, the DEA frontier is the piecewise connected line $x_1P_1P_3P_4R$.

However, the potential infeasibility problem is not addressed by dropping the convexity assumption because, as is obvious, the FDH production possibility set is a proper subset of the DEA production possibility set. This is shown in Figure 10.4. Suppose that the observed input–output pairs are $P_1(5,4), P_2(10,6), P_3(16,8), P_4(6,3)$, and $P_5(12,6)$. The corresponding FDH frontier is $AP_1BP_2CP_3Q$. While it includes all observed input–output bundles, the merged input–output bundle from P_4 and P_5 shown by the point $P_{45}(18,9)$ will lie above the FDH frontier, and its input-oriented efficiency cannot be determined.

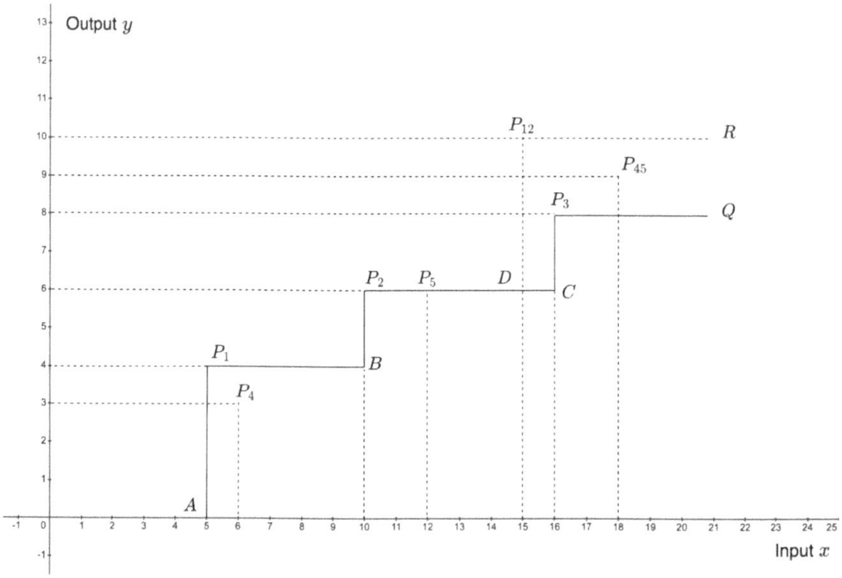

FIGURE 10.4 FDH with additivity.

One can overcome the infeasibility problem by assuming *additivity* in place of convexity. Additivity implies that if $\left(x^i, y^i\right)$ and $\left(x^j, y^j\right)$ are both feasible input–output bundles, then $\left(x^i + x^j, y^i + y^j\right)$ is also a feasible bundle. As argued by Bogetoft and Wang (1996, 2006), one can think of the merged unit as operating two distinct divisions within a single organization. These semi-autonomous sub-units can produce like the two distinct units as before and produce the combined output from their combined input.[8] When additivity is allowed, one can think of combining units P_1 and P_2 to hypothetically create a single unit $P_{12}\left(15,10\right)$ which would dominate the merged unit P_{45}.

The mixed integer programming problem for measuring input-oriented merger efficiency relative to an additive FDH frontier is

$$\begin{aligned} &ME_x^{FDH} = \min \theta^M \\ &s.t \sum_{j=1}^{N} \lambda_j x^j \le \theta^M x_T; \\ &\sum_{j=1}^{N} \lambda_j y^j \ge y_T; \\ &\lambda_j \in \{0,1\}; j = 1,2,\ldots,N; \end{aligned} \tag{10.31}$$

Note that additivity ensures that by setting λ_j equal to 1 for each firm j involved in the merger and 0 for all other units, one gets θ^M equal to 1 as an upper bound on the optimal value of the objective function in (10.31). Thus, if the technology is additive, merger will never be more input-using.

Tulkens (1993) proposed the idea of replication in the sense that if $\left(x^0, y^0\right)$ is observed, then one can consider replicating it and thus have two identical sub-units each with input–output, and thus $\left(2x^0, 2y^0\right)$ is also feasible. This is a special example of additivity where $\left(x^i, y^i\right) = \left(x^i, y^i\right) = \left(x^0, y^0\right)$. Generalizing this to k identical units, one gets the property

$$\left(x, y\right) \in T^{\in FDH} \Rightarrow \left(kx, ky\right) \in T^{\in FDH}, k = \text{integer}. \tag{10.32}$$

This is quite similar to CRS except that only integer scaling of input–outputs will be feasible. Tulkens (1993) described the FDH with replication as free replication hull (FRH).

Under the assumption of both additivity and replicability, the revised optimization problem for input-oriented merger efficiency becomes

$$
\begin{aligned}
& ME_x^{FDH} = \min \theta^M \\
& s.t. \\
& \sum_{j=1}^{N} \lambda_j x^j \leq \theta^M x_T; \\
& \sum_{j=1}^{N} \lambda_j y^j \geq y_T; \\
& \lambda_j = \text{integer}; j = 1,2,\ldots,N.
\end{aligned}
\tag{10.33}
$$

It needs to be underscored that although mathematically there is no upper bound on the number of replications (k) of any observed input–output bundle, it may be necessary to impose a realistic upper bound in any given context.

Ray and Hu (1997) used a version of this model to determine the technically optimal structure of the airlines industry in the United States. Green and Cook (2004) described the FDH with additivity as the free coordination hull (FCH).

10.4 SUBADDITIVITY OF THE COST FUNCTION AND POTENTIAL GAINS FROM MERGERS[9]

The role of the production technology in determining the optimal configuration of firms in an industry has been analyzed in detail by Baumol (1977), Baumol and Fischer (1978), and Baumol, Panzar, and Willig (hereafter BPW) (1982), among others. In particular, Baumol and Fischer (1978) considered the optimal number of firms that could collectively produce the total industry output at the minimum cost. Ray and Hu (1997) developed a nonparametric optimization model for the technically optimal organization of an industry that would result in the maximum radial scaling down of the aggregate input bundle without reducing the industry output. A different, though closely related, question is whether it is more cost efficient if the outputs of two or more existing firms are combined and produced by a single firm than when produced separately. The cost function is said to be subadditive at a given output (bundle) if the combined cost of producing any number of smaller output bundles adding up to a specified output bundle is higher than the cost of producing it as a single aggregate bundle.

Clearly, there will be incentives for firms to merge voluntarily when the cost function is subadditive at the combined output bundle.

In this chapter, we circumvent the effect of mergers on market demand by assuming that the target output remains unchanged whether or not the firms merged.[10] This is comparable to Baumol and Fischer (1978), who also assume that the aggregate demand in the market is not affected by the number of firms. However, while Baumol and Fischer (1978) search for the optimal number of firms to produce the industry output, the question in the context of a merger is whether the observed outputs of a number of firms can be produced at a lower cost by a single firm.

Also, BPW (1982) emphasizes the role of the startup cost. A key factor behind the possible subadditivity of the cost function is the fact that when produced by a single firm the combined output entails the startup cost only once while the startup cost is incurred by every firm that produces the smaller bundles adding up to the merged output. The implication is that it is an *ex-ante* comparison carried out at the planning stage to decide whether, for producing a target output bundle, a single production unit is more cost efficient than many. By contrast, in this chapter, evaluation of potential gains from mergers is an *ex-post* analysis in the sense that the merging firms are already in existence. Thus, each firm considered for a merger already has an observed output bundle along with a corresponding observed bundle of fixed inputs. When firms are merged, fixed inputs and, hence the fixed costs, are also combined side by side with the outputs.[11]

While using the basic concepts of sub-additivity and ray convexity of the cost function taken directly from BPW (1982), Ray and Sethia (2022) focus mainly on the short-run ray average cost and total costs specific for the product mix of the merged output bundle and the combined bundle of the fixed inputs. They provide an additive decomposition of the total gain from a merger into three separate components attributable to (i) trans-ray convexity of the technology, (ii) sub-additivity of the short-run ray cost curve, and (iii) increase in the bundle of the fixed inputs (broadly described in Ray and Sethia (2022) as 'capital expansion'). In this process, they show that subadditivity of the short-run ray cost function is neither sufficient nor necessary for gains from merger in the presence of fixed inputs.

All of the discussion in the preceding sections has been in terms of the potential for increasing output quantities or reducing input quantities through merger. Moreover, the focus has been on proportional scaling down of all inputs or scaling up of all outputs. Such radial technical efficiency analysis ignores input or output slacks. However, when slacks are present at the optimal solution of a DEA problem, the radial efficiency factor underestimates the potential for cost savings or revenue increasing through merger. In this chapter, we concentrate on the potential for cost saving and treat the output (and hence the revenue) as given.

As noted earlier, the production technology can also be defined in terms of a family of input requirement sets. Consider a target output bundle and the corresponding input requirement set

$$V\left(y^0\right)=\left\{x : x \text{ can produce } y^0\right\} \Leftrightarrow \left\{x : \left(x, y^0\right) \in T\right\}. \tag{10.34}$$

Let $p \in R^n_{++}$ be a strictly positive vector of input prices. Then, the minimum cost of producing the output bundle y^0 is

$$\begin{aligned} C\left(y^0, p\right) &= \min p'x \\ s.t. x &\in V\left(y^0\right). \end{aligned} \tag{10.35}$$

10.4.1 Subadditivity of the Cost Function: Long Run

Now consider breaking up the bundle y^0 into q smaller bundles $y^s \left(s=1,2,\ldots,q\right)$ such that $\sum_{s=1}^{q} y^s = y^0$. The cost function is subadditive at y^0 if for any $q>1$,

$$\sum_{s=1}^{q} C\left(y^s, p\right) > C\left(y^0, p\right). \tag{10.36}$$

As emphasized by BPW (1982), subadditivity is not a local concept (like scale economies) because one needs information about the cost of *all feasible output bundles smaller than* y^0. Clearly, there is a case in favor of merging several firms within the same industry or several units within the same firm if the cost function is subadditive at the aggregated output bundle. Note that in (10.35), all inputs are chosen optimally in order to minimize the total cost. Thus, (10.35) and (10.36) refer to the long-run cost function.

For simplicity of exposition, consider the potential gain (or loss) from merger between two firms – *A* and *B*. Suppose that their respective output bundles are y^A and y^B. Then, evaluated at the same input price vector $\left(p\right)$, the minimum cost of producing their outputs are $C_A = C\left(y^A, p\right)$ and

$C_B = C(y^B, p)$. The cost of producing their combined output $y^M = y^A + y^B$ is $C_M = C(y^M, p)$. Thus, the potential gain from merger is

$$\begin{aligned} g &= (C_A + C_B) - C_M \\ &= \left[C(y^A, p) + C(y^B, p)\right] - C(y^M, p). \end{aligned} \tag{10.37}$$

Next, define $\bar{y} = \frac{1}{2}(y^A + y^B)$, and $\bar{C} = \frac{1}{2}(C_A + C_B)$. Thus, the combined output is $y^M = 2\bar{y}$ and (10.37) can also be expressed as

$$g = \left[C(y^A, p) + C(y^B, p)\right] - C(2\bar{y}, p). \tag{10.38}$$

Now suppose that the cost-minimizing variable input quantities are x^A for y^A and x^B for y^B. That is both (x^A, y^A) and (x^B, y^B) are feasible input–output bundles. Let $\bar{x} = \frac{1}{2}(x^A + x^B)$. Therefore, by convexity of the set T, $(\bar{x}, \bar{y})$ is also a feasible input–output combination. That is $\bar{x} \in V(\bar{y})$. But this need not be the cost-minimizing input bundle for $\bar{y}$.

$$w'\left(\frac{x^A + x^B}{2}\right) = \frac{1}{2}\left[C(y^A, p) + C(y^B, p)\right] \geq C(\bar{y}, p). \tag{10.39}$$

Now, (10.38) can also be written as

$$\begin{aligned} g &= C(y^A, p) + C(y^B, p) - C(y^A + y^B, p) \\ &= C(y^A, p) + C(y^B, p) - 2C(\bar{y}, p) + 2C(\bar{y}, p) - C(y^A + y^B, p); \\ &= 2\left[\frac{C(y^A, p) + C(y^B, p)}{2} - C(\bar{y}, p)\right] + 2\left[C(\bar{y}, p) - \frac{C(2\bar{y}, p)}{2}\right]. \end{aligned} \tag{10.40}$$

For simplicity of exposition, we can identify the two terms on the right-hand side of (10.40) as:

$$I = 2\left[\frac{C(y^A, p) + C(y^B, p)}{2} - C(\bar{y}, p)\right] \tag{10.41a}$$

$$II = 2\left[C(\bar{y}, p) - \frac{C(2\bar{y}, p)}{2}\right] \tag{10.41b}$$

The condition $g > 0$ in (10.40) is what Baumol (1977) calls strict subadditivity for bundle $y^A + y^B$. The costs $C(y^A, p)$, $C(y^B, p)$, $C(y^A + y^B, p)$, and $C(\overline{y}, p)$ are, as per (10.35), optimal long-run total costs of producing bundles y^A, y^B, $(y^A + y^B)$, and $\overline{y}$, respectively, given the factor price vector p. The expression in the square brackets on the right-hand side of (10.41a) pertains to the trans-ray convexity of the technology and is nonnegative. The focal point of our discussion in the long run is the expression in the right-hand side square brackets of (10.41b) which could be positive or negative depending on whether or not the long-run cost function is ray subadditive at the combined output bundle. Even when this term is negative (that is, the costs are not subadditive at the relevant output bundle), the net effect can still be positive. So, although sufficient for positive gains from merger, ray subadditivity of the cost function is not a necessary condition for positive gains from merger in the long run.

10.4.2 Subadditivity of the Cost Function: Short Run

Now suppose that the input vector is partitioned as $x = (v, K)$ where v and K are, respectively, the sub-vectors of the variable and fixed inputs. Correspondingly, w and r are the price vectors of the variable and the fixed inputs.

Holding K fixed at K^0, one can define the conditional input requirement set

$$V(y^0 \mid K^0) = \left\{v : \left(v, K^0\right) \text{can produce } y^0\right\} \Leftrightarrow \left\{v : \left(v, K^0, y^0\right) \in T\right\}. \quad (10.42)$$

The minimum cost of producing the output bundle y^0 at input prices (w, r) with K held fixed at K^0 is

$$\begin{aligned} C\left(y^0, w, r, K^0\right) &= \min w'v + r'K^0 \\ s.t.\, v &\in V(y^0 \mid K^0). \end{aligned} \quad (10.43)$$

The condition for subadditivity of the short-run cost function defined in (10.43) at y^0 comparable to (10.36) above will be

$$\sum_{s=1}^{q} C\left(y^s, w, r, K^0\right) > C\left(y^0, w, r, K^0\right). \quad (10.44)$$

There are two complicating factors why (10.44) does not by itself provide a justification for merger across firms or consolidation within a firm. First, typically the q individual firms under consideration for merger will differ not only in their output bundles y^s but also in their bundles of fixed inputs K^s. Obviously, the short-run total cost curves are different when the fixed input bundles are different. Second, when the individual firms are merged, their fixed inputs are also aggregated to $K^M \equiv \sum_{s=1}^{q} K^s$. Even if all of the individual firms had the same bundle of fixed inputs K^0, the merged firm would have an aggregated bundle of fixed inputs $\left(K^M \equiv \sum_{s=1}^{q} K^s = qK^0 \right)$. For positive gains from merger, we need

$$g \equiv \sum_{s=1}^{q} C\left(y^s, w, r, K^s\right) - C\left(\sum_{s=1}^{q} y^s, w, r, \sum_{s=1}^{q} K^s \right) > 0. \qquad (10.45)$$

Clearly, the conditions (10.44) and (10.45) are different. In the rest of this section, we derive a set of conditions which are together sufficient for positive gains from merger. We also provide an additive decomposition of the gain (g) into three separate components relating, respectively, to (a) convexity of the production possibility set (b) subadditivity of the short-run cost curve corresponding to the aggregate bundle of fixed inputs $\left(K^M = \sum_{s=1}^{q} K^s \right)$ at the aggregate output bundle y^M, and (c) the difference between two short-run total cost curves at the average output bundle $\bar{y}$.

We continue with the example of merger between two firms – A and B, with output bundles y^A and y^B, and the fixed input quantities K^A and K^B. Evaluated at the same input price vector (w, r), the minimum short-run costs of producing their outputs given their observed levels of fixed inputs are $C_A = C\left(y^A, w, r, K^A\right)$ and $C_B = C\left(y^B, w, r, K^B\right)$. The cost of producing their combined output $y^M = y^A + y^B$ given the combined bundle of fixed inputs $K^M = K^A + K^B$ is $C_M = C\left(y^M, w, r, K^M\right)$. Thus, the potential gain from merger is

$$g = \left(C_A + C_B\right) - C_M = \left[C\left(y^A, w, r, K^A\right) + C\left(y^B, w, r, K^B\right) \right] - C\left(y^M, w, r, K^M\right). \qquad (10.46)$$

Next, define $\bar{y} = \frac{1}{2}(y^A + y^B)$, $\bar{K} = \frac{1}{2}(K^A + K^B)$, and $\bar{C} = \frac{1}{2}(C_A + C_B)$. Thus, the combined output is $y^M = 2\bar{y}$, and the combined fixed input is $K^M = 2\bar{K}$. Hence, (10.13) can also be expressed as

$$g = \left[C(y^A, w, r, K^A) + C(y^B, w, r, K^B)\right] - C(2\bar{y}, w, r, 2\bar{K}). \quad (10.47)$$

Now suppose that the cost-minimizing variable input quantities are $(v^A \mid K^A)$ for y^A and $(v^B \mid K^B)$ for y^B. That is both (v^A, K^A, y^A) and (v^B, K^B, y^B) are feasible input–output bundles. Let $\bar{v} = \frac{1}{2}(v^A + v^B)$. Therefore, by convexity of the set T, $(\bar{v}, \bar{K}, \bar{y})$ is also a feasible input–output combination. That is $\bar{v} \in V(\bar{y} | \bar{K})$. But this need not be the short-run cost-minimizing bundle of variable inputs. Hence,

$$w'\left(\frac{v^A + v^B}{2}\right) + r'\left(\frac{K^A + K^B}{2}\right) = \frac{1}{2}\left[C(y^A, w, r, K^A) + C(y^B, w, r, K^B)\right]$$
$$\geq C(\bar{y}, w, r, \bar{K}). \quad (10.48)$$

Now, (10.46) can also be written as

$$\begin{aligned} g &= \left[C(y^A, w, r, K^A) + C(y^B, w, r, K^B) - 2C(\bar{y}, w, r, \bar{K})\right] \\ &\quad + 2C(\bar{y}, w, r, \bar{K}) - C(2\bar{y}, w, r, 2\bar{K}) \\ &= 2\left[\frac{C(y^A, w, r, K^A) + C(y^B, w, r, K^B)}{2} - C(\bar{y}, w, r, \bar{K})\right] \quad (10.49) \\ &\quad + 2\left[C(\bar{y}, w, r, \bar{K}) - \frac{C(2\bar{y}, w, r, 2\bar{K})}{2}\right]. \end{aligned}$$

The expression inside the square brackets in the last term on the right-hand side of (10.49) can be further broken up as

$$\begin{aligned} & C(\bar{y}, w, r, \bar{K}) - \frac{C(2\bar{y}, w, r, 2\bar{K})}{2} \\ &= \left[C(\bar{y}, w, r, 2\bar{K}) - \frac{C(2\bar{y}, w, r, 2\bar{K})}{2}\right] \quad (10.50) \\ &+ \left[C(\bar{y}, w, r, \bar{K}) - C(\bar{y}, w, r, 2\bar{K})\right] \end{aligned}$$

Substituting (10.50) into (10.49), one gets

$$g = 2\left[\frac{C(y^A, w, r, K^A) + C(y^B, w, r, K^B)}{2} - C(\bar{y}, w, r, \bar{K})\right] + 2\left[C(\bar{y}, w, r, 2\bar{K}) - \frac{C(2\bar{y}, w, r, 2\bar{K})}{2}\right] + 2\left[C(\bar{y}, w, r, \bar{K}) - C(\bar{y}, w, r, 2\bar{K})\right] \quad (10.51)$$

It will be helpful to separately identify the three terms on the right-hand side of (10.51) as

$$I = 2\left[\frac{C(y^A, w, r, K^A) + C(y^B, w, r, K^B)}{2} - C(\bar{y}, w, r, \bar{K})\right] \quad (10.52a)$$

$$II = 2\left[C(\bar{y}, w, r, 2\bar{K}) - \frac{C(2\bar{y}, w, r, 2\bar{K})}{2}\right] \quad (10.52b)$$

$$III = 2\left[C(\bar{y}, w, r, \bar{K}) - C(\bar{y}, w, r, 2\bar{K})\right] \quad (10.52c)$$

At this point, a slight digression on scale economies is worthwhile. Consider, first, the single output case. Positive scale economies hold locally if a small percentage increase in the output level results in a smaller percentage increase in the total cost. In this case, the elasticity of cost with respect to output is less than 1. That is $\varepsilon_{Cy} = \frac{\partial \ln C}{\partial \ln y} < 1$. But that implies that the marginal cost is less than the average cost, and the average cost curve is falling at that output level. In the multiple output case, of course, there is no natural interpretation of an average cost. But one may consider the long-run *ray average cost*, $\text{RAC} = \frac{C(ty^0, p)}{t}$. At a given output bundle y^0, t equals unity so that the RAC and the total cost are the same at the unit scale. Now suppose that each output in the y^0 bundle is scaled up by a factor $t > 1$. If, as a result, the cost increases by a factor less than t, the scale elasticity of cost is less than 1. But also note that in this case, $C(ty^0, p) < tC(y^0, p)$. That is, $\frac{C(ty^0, p)}{t} < C(y^0, p)$. This implies that the ray average cost is falling as the scale of the output bundle is increased (without changing the output mix).

In the short run, as the output increases, the fixed cost is spread over a larger number of units, and the average fixed cost continues to fall indefinitely. However, with some inputs held fixed, the variable cost will eventually increase more than proportionately as the output increases. At that stage, increase in the average variable cost will more than offset the decline in the average fixed cost so that the short-run average cost increases. Hence, it is reasonable to assume the U shape of the average cost curve in the short run (whether or not it is U shaped in the long run). This argument applies just as well to the ray average cost in the multiple output case. That is, economies of scale holds locally if an output bundle ty^0 is located on the declining segment of the short-run ray average cost curve for a given bundle of the fixed inputs, $\frac{C(ty^0, w, r \mid K^0)}{t}$. But, as is shown below, even if it is located on the upward rising segment so that locally diseconomies of scale holds at ty^0, the short-run cost function may still be subadditive at that output scale t.

We now return to Equation (10.51) and explain each of the three terms on the right-hand side geometrically in Figures 10.5–10.7.

In this two-output example, the outputs are measured along the two axes labeled y_1 and y_2, and the third axis measures cost. The points y^A and y^B show the output bundles of the firms A and B. The rays through y^A and y^B in the output space show radial expansion/contraction of the output bundles of A and B. The curves $C(ty^A \mid K^A)$ and $C(ty^B \mid K^B)$ are two short-run ray total cost curves for fixed input bundles K^A and K^B, respectively. The corresponding fixed costs, FC_{y_A} and FC_{y_B}, are shown by the two vertical intercepts. Similarly, point $\bar{y}$ is the average output bundle, $C\left(t\bar{y} \middle| \bar{K}\right)$ is the short-run ray total cost curve for the fixed input bundle $\bar{K}$, and $FC_{\bar{y}}$ is the associated fixed cost. The points P, Q, and R show $C(y^A \mid K^A)$, $C(y^B \mid K^B)$, and $C\left(\bar{y} \middle| \bar{K}\right)$. The point S is the mid-point of the line segment PQ and is the average of $C(y^A \mid K^A)$ and $C(y^B \mid K^B)$. Convexity of the production possibility set ensures that point S lies above R. The first term on the right in (10.18) is twice the amount SR.

In Figure 10.6, we consider the question of subadditivity of the short-run ray cost function $C\left(t\bar{y} \middle| 2\bar{K}\right)$ at $t = 2$. Note that because we consider an unchanged output mix, we can measure the change in the scale along the horizontal axis and the corresponding short-run total cost up the vertical axis so that the usual two-dimensional diagram is adequate. The points P and Q on the same ray total cost curve show the short-run total cost of the bundles $\bar{y}$ (i.e., at t=1) and $y^M = y^A + y^B = 2\bar{y}$ (i.e., t = 2). At $\bar{y}$, t = 1, and the total and ray average cost are both equal to the slope of the line OP.

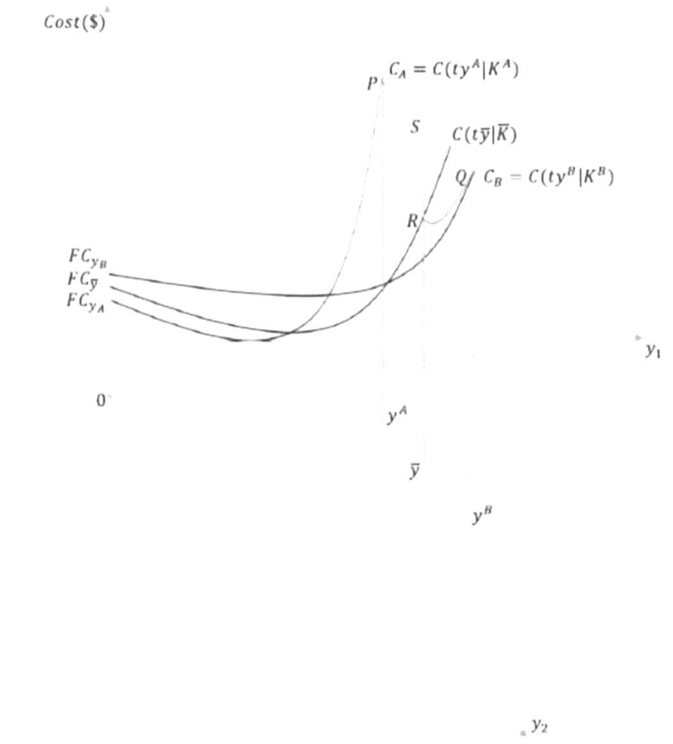

FIGURE 10.5 Trans-ray convexity of short-run cost curves.

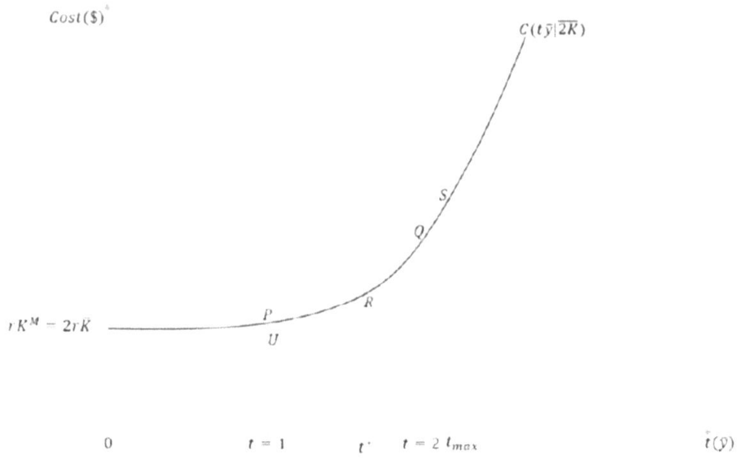

FIGURE 10.6 Subadditivity of the short-run ray cost curve.

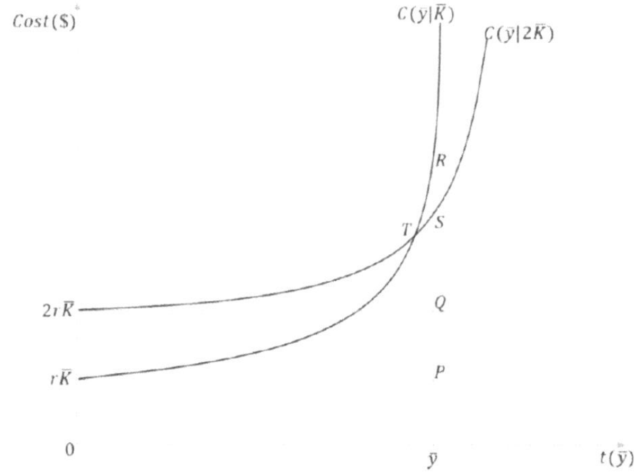

FIGURE 10.7 Capital expansion and trade off between fixed and variable costs.

By comparison, the ray average cost at $y^M = 2\overline{y}$ is measured by the slope of OQ. Two other points of interest in this diagram are R and S. At R, where the ray average cost reaches a minimum, $t = t^*$. This is the output scale which can be described as the cost-efficient production scale in the short run. Scale economies are present at all production scales up to t^*, where scale economies have been exhausted. Beyond that, diseconomies of scale are present. That is, ray average cost is rising at points to the right of R. Thus, in this example, diseconomies of scale are present at $y^M = y^A + y^B = 2y$. At the point S, the ray average cost has caught up to what it was at P. This is evident from the fact that the points P and S both lie on the same ray through the origin in Figure 10.2. So long at point Q lies to the left of S, ray average cost will be lower at Q than at P. We identify the scale at S as $t^{\max}$ For $t^* < 2 < t^{\max}$, subadditivity holds without economies of scale. Of course, if $t^* \geq 2$, economies of scale exist at y^M, and along this short-run ray total cost curve, subadditivity holds at y^M.

This discussion, of course, assumes that we are considering the merger of two units. In other words, we are comparing the cost of production of y^M with the total cost of producing y^M broken up into any two smaller bundles. What happens if y^M was broken up into q bundles (q>2)? Two changes need to be considered. First, for the same aggregate bundle y^M, $\overline{y} = \dfrac{y^M}{q}$ will be a smaller bundle for $q > 2$. Also, when a different set of firms are merged, the aggregated bundle of fixed inputs $K^M = \sum_{s=1}^{q} K^s$ will usually be different from $K^A + K^B$ so that the short-run ray total cost curve will also be different. However, so long as $q \leq t^{\max}$, ray average cost at $y^M = q\bar{y}$ is lower than the cost of the bundle $\bar{y}$. That is, the ray short-run cost curve for K^M is subadditive at y^M.

Finally, Figure 10.7 shows the tradeoff between the reduction in the variable cost and the increase in the fixed cost as the fixed input bundle is increased from $\bar{K}$ to $K^M = 2\bar{K}$. The two short-run ray total cost curves for $t\bar{y}$ are labeled $C(t\bar{y}|\bar{K})$ and $C(t\bar{y}|2\bar{K})$. At $t = 1$, *i.e.*, $(t\bar{y} = \bar{y}), C(\bar{y}|\bar{K}) = R\bar{y}$, and $C(\bar{y}|2\bar{K}) = S\bar{y}$ so that there is a net reduction of cost by *RS*. The increase in the fixed cost is $2r\bar{K} - r\bar{K} = PQ$, whereas the decline in variable cost is *PR-QS*. The net result is the reduction in short-run total cost by *RS*. In any specific case, whether the net change in cost will be positive (i.e., increase) or negative (i.e., decrease) will depend on whether the $\bar{y}$ bundle is located to the left or to the right of the point T where the two curves intersect.

10.4.3 The Nonparametric Models

In this section, we present the different optimization models using the nonparametric method of DEA to evaluate the various short-run ray total and average cost curves needed to measure the potential gains from merger of firms and also to additively decompose the total gain into the three components shown in Equations (10.52a–10.52c). We also formulate relevant DEA models for estimating merger gains in the long run.

Consider again the input–output data set $(x^j, y^j)(j = 1,2,\ldots,N)$. Furthermore, suppose that each input vector can be partitioned as $x^j = (v^j, K^j)$, where v and K are the sub-vectors of variable and fixed inputs. Under the assumption of free disposability of outputs and inputs along with convexity of the technology, one can empirically approximate the production possibility set as

$$\hat{T} = \left\{ \begin{array}{c} (v,K,y): v \geq \sum_{j=1}^{N} \lambda_j v^j; K \geq \sum_{j=1}^{N} \lambda_j K^j; \\ y \leq \sum_{j=1}^{N} \lambda_j y^j; \sum_{j=1}^{N} \lambda_j = 1; \lambda_j \geq 0; (j = 1,2,\ldots,N); \end{array} \right\} \quad (10.53)$$

10.4.3.1 Finding the Multi-Output Short-Run Total Cost

For any output bundle y^0, the short-run total cost at any given input price vector $\omega = (w,r)$ for the fixed input bundle K^0 can be obtained as

$$C(y^0, w, r \mid K^0) = \min w'v + rK^0$$
$$s.t. \sum_{j=1}^{N} \lambda_j y^j \geq y^0;$$
$$\sum_{j=1}^{N} \lambda_j v^j \leq v; \tag{10.54}$$
$$\sum_{j=1}^{N} \lambda_j K^j \leq K^0;$$
$$\sum_{j=1}^{N} \lambda_j = 1; \lambda_j \geq 0; \left(j = 1, 2, \ldots, N\right).$$

Using $\left(y^A, K^A\right), \left(y^B, K^B\right)$, and $\left(y^M, K^M\right) = \left(y^A + y^B, K^A + K^B\right)$ for $\left(y^0, K^0\right)$ in (10.20) we can get $C(y^M, w, r \mid K^M), C(y^A, w, r \mid K^A)$, and $C(y^B, w, r \mid K^B)$ from which one can measure the potential gain from merger of the units A and B defined above in (10.38).

The other costs $C\left(\bar{y}, w, r, 2\bar{K}\right)$ and $C\left(\bar{y}, w, r, \bar{K}\right)$ can also be similarly obtained using $\bar{y} = \dfrac{y^A + y^B}{2}, \bar{K} = \dfrac{K^A + K^B}{2}$, and $2\bar{K} = K^A + K^B = K^M$ for parameters in (10.49), as appropriate.

10.4.3.2 Finding the Short-Run Cost-Efficient Production Scale

The cost-efficient production scale of any output bundle y^0 is $t = t^*$ where the short-run ray average cost for a bundle of fixed inputs K^0, $RAC(ty^0, w, r \mid K^0) = \dfrac{w'v + rK^0}{t}$ is minimized subject to the condition that $\left(v, K^0, ty^0\right) \in T$. Under variable returns to scale, the relevant DEA optimization problem is

$$\min \frac{w'v + r'K^0}{t}$$
$$s.t. \sum_{j=1}^{N} \lambda_j y^j \geq ty^0;$$
$$\sum_{j=1}^{N} \lambda_j v^j \leq v; \tag{10.55}$$
$$\sum_{j=1}^{N} \lambda_j K^j \leq K^0;$$
$$\sum_{j=1}^{N} \lambda_j = 1; \lambda_j \geq 0; \left(j = 1, 2, \ldots, N\right).$$

This, clearly, is a non-linear programming problem. However, following Ray, Walden, and Chen (2021), one can define the variables

$$\mu_j = \frac{1}{t}\lambda_j, u_j = \frac{1}{t}v_j, \text{and } z^0 = \frac{1}{t}K^0. \tag{10.56}$$

Then,

$$\sum_{j=1}^{N}\mu_j = \frac{\sum_{j=1}^{N}\lambda_j}{t} = \frac{1}{t} \Rightarrow t = \frac{1}{\sum_{j=1}^{N}\mu_j}. \tag{10.57}$$

It may be noted that although K^0 is a parameter vector, $z^0 = \frac{1}{t}K^0$ is a vector of choice variables, because t is a choice variable. The non-linear problem can be written as the following linear programming problem:

$$\begin{aligned}
&\min w'u + r'z^0\\
&\text{s.t.} \sum_{j=1}^{N}\mu_j y^j \geq y^0;\\
&\sum_{j=1}^{N}\mu_j v^j \leq u;\\
&\sum_{j=1}^{N}\mu_j K^j \leq z^0;\\
&z^0 = \sigma K^0;\\
&\sum_{j=1}^{N}\mu_j = \sigma; \mu_j \geq 0; (j = 1,2,\ldots,N).
\end{aligned} \tag{10.58}$$

From the optimal solution of the LP problem in (10.58), we can get the cost-efficient scale as $t^* = \frac{1}{\sigma^*}$.

10.4.3.3 Finding the Maximum Scale for Subadditivity

An interesting question at this point is: what is the largest scale factor, $t^{\max}$, from any given output bundle y^0 such that $\text{RAC}(t^{\max} \mid y^0, w, r, K^0) = \frac{C(t^{\max}y^0, w, r, K^0)}{t^{\max}} \leq C(y^0, w, r, K^0)$? We have already seen that if the scale of the average output bundle of the merging firms is greater than the short-run cost-efficient scale t^* of the relevant output mix, then the merged output bundle (which is a radial expansion of the average bundle) cannot have a lower ray average cost. But even if the average bundle is smaller than its cost-efficient scale t^* ray average cost at the merged output bundle will be higher unless the number of merged units, q, is smaller than $t^{\max}$. The critical points are

$$C(y^0, w, r \mid K^0) = \frac{C(t^{\max} y^0, w, r \mid K^0)}{t^{\max}}, \tag{10.59}$$

and

$$\frac{C(ty^0, w, r \mid K^0)}{t} \leq C(y^0, w, r \mid K^0) \text{ for } 1 \leq t \leq t^{\max} \tag{10.60}$$

In order to determine $t^{\max}$ for any output bundle y^0 for the short-run cost function corresponding to K^0, we first solve the LP problem is (10.19) to determine $C(y^0, w, r \mid K^0) = C^*(y \mid K^0)$.

Next we solve the problem

$$\begin{aligned} t^{\max} &= \max t \\ s.t. &\sum_{j=1}^{N} \lambda_j y^j \geq t y^0; \\ &\sum_{j=1}^{N} \lambda_j v^j \leq v; \\ &\sum_{j=1}^{N} \lambda_j K^j \leq K^0; \\ &tC^*(y^0 \mid K^0) - w'v = rK^0; \\ &\sum_{j=1}^{N} \lambda_j = 1; \lambda_j \geq 0; (j = 1, 2, \ldots, N). \end{aligned} \tag{10.61}$$

10.4.4 Finding the Multi-Output Long-Run Total Cost

In the long run, we no longer partition the input vector into the vector of variable and fixed inputs. All inputs are variable, denoted by x^j and the associated input price vector is p.

Therefore, the corresponding DEA problem for estimating long-run costs of producing the output bundle y^0 given the input price vector p becomes[12]

$$
\begin{aligned}
& C\left(y^0, p\right) = \min p'x \\
& s.t. \\
& \sum_{j=1}^{N} \lambda_j y^j \geq y^0; \\
& \sum_{j=1}^{N} \lambda_j x^j \leq x; \\
& \sum_{j=1}^{N} \lambda_j = 1; \lambda_j \geq 0; \left(j = 1, 2, \ldots, N\right).
\end{aligned}
\tag{10.62}
$$

The costs $C\left(y^A, p\right)$, $C\left(y^B, p\right)$, $C\left(\bar{y}, p\right)$, and $C\left(y^M, p\right)$ are obtained by solving (10.28), respectively, for y^A, y^B, $\bar{y}$, and $y^M = y^A + y^B$ in place of y^0. Using these costs, we can estimate the individual elements in 8a and 8b.

10.4.5 Finding the Long-Run Cost-Efficient Production Scale

The most cost-efficient scale size for output bundle y^0 in the long run is the scale $t = t^*$, where its long-run ray average cost, defined as $\text{RAC}\left(p_0, ty_0\right) = \frac{C\left(p_0, ty_0\right)}{t}$ reaches a minimum.

To find the optimal scale t^*, we define $\mu_j = \frac{1}{t}\lambda_j$, the vector $q = \frac{1}{t}x$, and $\sigma = \frac{1}{t}$ and solve the linear programming problem

$$
\begin{aligned}
& \min p'q \\
& s.t. \\
& \sum_{j=1}^{N} \mu_j y_j \geq y^0; \\
& \sum_{j=1}^{N} \mu_j x_j \leq q; \\
& \sum_{j=1}^{N} \mu_j = \sigma; \sigma, \mu_j \geq 0, j = 1, 2, \ldots, N.
\end{aligned}
\tag{10.63}
$$

From the optimal solution of (10.63), we get $t^* = \frac{1}{\sigma^*} = \frac{1}{\sum_{j=1}^{N} \mu_j^*}$. We solve (10.63) using $\overline{y}$ in place of y^0 to obtain the efficient production scale for the average output bundle of the merging firms.

10.4.6 Finding the Maximum Scale for Subadditivity in the Long Run

To determine $t^{\max}$ for an output bundle y^0 in the long run, we first determine $C\left(y^0, p\right) = C^*\left(y^0\right)$, from (10.35) and then solve the following LP:

$$\begin{aligned}
&\max t \\
&s.t. \\
&\sum_{j=1}^{N} \lambda_j y_j \geq t y^0 \\
&\sum_{j=1}^{N} \lambda_j x_j \leq x; \\
&p'x = tC^*\left(y^0\right); \\
&\sum_{j=1}^{N} \lambda_j = 1; \lambda_j \geq 0, j = 1, 2, \ldots, N.
\end{aligned} \tag{10.64}$$

Since we are interested in the $t^{\max}$ for $\overline{y}$, we replace y^0 and $C^*\left(y^0\right)$ by $\overline{y}$ and $C^*\left(\overline{y}\right)$ in the above problem.

10.5 FREE DISPOSAL HULL ANALYSIS AND MERGER EFFICIENCY

10.5.1 Cost Minimization With FDH Analysis

The radial input-oriented technical efficiency in either DEA or FDH analysis ignores the presence of any slacks in individual inputs. However, even when inputs are freely disposable, they are not typically acquired without cost. Hence, any point on the FDH frontier that lies on a flat segment cannot be cost efficient. If we consider the frontier $AP_1BP_2CP_3Q$ in the one-output one-input example (without additivity) in Figure 10.4, only P_1, P_2, and P_3 can be cost efficient. At a point like P_5, there is room for reducing the input from 12 to 10 without lowering the output from 6. Hence, at any input price, the cost can be reduced.

For any output bundle $y^0 \in T^{FDH}$, the cost minimization problem at input price w^0 can be formulated as

$$\begin{aligned} C^{FDH}\left(w^0, y^0\right) &= \min w^{0'} x \\ s.t. \sum_{j=1}^{N} \lambda_J x^j &\le x; \\ \sum_{j=1}^{N} \lambda_J y^j &\ge y^0; \\ \sum_{j=1}^{N} \lambda_J &= 1; \lambda_j \in [0,1\}. \end{aligned} \tag{10.65}$$

Several points are to be noted. First, as in any cost minimization problem, there cannot be any slacks in the input constraints in (10.65).

Furthermore, the binary restriction on the λs implies that the optimal bundle will be one of the FDH efficient points in the data set. Finally, the output constraints ensure that at the reference point, the output must be at least as large as y^0. Thus, all that one needs to do is to consider only those $\left(x^j, y^j\right)$ in the data where $y^j \ge y^0$ and evaluate their corresponding input bundles x^j at the input price w^0 to compute $\tilde{C}_j = w^{0\prime} x^j$. The optimal value in (10.65) would then be the minimum of $\tilde{C}_j$. [13] The corresponding FDH cost efficiency will be

$$CE^{FDH}\left(w^0, y^0\right) = \min_j \frac{w^{0\prime} x^j}{w^{0\prime} x^0}. \tag{10.66}$$

10.5.2 Cost Efficiency of Mergers

In the context of mergers, the constituent firms may be facing different input prices (due to regional differences or other contractual factors), and it is important to decide beforehand the input price vector that would apply for the merged entity. Suppose that the post-merger input price is $\bar{w}$. For each of the F firms being merged, we first solve a problem like (10.65) to determine $C_f^{FDH}\left(\bar{w}, y^f\right), f = 1,2,\ldots,F$. Next, problem (10.65) is solved one more time for the combined output $y^T = \sum_{f-1}^{F} y^f$ to find $C^{FDH}\left(\overline{w}, y^T\right)$. The FDH cost efficiency of the merger would then be

$$ME_C^{FDH} = \frac{C^{FDH}\left(\overline{w}, y^T\right)}{\sum_{f=1}^{F} C^{FDH}\left(\overline{w}, y^f\right)}. \tag{10.67}$$

If the combined output becomes infeasible for the simple FDH problem in (10.66), one can assume additivity and modify the problem in (10.66) by removing the constraint $\sum_{j=1}^{N} \lambda_j = 1$. This would allow creating a benchmark with combined input–outputs of multiple firms.

10.6 CONCLUSION

Merger is economically beneficial if the combined output of firms entering into merger can be produced with less input or their combined input can produce more output. The sources of potential gain are convexity of the technology and benefits of a bigger size. If the convexity assumption is dropped, one can rely on free disposability alone so long as the post-merger output lies within the FDH of the observed input–output bundles. To overcome the infeasibility problem, in DEA one can assume constant (or, at least non-decreasing) returns to scale. In FDH analysis, one needs to assume additivity (with or without replication).

This chapter concentrates mainly on horizontal mergers (even with multiple outputs). A parallel strand in the literature relates to possible economies of scope where two or more firms, each specializing in a different single output, are combined to produce a diversified output bundle. Another related problem is one of reallocation of outputs and inputs across a number of firms to minimize their total cost or maximize the total revenue.[14]

We end with a note of caution. An important determinant of the success or failure of a merger is the compatibility of management cultures in the individual units. When a public sector firm with quite rigid bureaucratic management is merged with a private sector firm with more autonomy in decision making at lower levels, merger is unlikely to yield the expected gains even when scale economies suggest otherwise. A classic example of such cultural incompatibility is the failed merger in 1998 between Chrysler, which had an entrepreneurial style of management with Daimler-Benz with a hierarchical management style. Ultimately in 2009, Daimler sold its 19.9% stake in Chrysler, and the merger was dissolved.

NOTES

1 However, in 2002, travelers split from the CitiGroup and was ultimately sold to MetLife.

2 Blancard, Boussemart, Chavas, and Leleu (2016) consider both merger and breakup of firms. They also use the non-convex Free Coordination Hull (FCH) analysis, which assumes additivity.

3 For a merger between k firms, so long as $f\left(k\bar{x}\right) > kf\left(\bar{x}\right)$ holds the first term will be positive. Unlike CRS, it need not be true at some other input level $x \neq \bar{x}$ or some scale factor other than k even for $\bar{x}$.

4 A previous version of their paper has been available at least a decade earlier. See, for example, Bogetoft and Wang (1996). The chapter in Ray (2004) on which the previous section is based cites their 1996 working paper.

5 Vectors are identified by superscripts.

6 Bogetoft and Wang (2006) also include a technical efficiency component. However, such improvement in efficiency can be independently attained by the constituent firms and do not arise out of the merger.

7 For a detailed discussion of non-convexity, see Briec, Kerstens, and Van den Woestyne (2022).

8 This, however, ignores the possible diseconomies of a merger arising out of more complex internal organization and/or mutually incompatible management cultures of the constituent units.

9 This section is manly from Ray and Sethia (2022).

10 The post-merger output will usually differ from the combined pre-merger outputs of the merging firms due to other factors. We only assume that such changes are not caused by merger itself.

11 Another difference is that change in the fixed costs due to changes in fixed inputs shifts the variable cost function. By contrast, in BPW, variable and marginal costs remain unaffected by startup costs.

12 For the long run average cost curve to be U-shaped under convex technology we need C(0) > 0. Since in DEA, the production frontier under VRS is the free disposal convex hull of the observed input-output bundles only and since $\left(x^j, y^j\right) = \left(0,0\right)$ is never observed for any j = 1,2,., N, there is an implied minimum scale below which the production frontier is not defined.

13 This is the Weak Axiom of Cost Minimization of Varian (1984). See in this context, the discussion in Ray (2004, chapter 10).

14 Ray, Chen, and Mukherjee (2008) considered optimal allocation of the target output across different plants located in different regions to take advantage of input price differences. Ray (2016) considered reallocation of output across different units of a multi-unit firm within a specific market segment to reduce cost without reducing the aggregate output for that region.

REFERENCES

Baumol, W. 1977 "On the proper cost tests for natural monopoly in a multiproduct industry". *The American Economic Review*, 67(5), 809–822.

Baumol, W. and D. Fischer 1978 "Cost-minimizing number of firms and determination of industry structure." *The Quarterly Journal of Economics*, 92(3), 439–467.

Baumol, W., J. Panzar and R. Willig 1982 *Contestable Markets and the Theory of Industry Structure*, (Harcourt Brace Jovanovich, Inc., New York).

Blancard, S., Boussemart, J.P., Chavas, J.P. and Leleu. H. 2016 "Potential gains from specialization and diversification further to the reorganization of activities." *Omega*, 63, 60–68.

Bogetoft, P. and Wang D. 1996 "Estimating the potential gains from mergers." Paper Presented at the *Third Georgia Productivity Workshop held in Athens, GA*.

Bogetoft, P. and Wang, D. 2006 "Estimating the potential gains from mergers." *Journal of Productivity Analysis*, 23(2), 145–171.

Briec, W., K. Kerstens and I. Van den Woestyne 2022 "Nonconvexity in production and cost functions: An exploratory and selective review", in S.C. Ray, R. Chambers, and S. Kumbhakar (eds) *Handbook of production economics* (Springer Nature) pp. 722–754.

Deprins, D., Simar, L. and Tulkens, H. (1984) "Labor-efficiency in post offices", in M. Marchand, P. Pestieau, and H. Tulkens (ed) *The performance of public enterprises: concepts and measurement* (Elsevier Science Publications B. V, North Holland), pp. 243–267.

Green, R. and Cook, W. 2004 "A free coordination hull approach to efficiency measurement." *Journal of the Operational Research Society*; 55(10): 1059–1063.

Maindiratta, A. 1990 "Largest size-efficient scale and size efficiencies of decision making units in data envelopment analysis." *Journal of Econometrics*, 46, 39–56.

Ray, S., J. Walden and L. Chen 2021 "Economic measures of capacity utilization: A nonparametric short-run cost function analysis." *European Journal of Operational Research*, 293(1), 375–387.

Ray, S.C. 2004 *Data envelopment analysis: Theory and techniques for economics and operations research*. (Cambridge University Press, Cambridge, UK).

Ray, S.C. 2016 "Cost efficiency in an Indian bank branch network: A centralized resource allocation model." *Omega*, 65, 69–81.

Ray, S.C., Chen, L. and Mukherjee,K. 2008. "Input price variation across locations and a generalized measure of cost efficiency." *International Journal of Production Economics*, 116(2), 208–218.

Ray, S.C. and Hu, X. 1997 "On the technically efficient organization of an industry: A study of U.S. airlines." *Journal of Productivity Analysis*, 8(1), 5–18.

Ray, S.C. and Sethia, S. 2022 "Nonparametric measurement of potential gain from mergers: An additive decomposition and application to Indian bank mergers." *Journal of Productivity Analysis*, 57, 115.

Tulkens, H. 1993 "On FDH efficiency analysis: some methodological issues and applications to retail banking, courts, and urban transit." *Journal of Productivity Analysis*, 4, 183–210.

Varian, H. R. 1984 The nonparametric approach to production analysis." *Econometrica* 52(3), 579–597.

CHAPTER 11

Data Envelopment Analysis for Mergers and Acquisitions Transactions

Avenues of Research Toward Efficiency Gains

Said Gattoufi, Nabil Ktifi, and Mokhtar Laabidi

Laboratoire SMART, Institut Supérieur de Gestion de Tunis, Université de Tunis, Tunis, Tunisia

11.1 MERGERS AND ACQUISITIONS

11.1.1 Introduction

Three fundamental principles are often considered in the management of modern businesses namely survival, profitability, and growth. In their interactions with their environment, businesses define the appropriate strategies and operational actions to observe these fundamental principles. As suggested by Porter (1985b), the definition of strategies consists of defining the vision and mission that guides businesses in reaching their goals in line with the nature of activities and their internal environment. Moreover, the strategies take into consideration their external

DOI: 10.1201/9781003345206-11

environment constituted by their competitors as well as their collaborators and partners. The largely adopted SWOT analysis explained in Puyt (2023) helps in developing strategies that are based on strengths and weaknesses identified in their internal environment as well as the opportunities and threats identified in their external business environment.

As reaction to the increasingly turbulent business environment characterized by deep mutations in all business sector generated by the global technological developments and the recent challenges, businesses around the world develop what they believe are the appropriate strategies to face challenges and realize significant gains in profitability, effectiveness, and efficiency in line with economic theory like Varian and Melitz's (2024) profit maximization and geopolitical changes that are guiding public policies and regulations. The recent pandemic Covid-19, global warming, and sustainable development consideration are changing the business environment and calls for a wise solution to overcome the challenges and realize adjusted objective for the medium and long run. Restructuring of activities and partnership and business reengineering are among the actions that are increasingly adopted. Businesses are keener to realize mergers and acquisitions (M&As) to strengthen their resilience and hence fortify survival chances, particularly as a reaction to the pandemic that reinforced the need for actions that support their resilience.

A natural way to identify targets for "friendly" M&A is to follow their value chain as well as their networks of alliances and partnerships. Obeying the well-known pragmatic adage "if you can't beat them, join them", M&A actions can be aggressive by takeovers on competitors. These operations can be seen as horizontal or vertical integrations. The formation of the supply chain through the value chain and its management can be an important factor in identifying M&A operations elements.

Moreover, circular economy and environmental obligations encouraged by public policies around the globe are generating a wide range of opportunities by extending and redesigning value chains for all kind of businesses and encouraged, or forced by regulations to redesign their business processes relying on spectacular information technology developments. McKinsey and Company (2024) report a surge in cross-sectorial M&A during the second half of 2023 that is expected to increase during 2024 to cover a wide range of business sectors and particularly cross sectorial M&A.

However, there is a need to assess the gains realized by these strategic actions to shed light for the future possible endeavor about the success factors and more importantly from the failure in past M&A operations. This assessment can take different forms. Our interest in this chapter goes to the assessment of efficiency gains realized through M&A using data envelopment analysis (DEA).

As emphasized in the related academic literature and professional reports analyzed for the purpose of the current book chapter, M&As are critical strategies that organizations employ to enhance their market position and operational efficiencies. However, there was always a need to assess the degree of effectiveness and efficiency realized through these strategic moves and conclude whether they were wise shifts or catastrophic drifts or somewhere in between. The existing literature dedicated to M&A, academic or professional, is rich in approaches that are claimed to be useful in this assessment. Frequently, the concept of efficiency gains and that of earnings management resulting from M&A are discussed and adopted using formal mathematic models and a variety of parametric and non-parametric approaches were suggested.

Among the non-parametric approaches, DEA and several of its variants gained the highest popularity and claimed to be a formal mathematical modeling that helps in assessing the efficiency gains realized through M&A. The assessment provides a benchmarking of the involved economic units, called decision making units (DMUs), pre- and post-M&A. More recent developments of the DEA proposed Inverse DEA as a tool that provides recommendations regarding inputs and outputs to reach a predefined efficiency target. The authors present these concepts with a focus on DEA as a tool to assess the efficiency gains realized through M&A. A bibliometric analysis is conducted to define the main features of the literature dedicated to DEA for M&A.

We present in the first part of this chapter the basic concepts called in this chapter followed by a bibliometric analysis of the recent scholarly publications related to M&A in different business sectors. The second part is devoted to presenting some theoretical concepts, away from mathematical complications, and then analyzing the DEA literature related with M&A by bibliometric analysis. The chapter ends with suggesting some research topics the authors believe are promising to the subject of assessing the gains realized through M&A using intelligent variants of DEA based on recent developments in data science and business analytics.

These research topics will be suggested in line with recent trending economic concepts like circular economy and sustainable development through value chain.

11.1.2 Theoretical Foundation and Historical Facts Value Chain

The concept of the value chain was developed and popularized by Porter (1985a). Johnson and Scholes (2008) describe the value chain as an interconnected set of value-generating activities, extending from the acquisition of raw materials from suppliers to the delivery of the final product to the customer, regardless the sector of activity of the company.

The value chain is broken down into main activities such as internal logistics, production, external logistics, marketing, sales, and services. This decomposition into different value activities makes it possible to understand the cost dynamics and the current and potential sources of differentiation, which makes the value chain a strategic tool as also emphasized by Porter (1985b) by highlighting the importance of competitive advantage in the economic framework and the possibilities of synergies between different companies.

The value chain is a powerful tool for strategic analysis and decision making, allowing companies to better understand their internal processes, optimize their resources, and create a sustainable competitive advantage in the market, and hence are fundamental concepts for companies and businesses and a starting point in defining the nature of expansion through M&A (or split) for the future. Thus, identify targets in defining the vision and mission as well as the targets to be reached through well-defined and justified strategies.

- Understanding of key activities: It allows a detailed analysis of the different activities involved in creating value within the company, from production to marketing to logistics, which helps identify strengths and weaknesses of the company.
- Cost Management: By understanding the costs associated with each activity in the chain, businesses can identify areas where savings can be made or investments to improve operational efficiency.
- Identification of strategic levers: By visualizing activities sequentially, the value chain allows the identification of activities where strategic changes can have the greatest impact on the overall performance of the company.

11.1.3 Mergers and Acquisitions

11.1.3.1 Definition and Historical Facts

M&A has always been an interesting subject for academicians, professionals, and policy makers. There are so many factors involved in an M&A process, and each factor requires detailed analysis. The importance of the topic can be gauged from the recent global economic slowdown and crisis, which has resulted in increased M&A activities. The unexpected developments in global financial markets have affected different industries in different ways. Some industries have experienced a downturn and consolidation in the form of M&A was evident. On the other hand, there were industries which were doing well and hence took advantage of the depression in the market and acquired companies at attractive prices. Yet, there were others that became targets for acquisition due to financial distress. Anyways, it provides a very rich ground for doing analysis of various factors and types of M&A activities, their success, and impact on the firms and the industry concerned.

M&As are in fact crucial business strategies to drive business growth and evolution. These operations and transactions allow smaller or less specialized companies to explore new potential markets or areas of activity to optimize their profits. Merger is generally defined as an operation where one entity is merged with another in such a way that at least one of the initial entities loses its own identity, established the distinction between merger and acquisition. In contrast, an acquisition is often seen as one company acquiring a majority share in another without necessarily merging the assets of the two companies involved. According to Alao and Kuje (2010), merger is the alliance of two or more organizations to form a single unified organizational entity. Georgios and Georgios (2011) describe a merger as the coming together of two or more companies to form one, whereas in the case of an acquisition, a larger and financially strong company buys a small company. The work of Durga et al. (2013) defines M&As as activities, including buyouts, restructurings, or takeovers that alter the ownership structure of companies.

The inception of the modern forms of M&A in line with the modern economics and business models goes back to the 18th century and took places in United States, according to several publications like Focarelli et al. (2002) and Goyal and Joshi (2011). There is an agreement in the literature that M&A went through six waves detailed hereafter, not including the recent ongoing and expected waves that had been triggered by the recent international financial crisis and more recently by the global pandemic and recent orientation toward more sustainable and responsible economic activity, boosted by the international awareness about global warming and the public policies encouraged by international organizations.

First wave: The first wave of M&A began in 1897 and extended until 1904. During this period, these practices emerged in companies and organizations seeking to capitalize on their production by becoming unique sellers in the market, including in sectors such as railways, lighting, and electricity. These transactions mainly took the form of horizontal mergers and took place in key industries. However, most of the deals initiated during this early merger and acquisition period turned out to be unsuccessful, as they failed to achieve the set of predefined goals.

Second wave: The second period of M&A began in 1916 and extended until 1929. The main objective of this period was to induce companies to engage in merger and acquisition transactions aimed at benefiting from an oligopoly rather than a monopoly. This period was marked by the expansion of high technology as well as advances in railroads and transportation. M&As in this period were mainly horizontal or conglomerate in nature according to Golubov et al. (2013). The companies and organizations involved in these operations were mainly major producers of ores and minerals, food products, oil and fuel, transportation, and chemicals, among others. Banks have played a crucial role in facilitating these transactions, with investment banks providing loans to investors with flexible payment terms. This wave of M&A ultimately caused the stock market to collapse in 1929.

Third wave: The third wave of M&A took place between 1965 and 1969, mainly characterized by transactions of a conglomerate nature. These M&A deals were generally financed by owners' capital, and banks appeared to play a less important role. However, toward the end of this period, the trend began to reverse when consolidations of various companies and organizations began to show unsatisfactory results in 1968 as reported in Fatima and Shehzad (2014).

Fourth wave: The fourth wave of mergers (1981–1989) was marked by a notable role of hostile mergers. During the 1980s, hostile mergers became an acceptable form of business expansion. Commercial invasion activity has become a highly speculative action, seen as a means of making short-term profits. Takeover bids (takeover bids) in this wave were considered either friendly or hostile, depending on the response of the target company's board of directors. If the board of directors approved the takeover bid, it would be considered friendly, otherwise it was considered hostile. According to Golubov, Petmezas and Travlos (2013), mergers initiated in the oil and gas, pharmaceutical, banking, and airline sectors mainly characterized this fourth wave.

The fifth wave: The fifth wave began in 1992 and lasted until 2000, inspired by the global stock market boom and the deregulation that accompanied it. This wave mainly affected the banking and telecommunications sectors, with transactions supported to some extent by equity rather than debt financing (Kouser and Saba, 2011).

The sixth wave: Mergers in the metals, oil and gas, utilities, telecommunications, banking, and health care industries characterized the sixth wave of mergers, which took place from 2003 to 2007. This wave was fueled by increasing globalization and government support from some countries like France, Italy, and Russia to create strong national and global champions. Private equity investors have played a crucial role, accounting for a quarter of the general buyout movement, aided by the availability of credit at low interest rates. Cash transactions were significantly more prevalent during this period (Alexandridis et al., 2012).

11.1.3.2 Types of Mergers

From an economic perspective, M&A can be categorized based on the level of economic activity. These mergers come in three main types: horizontal mergers, vertical mergers, and conglomerate mergers. Here are detailed explanations of these types, based on the observations of Weston and Jawien (1999).

- **Horizontal mergers**: Horizontal mergers involve two companies that operate and compete in the same sector of activity. Proponents of this type of merger argue that they can generate economies of scale. However, a critical problem associated with horizontal mergers is market concentration, where the number of companies potentially decreases, leading to monopoly risk.

- **Vertical mergers**: Vertical mergers occur between companies at different stages of the production chain. This often happens when the companies involved have a buying and selling relationship.
- **Conglomerate mergers**: Conglomerate mergers occur when the companies involved operate in unrelated sectors. In other words, there is no direct buying or selling relationship between the merging companies.

The acquisition of the target company can be done in two ways by statutory merger or by acquisition of assets. In the case of a statutory merger, the acquiring company legally takes control of the target company by exchanging its shares for those of the acquired company, leading to the legal dissolution of the latter. In the asset acquisition method, the acquiring company purchases the assets of the target company and becomes responsible for its liabilities. Then, the target company is liquidated by distributing its assets before being dissolved.

11.1.4 Bibliometric Analysis of the M&A Literature

11.1.4.1 Methodology of the Bibliometric Analysis

As reported in their website, https://www.bibliometrix.org/home/, bibliometrics is an open-source R tool designed for conducting comprehensive science mapping analysis of scientific literature. Developed in the R programming language, it is characterized by its flexibility and ease of integration with

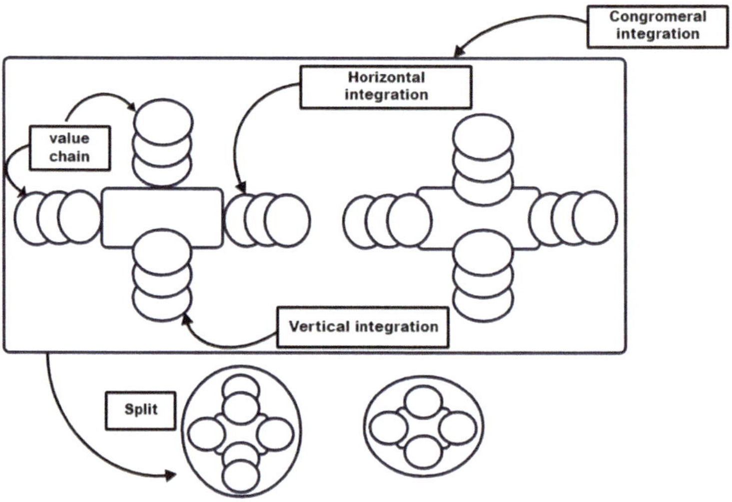

FIGURE 11.1 Different types of M&A.

various statistical and graphical packages. This adaptability is crucial given the dynamic nature of bibliometrics as a science. Bibliometrics' architecture allows for rapid updates and integrations, catering to the needs of an extensive and active community of developers and researchers.

11.1.5 Description of the of M&A Publications During 2018–2024 Dataset

11.1.5.1 Content Description

The dataset of papers is obtained from SCOPUS, available on request from the authors, and covers research articles and documents related to M&A, consolidations, and takeovers. It spans from 2018 to 2024 and comprises 1605 documents from various sources, including journals and books. It shows a negative annual growth rate of –8.64%, indicating a decline in the number of documents over time.

On average, each document is approximately 3.09 years old, with an average of 10.08 citations per document. The dataset contains 525 unique sources and includes keywords provided by authors and indexed by SCOPUS, with 4472 Keywords Plus (ID) and 4589 Author's Keywords (DE). It involves 3597 authors, with 200 single-authored documents and an average of 2.79 co-authors per document. Additionally, about 35.76% of documents have international co-authorships. Most documents are categorized as articles, with 1589 articles and 16 articles that are also tagged as "Book chapter."

Table 11.1 provides a comprehensive overview of scholarly research on mergers, acquisitions, consolidations, and takeovers from 2018 to 2024 indexed in SCOPUS and included in this bibliometric analysis.

11.1.5.2 Sources Analysis

As shown in Table 11.2, the dataset comprises a collection of articles from various academic journals, representing a diverse array of disciplines but with a strong focus on finance, management, and engineering. Notable journals included are the *Journal of Cleaner Production* with 61 articles, *Journal of Corporate Finance* contributing 44 articles, and *Management Science* with 38 articles. Other significant contributors include *International Journal of Industrial Organization*, *Long Range Planning*, and *Strategic Management Journal*, each offering a considerable number of publications.

Table 11.2 features also work for specialized journals such as *IEEE Transactions on Engineering Management* and industry-specific studies

TABLE 11.1 Information About the Dataset of References Considered in the Bibliometric Analysis

Description	Results
Timespan	2018:2024
Sources (Journals, Books, etc.)	525
Documents	1605
Annual Growth Rate %	−8.64
Document Average Age	3.09
Average Citations per Doc	10.08
Document Contents	
Keywords Plus (ID)	4472
Author's Keywords (DE)	4589
Authors	
Authors	3597
Authors of Single-Authored Docs	200
Authors Collaboration	
Single-Authored Docs	223
Co-Authors per Doc	2.79
International Co-Authorships %	35.76
Document Types	
Article	1589
Book Chapter	16

like those in *Journal of Construction Engineering and Management*. This collection serves as a rich resource for studying trends and methodologies across finance, strategic management, industrial organization, and engineering management, providing a comprehensive view of contemporary research in these fields.

11.1.5.3 Conceptual Analysis

The thematic map in Figure 11.2 categorizes various themes within a research domain based on their development degree (density) and relevance degree (centrality). This helps in understanding the status and importance of each theme within the research landscape.

Quadrants Analysis

1. **Niche Themes (Top-Left Quadrant)**: These themes, such as merging, horizontal merger, and restructuring, are highly developed but have low relevance. They are well-studied but not central to the main discourse of the research field, often appealing to specialized groups.

TABLE 11.2 Top 20 Most Relevant Sources

Sources	Articles
Journal of Cleaner Production	61
Journal of Corporate Finance	44
Management Science	38
International Journal of Industrial Organization	37
Long Range Planning	32
Strategic Management Journal	32
Technological Forecasting and Social Change	29
Finance Research Letters	23
Journal of Banking and Finance	22
International Review of Financial Analysis	20
Journal of Business Research	19
Journal of Financial Economics	19
Emerald Emerging Markets Case Studies	18
IEEE Transactions on Engineering Management	18
Accounting and Finance	17
Pacific Basin Finance Journal	15
Managerial Finance	14
Review of Quantitative Finance and Accounting	14
Journal of Construction Engineering and Management	13
Journal of Management in Engineering	13

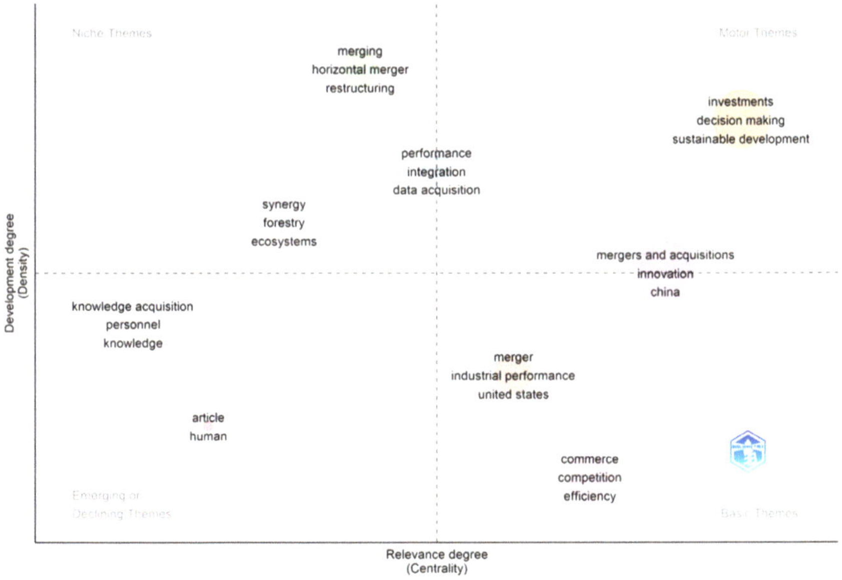

FIGURE 11.2 Thematic map of research topics in M&As.

2. **Motor Themes (Top-Right Quadrant)**: Examples include investments, decision making, and sustainable development. These themes are both highly developed and highly relevant, driving innovation and new discoveries within the field.

3. **Emerging or Declining Themes (Bottom-Left Quadrant)**: Themes like knowledge acquisition, personnel, and human are characterized by low development and low relevance. They may be in the early stages of research or losing relevance over time.

4. **Basic Themes (Bottom-Right Quadrant)**: These themes, such as commerce, competition, and efficiency, are highly relevant but underdeveloped. They form the foundation of the field, supporting more complex research.

Figure 11.3 ends our bibliometric analysis of the M&A literature. The dendrogram represents a hierarchical clustering of publications related to M&A. Upon examining the structure and the formation of clusters, key observations, and insights emerge.

The lower-level clusters or leaves represent individual publications or very closely related studies. These might include detailed case studies of specific M&A transactions, empirical analyses of M&A outcomes in particular industries, or explorations of the role of organizational culture in successful mergers. The fine granularity at this level suggests a high degree of similarity in the research questions, methodologies, and findings of the grouped publications.

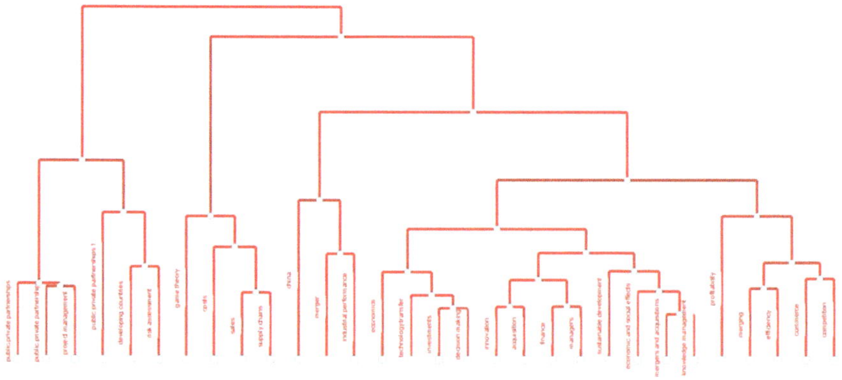

FIGURE 11.3 Dendrogram of research topics clusters in M&As.

A closer examination of the labels can provide additional insights into the thematic structure of M&A research. For instance, clusters with labels such as "post-merger integration," "cross-border M&A," and "corporate governance" indicate focused research topics. Publications on post-merger integration might further cluster into sub-topics such as cultural integration, synergies realization, and integration planning. Cross-border M&A clusters might be divided into studies examining the impact of different regulatory environments, cultural differences, and international finance considerations.

Understanding these clusters can be highly beneficial for researchers and practitioners. For researchers, the dendrogram can help identify gaps in the literature where further study might be needed, such as under-researched aspects of M&A or emerging trends that have not yet been extensively covered. It also aids in recognizing the most influential works and key publications within specific subfields, providing a roadmap for further reading and exploration.

11.2 DEA IN M&A PROCESS EVALUATION

11.2.1 Data Envelopment Analysis

Unlike many analysis methods that rely on parametric models with specific assumptions about data distribution, such as financial ratio analysis or discriminant analysis, DEA is a non-parametric benchmarking approach based on analyzing the economic efficiency of a homogeneous set of economic units by assessing their relative efficiency and producing an efficiency score of each to rank them based of their individual efficiency scores considered as a measure of goodness in transforming inputs into outputs. The inception of DEA was initially coined by Charnes et al. (1978) and finds its origins in Farrell (1957) who suggested the existence of an efficient frontier for each production process in his pioneering paper about efficiency analysis that was explained and commented in Førsund (2015).

Mathematical models and analytical elements as well as most important subsequent developments and extensions with comprehensive explanation of the underlying DEA theory can be found in Cooper et al. (2007) and extended literature surveys can be found in Gattoufi et al. (2004a, 2004b, 2004c), Emrouznejad and Yang (2018), as well as in a recent paper produced by Emrouznejad et al. (2023) focusing on the Inverse DEA.

The advantage of DEA resides in the fact that it is not stochastic, and it does not require a priori assumptions about the relationship between inputs and outputs unlike parametric approaches for efficiency analysis that impose a functional Cob-Douglas type form of the production function estimated using data about inputs and outputs of the production process. However, DEA approach is very sensitive to data anomalies, particularly outliers. Moreover, the rule of thumb limiting the number of factors (inputs and outputs) included in the analysis to the third of economic units is preventing the analysis of small datasets of units.

The DEA approach works by modeling the inputs and outputs of entities as vectors in a multidimensional space, then identifying efficiency boundaries by wrapping the data. By reference to Figure 11.4, entities located on this boundary (efficient isoquant in Figure 11.4) are considered efficient, while those above the boundary can be considered potentially improvable in terms of resources usage. DEA approach for efficiency analysis identifies efficiency frontiers by enveloping the best possible performance, which allows the performance of entities to be compared against these frontiers and defines targets for those above the boundaries.

DEA is widely used in fields such as operational management, finance, economics, and operations research to evaluate the effectiveness and efficiency of production, distribution or service processes and producing benchmarks within a homogeneous set of economic units.

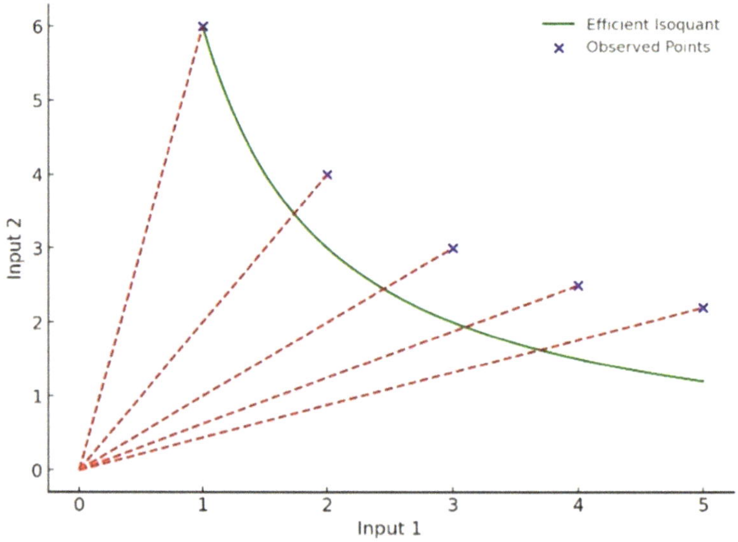

FIGURE 11.4 Farrell's input orient efficient frontier under constant return to scale assumption.

11.2.2 DEA for M&A Processes

DEA is a linear programming model-based technique used to evaluate the relative efficiency of decision making units. In the context of M&A, efficiency here refers to the optimal use of resources to enhance productivity. In layman's terms, DEA can be used to evaluate how well firms transform their inputs to outputs. DEA finds its application in M&A at all stages. It can be used at the pre-merger phase to assess the efficiency of candidate firms and to predict the post-merger prospects of efficiency. At the transition phase, DEA can be used to identify the impact of mergers or acquisitions on the efficiency of participating firms and to suggest ways to improve it. It can be used to compare efficiency levels of firms to identify the potential benefits of forming a consortium.

The objective of DEA is to provide a structured method for evaluating the relative efficiency of decision making units (DMUs) that represent separate business entities like Charnes et al. (1978), Gattoufi et al. (2004a), Barros et al. (2018), among others. This could be an individual plant, sales office, subsidiary company, or joint venture – any part of an overall organization engaged in a well-defined activity with the capability of exercising managerial control over its inputs and outputs.

A typical DMU operating in a private sector industry targeted by M&A activity will face increased pressure to produce better results from a given set of resources. On the one hand, it will be subjected to immediate competitive forces compelling it to raise productivity. On the other hand, it may face an internal assessment of its cost-effectiveness relative to other parts of the organization with a view to outsourcing or offshoring the activity in question. At the extreme, the DMU may find its activity is no longer viable and face disposal or closure. All these scenarios involve a comparison of the DMU's input/output performance with that of others inside or outside the host organization. DEA provides a means of making these comparisons in a systematic and unbiased manner by focusing on the relative efficiency of each DMU.

In contrast to statistical techniques based on econometric or parametric programming, estimation of a production frontier is not required. This can be a major advantage where the data to be analyzed is not well-structured or where a true understanding of DMU performance is not yet available. The authors believe, based on their review of a number of publications among the abundant literature, to date estimated to be over 40,000 publications as reported in the website DEAZone maintained by Emrouznejad (www.deazone.com), that DEA can have a valuable and crucial role in the process of M&A decision making as confirmed in Gattoufi et al. (2014b), Amin et al. (2017a) and Amin et al. (2017b), among others.

DEA approach analyses the comparative efficiency of each DMU and identifies sources of inefficiency. It can create opportunities for M&A activity by identifying which DMUs will be more efficient if they join with other DMUs. This is directly analogous to the firms involved in M&A. One of the models used to measure efficiency in DEA is the basic and original model suggested in Charnes et al. (1978). The output from this model can be used to find targets for M&A activity. A target is found for a DMU if it is relatively inefficient in its current state and the target represents the level of inputs (or outputs) needed for the DMU to become efficient in comparison to the others. If the output is less than 1, it is suggestive that the DMU should undertake a reduction in inputs, and if it is greater than 1, a DEA analysis can be re-run using the Charnes-Cooper-Rhodes original DEA model developed in Charnes et al. (1978) to identify the level of outputs needed to achieve efficiency.

A DMU (in this context), a firm's efficiency, is found by comparing the weighted sum of its inputs to the weighted sum of its outputs. If targets are found where more than one DMU should perform them, there is potential to create mergers between those DMUs so that a single DMU would be able to perform the target more efficiently. This is exactly the situation in M&A activity or any alike transactions considered to be realized by a company for a better performance.

11.2.3 The Role of DEA in M&A

Essentially, DEA measures the efficiency of DMUs relative to a frontier. Given the relatively low operational efficiencies in M&A, DEA can be applied through benchmarking an acquiring firm against the most efficient firms in the industry. DEA is appealing to academics and researchers due to the ability to control all variables and determine effects solely from efficiency differences. DEA's control of variables can be compared to the event studies methodology, in which case the Abnormal Mean Difference Method (AMDM) is used to assess the effectiveness and efficiency of the implemented M&A. Industries with highly leveraged capital structures and reporting requirements are more suited to regression frameworks and inferential statistics. This is not to say DEA is restricted as a deterministic technique. There are instances of stochastic frontier analysis being used to assess relative efficiency. Though DEA is non-parametric analysis, unlike regression analysis, which is a parametric estimation techniques, they were successfully combined in many contexts. DEA and regression are used in sequence and perfect complementarity. As another beneficial application, DEA is useful for the firm attempting to improve efficiency over multiple periods to maximize a certain output.

In the context of M&A, DEA efficiency scores can be used as an ancillary technique with the estimated coefficients conforming output/input changes. By estimating cost, revenue, or profit efficiency, the M&A can also be analyzed using DEA through the process of returning to profitability post integration, the key issue in M&A being whether these scores will improve. There were several publications in different sectors like Oukil (2023), Amin & Al-Muharrami (2018), Amin & Ibn Boamah (2020), and Lozano and Villa (2010), among others. DEA was combined with other approaches in some publications like in Amin et al. (2019) where DEA was combined with goal programming for target selection in M&A transactions and in Barros & Wanke (2015) combining DEA with neural networks to analyze efficiency and Wanke et al. (2016, 2017).

11.2.4 Efficiency Gains and Earning Management in M&A Transactions

Efficiency gains and earnings management are two critical concepts in financial performance and corporate governance. Efficiency gains refer to improvements in productivity and cost-effectiveness, while earnings management involves the manipulation of financial reports to achieve desired financial outcomes. Efficiency gains can potentially reduce the need for earnings management by improving actual financial performance. However, in some cases, companies may still engage in earnings management to meet short-term financial targets or expectations.

On the one hand, efficiency gains in M&A refers to the increased productivity and reduced costs that result from combining two or more companies and combining their resources. These gains can arise from various sources, including cost synergies, revenue synergies, and improved operational efficiencies. They can significantly enhance the value of the combined entity by reducing costs, increasing revenues, and improving operational efficiencies. However, realizing these gains requires careful planning, effective integration, and ongoing management to address potential challenges and risks.

According to the existing literature, like Amin and Ibn Boamah (2021), the efficiency gains can be noticed in several aspects, among them we cite:

- **Cost Synergies**: These refer to the reduction in operating costs that can be achieved by merging two companies. These can be realized through:
  - Economies of Scale: Larger operations can reduce per-unit costs by spreading fixed costs over a larger output.
  - Consolidation of Facilities: Reducing the number of production plants, warehouses, or office locations.
  - Headcount Reduction: Eliminating redundant positions to reduce labor costs.
  - Procurement Savings: Combining purchasing functions to negotiate better terms with suppliers.
- **Revenue Synergies**: These are represented by the increased revenues that result from the merger, achieved through:
  - Cross-Selling Opportunities: Selling complementary products to each company's customer base.
  - Market Expansion: Entering new geographic markets or customer segments.
  - Enhanced Product Offerings: Combining R&D efforts to develop new products.
- **Operational Efficiencies**: Operational efficiencies can be achieved by streamlining processes and adopting best practices from each company:
  - Process Standardization: Harmonizing processes and systems across the merged entity to improve efficiency.
  - Technology Integration: Leveraging shared technology platforms to reduce IT costs and improve data management.
  - Best Practices Adoption: Implementing the most effective methods and practices from each company.
- **Strategic Realignment**: M&A can lead to strategic realignment that enhances efficiency:
  - Focus on Core Competencies: Divesting non-core assets to focus on more profitable business areas.

- Innovation and R&D: Combining R&D capabilities to accelerate innovation and reduce time-to-market for new products.

Emrouznejad and Amin (2023), Emrouznejad and Yang (2018), and Emrouznejad et al. (2023) report an impressive body of recent published research that significantly enriched DEA literature and widened its body of knowledge for decision making related with restructuring companies, particularly in banking. These reviews of recent works provide a comprehensive exploration of the applications of DEA and its inverse version as well as its version with negative data methodologies to mergers, acquisitions, consolidation, and restructuring in general particularly in banking. The overarching themes and contributions related with M&A can be synthesized as follows:

1. **Efficiency Gains in M&A**: Exploring the potential efficiency gains that can be achieved through M&A, particularly in the banking sector.
2. **Strategic Resource Allocation**: The focus on optimal resource allocation in mergers and partnerships as well as target selection using DEA to guide decision making processes and ensure that combined resources are utilized effectively to achieve maximum efficiency.
3. **Handling Complex Data**: Refined DEA models to handle various types of data, including negative and ordinal data, which are common in real-world scenarios have been developed and applied. This allows for more accurate and flexible efficiency assessments in diverse sectors.
4. **Target Setting and Goal Programming**: Combining inverse DEA with goal programming provides robust frameworks for setting realistic and achievable targets in mergers, helping organizations balance multiple objectives such as cost reduction and operational efficiency.
5. **Restructuring and Consolidation**: Developing restructuring strategies and analyzing the impact of consolidations (both minor and major) on organizational efficiency. Pioneering models have been developed to offer valuable insights about the actual and future performance of firms undergoing significant structural changes and compare that with pre-restructuring performance in terms of economic efficiency.

6. **Applications Beyond Banking**: Recent developments have seen a diversification of applications exemplifying the new models, though most of the applications have been focused on the banking sector. This confirms their applicability and their valuable added value as decision making aid tool across various industries, including manufacturing and technology, showcasing the versatility of DEA in different contexts and different economic sectors.

Overall, recent contributions, like Gattoufi et al. (2014b), Amin et al. (2017a, 2017b, 2018, 2019), and more recently Ibn Boamah and Amin (2024), Emrouznejad and Amin (2023), Amin and Ibn Boamah (2023), Mahla et al. (2023), Amin and Ibn Boamah (2020), Amin and Al-Muharrami (2018), Soltanifar et al. (2023, 2024), as well as Amin and Ibn Boamah (2021) report about how recent DEA literature have significantly advanced the theoretical and practical applications of DEA and its inverse DEA variant in strategic business decisions, particularly in the areas of mergers, acquisitions, and restructuring. They provide valuable tools and insights for academics and practitioners aiming to optimize efficiency and performance in organizational transformations.

In particular, it is important to mention the recent trending research topic in DEA literature for efficiency gains assessment which is Inverse DEA (InvDEA). It is a variation of the traditional DEA method used to evaluate the efficiency of decision making units (DMUs). While traditional DEA seeks to identify the most efficient DMUs by comparing inputs and outputs, inverse DEA focuses on determining the necessary changes in inputs and outputs to achieve a specific efficiency score. This approach can be particularly useful in understanding efficiency gains required in various scenarios, such as M&A and analyze the potential efficiency improvements that can be realized and to assess how changes in operations can lead to efficiency gains.

Inverse DEA can be applied in M&A transactions to assess pre-merger efficiency and estimate post-merger efficiency to determine the necessary adjustments in inputs and outputs for the merged entity to achieve target efficiency and to plan integration strategies to identify areas where efficiency gains can be achieved through the integration of operations, resources, and processes. Inverse DEA consists of:

1. Defining Inputs and Outputs: Select the relevant inputs (e.g., labor, capital) and outputs (e.g., revenue, production volume) for the DMUs involved.

2. Setting Target Efficiency: Determine the desired efficiency score for the merged entity.
3. Solving Inverse DEA Model: Use linear programming to calculate the necessary changes in inputs and outputs to achieve the target efficiency.

Hence, the abundant recent publications in inverse DEA and the increasing interest in this technique confirm it as a valuable tool for analyzing and planning efficiency gains in M&A. By identifying the necessary changes in inputs and outputs, companies can develop strategies to achieve desired efficiency levels and realize the full potential of M&A synergies.

Earnings management, on the other hand, was discussed in the literature in different contexts and industries using a variety of measures, parametric and non-parametric, particularly in banking. Mainly, the tools used are financial analysis to discuss corporate governance issues. Ben-Amar and Missonier-Piera (2008) explore the phenomenon of earnings management by Swiss firms targeted for friendly takeovers. Fakhfakh and Nasfi (2012) delve into the complex realm of earnings management within acquiring firms during M&A in French companies. Vo and Chu (2019) investigate the relationship between foreign shareholdings and earnings quality in Vietnamese non-financial listed firms. Valiya Purayil and Lukose (2022) analyzed the intricate dynamics of how cross-border acquisitions influence earnings management practices. Gonçalves and Coelho (2019) present a comprehensive analysis of earnings management practices in European firms within the context of M&A transactions. An interesting publication produced by Proença et al. (2023) combined efficiency analysis with earning management. The paper investigates the impact of earnings management on the efficiency of Eurozone banks, examining its chronological evolution until the implementation of International Financial Reporting Standard 9. The research claims that its findings help stress the importance of adequately defining earnings management, namely for the purpose of analyzing its effect on banking efficiency.

As conclusion from comparing these two concepts, efficiency gains and earnings management are, as confirmed in the literature, two related critical concepts in financial performance and corporate governance. Efficiency gains refer to improvements in productivity and cost-effectiveness, while earnings management involves the manipulation of financial reports to achieve desired financial outcomes.

Combining efficiency gains assessment with earning management constitutes a comprehensive analysis tool to assess and evaluate the benefits of restructuring transactions, including split, rarely studied, takeovers, and M&A, and hence provide a useful tool for decision makers in appreciating the synergies realized through restructuring endeavors. Hence, understanding the interplay between efficiency gains and earnings management is crucial for investors, regulators, and corporate managers. However, while efficiency gains generally lead to better financial performance and reduce the need for earnings management, vigilance is necessary to ensure that any improvements in reported earnings are genuinely reflective of operational enhancements and not the result of manipulative practices.

11.2.5 Bibliometric Analysis of the DEA in M&A Literature

11.2.5.1 Description of the Dataset of the DEA in M&A Literature Published in SCOPUS

11.2.5.1.1 Content Description The dataset, downloaded from SCOPUS, that covers research from 1992 to 2024 on the topics of "M&A" and "DEA" is available on request from the authors. It includes 217 documents collected from 159 different sources such as journals and books, demonstrating a robust annual growth rate of 6.27% in this field. The documents have an average age of 8.65 years and receive an average of 22.78 citations each, underscoring their significant impact within the academic community.

The dataset is rich in keywords, featuring 490 "Keywords Plus" and 529 "Author's Keywords," which illustrates the wide range of research focuses within the scope of M&A and efficiency analysis. From the authorship perspective, the dataset lists 501 different authors, with 25 documents written by single authors. The collaboration data shows an average of 2.72 co-authors

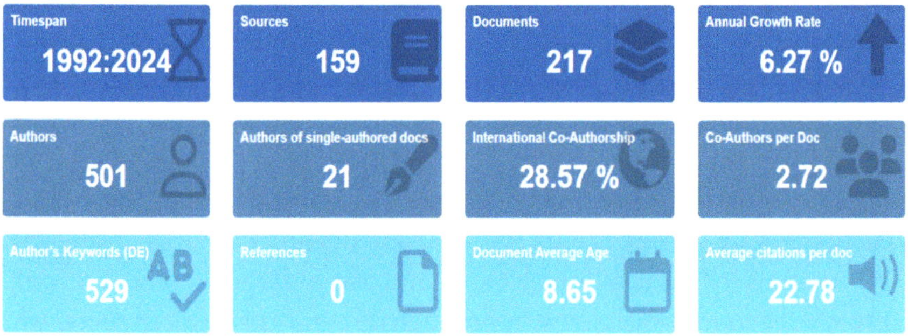

FIGURE 11.5 DEA in M&A DEA Dataset Description.

per document and indicates that 28.57% of the papers involve international collaborations, highlighting a considerable degree of global cooperation.

Document types in this collection vary, with 199 articles, eight book chapters, and nine reviews, showcasing a variety of research outputs from detailed analyses to comprehensive overviews. This dataset is a valuable resource for bibliometric analysis, offering insights into the evolution and current trends in the field of business efficiency and strategic corporate actions.

Table 11.3 provides a detailed analysis of the average citations number per year for articles published over a range of years, focusing on the

TABLE 11.3 Average Citations Number per Year of DEA in M&A Publications

Year	MeanTCperArt	*N*	MeanTCperYear	CitableYears
2023	4.24	25	2.12	2
2022	6.88	17	2.29	3
2017	17.27	15	2.16	8
2020	11.33	15	2.27	5
2021	6.67	15	1.67	4
2013	18.93	14	1.58	12
2018	12.38	13	1.77	7
2016	25.36	11	2.82	9
2008	36.78	9	2.16	17
2010	65.12	8	4.34	15
2012	66.62	8	5.12	13
2015	20.25	8	2.02	10
2009	54	7	3.38	16
2014	18.57	7	1.69	11
2024	0.43	7	0.43	1
2011	10.17	6	0.73	14
1999	54.2	5	2.08	26
2006	95	5	5	19
2019	9.6	5	1.6	6
2004	56.75	4	2.7	21
2001	15	3	0.62	24
2003	35.33	3	1.61	22
2005	16	2	0.8	20
2007	7.5	2	0.42	18
1992	34	1	1.03	33
1993	11	1	0.34	32
2000	103	1	4.12	25

effectiveness and impact of scholarly works within their respective fields. The dataset lists years from 1992 to 2024, with the number of articles (*N*) published each year and the average total citations those articles have received (MeanTCperArt). It also calculates the mean total citations per year (MeanTCperYear) and the number of years since publication (CitableYears), providing insights into the citation dynamics over time.

In recent years, articles from 2023 received an average of 4.24 citations, with a mean of 2.12 citations per year over two years. The year 2022 shows articles being cited 6.88 times on average, with a slightly higher yearly rate of 2.29 over three years. Looking back further, articles from 2012 stand out, with each article averaging 66.62 citations and an impressive 5.12 citations per year over thirteen years, indicating high relevance and impact.

The dataset also highlights peak citation years, such as 2006, where each article garnered an exceptional average of 95 citations, translating to a yearly rate of 5.00 over 19 years, the highest yearly citation rate in the dataset. Similarly, in 2000, each article accumulated 103 citations on average, which is the highest total citation per article across the dataset, with a rate of 4.12 citations per year over 25 years.

This citation analysis offers a comprehensive view of the scholarly impact over time, useful for assessing the longevity and influence of research in academic and professional fields.

11.2.5.1.2 Sources Analysis Table 11.4 provides a list of the most prolific journals in this dataset. It includes the *European Journal of Operational Research*, *Journal of Banking and Finance*, and *Journal of the Operational Research Society*, each contributing six articles. These journals are followed closely by the *IMA Journal of Management Mathematics* with five articles and both the *Annals of Operations Research* and *Journal of Air Transport Management* contributing four articles each.

Other notable sources include *Managerial and Decision Economics*, *Applied Economics*, *Banks and Bank Systems*, *Cogent Economics and Finance*, *Expert Systems with Applications*, *Journal of Financial Services Research*, *Journal of Productivity Analysis*, and *Research in International Business and Finance*, each providing three articles to the dataset.

The collection also features contributions from journals that have published two articles each, such as *Actual Problems of Economics*, *Advances in Airline Economics*, *Agricultural Finance Review*, *Benchmarking*, *Cogent Business and Management*, and *Computers and Industrial Engineering*. This diverse array of sources highlights the interdisciplinary approach to

TABLE 11.4 Top 20 Journals in M&A Publications Number

Sources	Articles
European Journal of Operational Research	6
Journal of Banking and Finance	6
Journal of the Operational Research Society	6
IMA Journal of Management Mathematics	5
Annals of Operations Research	4
Journal of Air Transport Management	4
Managerial and Decision Economics	4
Applied Economics	3
Banks and Bank Systems	3
Cogent Economics and Finance	3
Expert Systems With Applications	3
Journal of Financial Services Research	3
Journal of Productivity Analysis	3
Research in International Business and Finance	3
Actual Problems of Economics	2
Advances in Airline Economics	2
Agricultural Finance Review	2
Benchmarking	2
Cogent Business and Management	2
Computers and Industrial Engineering	2

research on mergers, acquisitions, and efficiency analysis, indicating a broad interest in and application of these topics across different fields and industries.

11.2.5.1.3 Authors Analysis Table 11.5 highlights the contributions of the most relevant authors in the field of mergers, acquisitions, and DEA. Among these, Amin GR stands out with the highest number of articles, having authored seven papers with a fractional contribution of 2.75, indicating significant influence in this research area. Following closely, Emrouznejad A has contributed to six articles with a fractionalized score of 1.70, showcasing consistent scholarly output. Gattoufi and Oukil contributed four publications with many citations, putting them among leading researchers in the domain.

Amin is heading the list with seven papers followed by Emrouznejad with six publications. Sufian F and Wanke P are also notable contributors, each with five publications. This rich array of authors highlights the dynamic and collaborative nature of research in M&A, and efficiency

TABLE 11.5 Top 20 Most Relevant Authors of DEA in M&A Publications Listed in SCOPUS

Authors	Articles	Articles Fractionalized
Amin GR	7	2.75
Emrouznejad A	6	1.7
Sufian F	5	3.08
Wanke P	5	1.42
Barros CP	4	1.25
Cummins JD	4	1.83
Gattoufi S	4	1.33
Lambkin M	4	1.67
Oukil A	4	1.7
Rahman M	4	1.67
Ghiyasi M	3	0.83
Ibn Boamah M	3	1.5
Lozano S	3	1.5
Soltani AA	3	0.7
Wanke PF	3	1.5
Wu J	3	1.03
Azad Mak	2	0.58
Bencheikh A	2	0.45
Chiu Y	2	0.45
Correa HL	2	0.5
Dranev Y	2	0.75
Girardone C	2	0.83
Glass JC	2	0.5
Halkos GE	2	0.83
He Y	2	0.83

analysis, pointing to a well-connected community of scholars dedicated to advancing knowledge in this domain.

Table 11.6 illustrates the application of Lotka's Law, which describes the frequency of publication by authors in each field, showing that a small number of authors produce a large portion of the work. The data comprises authors categorized by the number of documents they have written. The vast majority, 444 authors, have written just one document each, accounting for 88.6% of the total authors. This is followed by 41 authors who have written two documents each (8.2%), and smaller numbers continue to produce fewer documents: six authors each have written three or four documents (1.2% each), two authors have written five documents (0.4%), and one author each has written six or seven documents (0.2%

TABLE 11.6 Author Productivity Through Lokta's Law

Documents Written	No. of Authors	Proportion of Authors
1	444	0.886
2	41	0.082
3	6	0.012
4	6	0.012
5	2	0.004
6	1	0.002
7	1	0.002

each). This distribution confirms Lotka's Law, highlighting the concentration of publication output among a smaller number of prolific authors within the dataset.

11.2.5.1.4 Documents Analysis As depicted in Figure 11.6, the dataset encompasses a selection of the most globally cited documents in the field of applying DEA in a context of M&A particularly focusing on operational efficiency and financial analysis. It highlights key papers from various academic journals that have significantly influenced scholarly discourse over the years. The most cited paper in this collection is by Barros CP from 2012, published in Omega, with 279 total citations and an average of 21.46 citations per year, reflecting its considerable ongoing relevance and impact in the field. Other notable works include Casu B's 2006 study in Manchester School with 246 total citations and De Young R's 2009 paper in *Journal of Financial Services Research* with 222 total citations.

Table 11.6 covers a range of topics, including bank, finance, economic modeling, and operational research, with papers typically receiving citations that reflect their normalized impact and relevance over time. Cummins JD appears multiple times in the list, indicating a significant contribution to the literature on banking and financial stability. Each entry in the dataset is evaluated on total citations, annual citation rate, and a normalized citation count, providing a comprehensive insight into the influence and longevity of the research within the academic community. This collection serves as a valuable resource for understanding the evolution of economic and business strategies through scholarly work.

This word cloud derived from the titles of documents in a dataset focuses on "M&A" and "DEA." The term "efficiency" leads with a frequency of 117, highlighting its prominence in the research. This is followed by specific terms like "mergers" (48) and "acquisitions" (38), which are closely linked

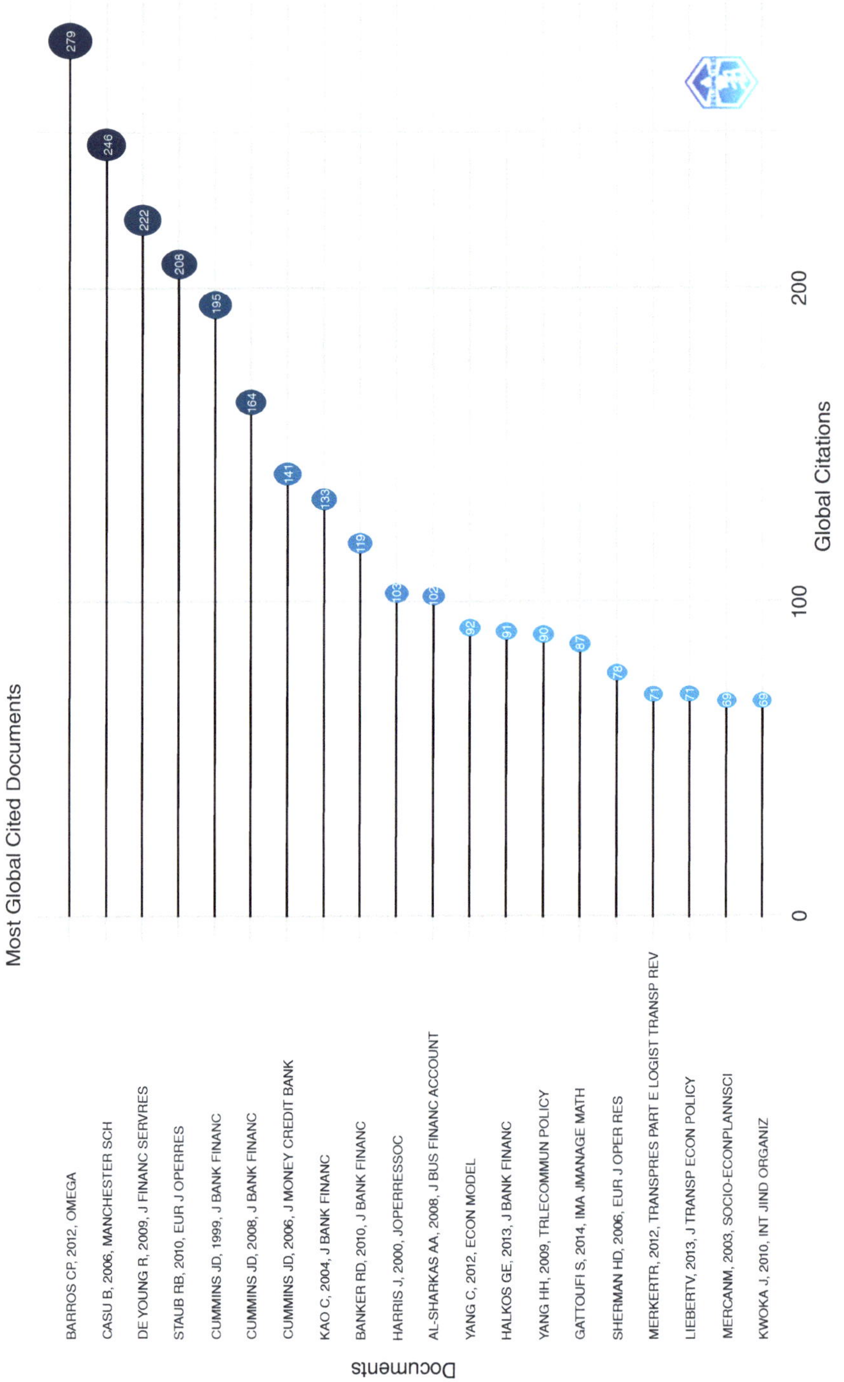

FIGURE 11.6 Most cited DEA in M&A publications referenced in SCOPUS.

FIGURE 11.7 Word cloud of the most cited titles' words in M&A publications.

to "analysis" (43). The presence of "DEA" (35) underscores the analytical methodology used in studying these topics. Other significant terms include "bank" and "banking" (a combined frequency of 53), pointing to a strong focus on the financial sector. Terms such as "performance," "evidence," and "industry" reflect the diverse areas of application and study within the dataset. Additionally, words like "productivity," "gains," and "restructuring" suggest specific outcomes and processes being explored in the context of M&A. This word cloud represents a concentrated look at key themes and sectors within the dataset, indicating a rich field of study in both corporate strategy and efficiency analysis.

11.2.5.2 Knowledge Synthesis

11.2.5.2.1 Conceptual Structure This paragraph describes the conceptual structure of a co-occurrence network, shown in Figure 11.8, derived from a dataset focusing on efficiency and mergers in various sectors. The network maps out key concepts such as "bank mergers," "cost benefit analysis," and "DEA," assigning them to clusters based on thematic similarities. For instance, terms related to electric utilities and distribution are grouped together in cluster 5, while a broad range of efficiency-related terms, including "efficiency," "merging," and "decision making," are found in cluster 4.

The term "M&A" anchors cluster 3, demonstrating significant betweenness, which highlights its role as a central connector within the network. Notably, "DEA" stands out in cluster 2 with the highest betweenness, suggesting its pivotal importance in linking diverse research areas. The network also captures less discussed yet relevant terms in cluster 1 like "banking" and "airline industry," showing their specific but critical roles in the broader discussion. Each node's influence is quantified through

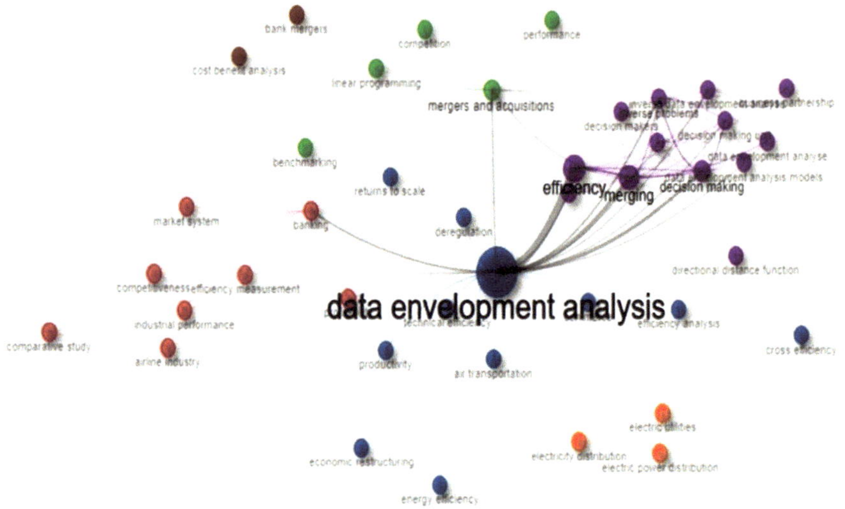

FIGURE 11.8 Co-occurrence of key concepts assigned to clusters based on thematic similarities.

metrics such as Betweenness, Closeness, and PageRank, providing a numerical representation of each term's impact and connectivity within the network.

The provided data is a thematic map from a bibliometric analysis, focusing on the conceptual structure of various research themes related to efficiency and industry studies. It includes a detailed breakdown of key terms and their associations within different clusters. Each term is analyzed for its occurrences and centrality measures, reflecting its significance and connectivity within the research landscape.

Key terms such as "DEA" and "M&A" anchor major clusters, with "DEA" being particularly central, showing high betweenness, closeness, and PageRank centrality scores. Other important terms include "benchmarking," "linear programming," and "efficiency measurement," each forming distinct clusters like "benchmarking" and "DEA." These clusters also incorporate more specific topics like "technical efficiency" and "profitability," indicating a diverse range of focused research areas.

Additionally, industry-specific terms like "airline industry" and "banking" highlight the application of these analytical techniques in particular sectors, with "banking" emerging as a standalone cluster. This thematic map not only showcases the central topics within the field but also illustrates the dynamic interrelations and importance of these terms in ongoing research.

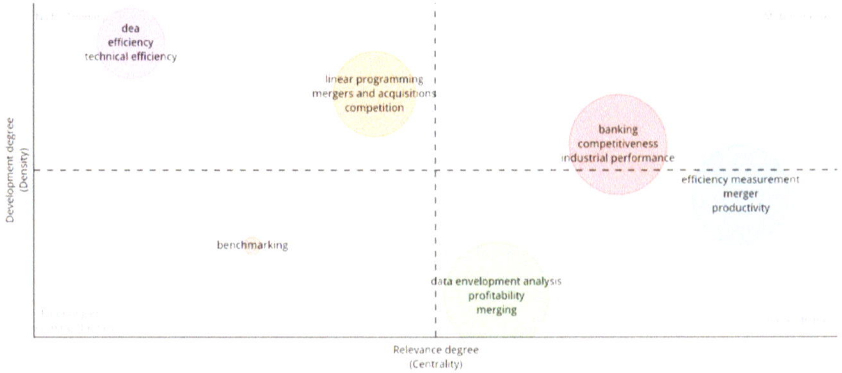

FIGURE 11.9 Thematic clusters of key concepts based on relevance and development degree.

11.2.5.2.2 Intellectual Structure The dendrogram shown in Figure 11.10 summarizes and synthesizes the bibliometric analysis. It presents a hierarchical clustering of publications related to M&A and DEA. This detailed analysis focuses on how DEA and its variants, including Inverse DEA, are utilized in assessing M&A performance. The dendrogram reveals key research themes and their implications, offering insights into the various methodologies and sector-specific applications.

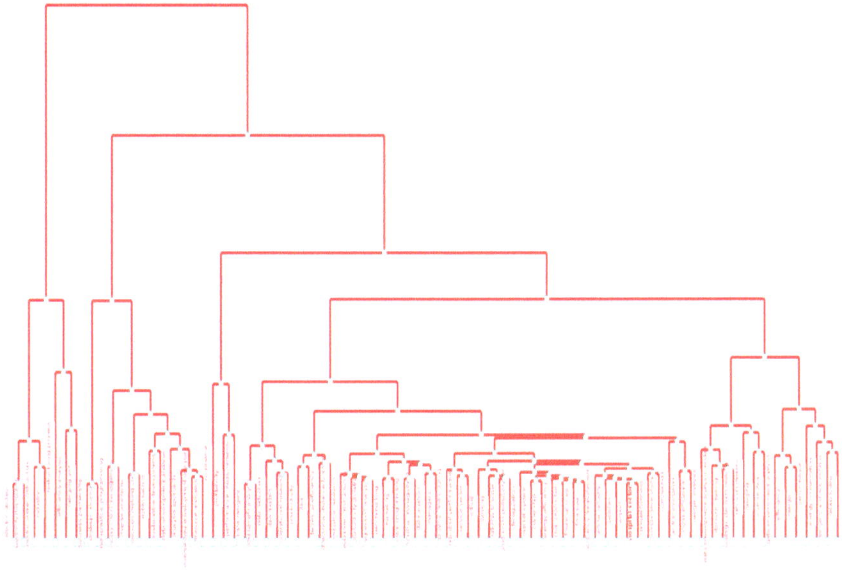

FIGURE 11.10 Dendrogram of research topics in DEA applied to M&A.

11.2.5.2.2.1 DEA in M&A The dendrogram reveals that DEA and its variants play a crucial role in evaluating the performance of M&A transactions. At the highest level, the literature can be divided into broad themes: DEA methodologies tailored for M&A performance assessment and sector-specific applications of DEA across different industries. Within these themes, we see the application of traditional DEA, advanced DEA techniques, and specific methodological innovations, including Inverse DEA.

11.2.5.2.2.2 Traditional DEA Models Traditional DEA models are widely used to measure the operational efficiency and productivity of firms involved in M&A. These studies typically analyze inputs such as capital, labor, and technology, and outputs like revenue and market share, to determine whether M&A leads to efficiency gains and improved performance. Traditional DEA models serve as the foundation for understanding how resource utilization and operational practices are impacted by M&A.

11.2.5.3 Advanced DEA Models

In addition to traditional models, advanced DEA techniques such as network DEA, dynamic DEA, and fuzzy DEA are increasingly applied to capture the complexities of M&A transactions:

1. **Network DEA**: This approach allows researchers to assess the efficiency of interconnected processes within merged entities. Network DEA provides a more nuanced analysis by evaluating different divisions or subsidiaries within a company, highlighting how internal efficiencies contribute to overall performance.
2. **Dynamic DEA**: Dynamic DEA models evaluate performance over time, considering the changing nature of inputs and outputs across different periods. This method is particularly useful for assessing the long-term impact of M&A on firm efficiency, revealing trends and patterns that static models might miss.
3. **Fuzzy DEA**: Fuzzy DEA addresses uncertainty and imprecision in M&A data, offering more robust efficiency assessments in volatile markets. This approach is beneficial for evaluating M&A transactions in industries characterized by high variability and unpredictability.

4. **Inverse DEA**: Inverse DEA, a specific variant of DEA, is particularly valuable in determining the required input levels or feasible output levels given a set of efficiency targets. In the context of M&A, Inverse DEA is used to understand how resource adjustments can improve performance. It helps identify the necessary changes in inputs (e.g., capital, workforce) to achieve desired efficiency levels and optimize resource utilization.

Sector-Specific Applications: The dendrogram also highlights the extensive application of DEA in evaluating M&A performance across various industries. Each sector presents unique challenges and opportunities for applying DEA methodologies.

Banking and Finance: In the banking sector, DEA is used to assess the financial stability and operational efficiency of financial institutions post-M&A. Studies focus on immediate financial outcomes, changes in operational practices, and long-term stability. Inverse DEA, particularly, helps determine how banks can adjust their resource inputs to improve performance and achieve efficiency targets.

Healthcare: In the healthcare sector, DEA evaluates hospital mergers, focusing on cost efficiency and quality of care. Studies assess how resource allocation changes post-merger impact patient outcomes and service delivery. Inverse DEA assists in optimizing resource inputs, such as medical staff and equipment, to enhance efficiency.

Manufacturing: DEA is applied to assess productivity gains and supply chain integration following M&A in the manufacturing sector. Research evaluates how manufacturing firms can optimize resource inputs to achieve desired efficiency levels and improve production processes. Inverse DEA provides insights into necessary adjustments to maintain or enhance efficiency.

Energy Sector: In the energy sector, DEA evaluates the efficiency of traditional and renewable energy companies post-M&A. Studies focus on overall efficiency and sustainability, considering factors such as resource utilization and environmental impact. Inverse DEA helps energy companies adjust their inputs to improve efficiency and meet sustainability goals.

Comparative and Longitudinal Studies: Another significant theme in the dendrogram is the comparison of pre- and post-M&A efficiency using DEA and its variants. These studies offer valuable insights into how M&A transactions affect firm performance over time:

Comparative Studies: Comparative studies benchmark the efficiency of firms against industry standards or competitors, using DEA to identify strengths and weaknesses. Cross-country analyses are common, evaluating how different regulatory environments impact M&A efficiency. Inverse DEA is used to pinpoint the necessary resource adjustments to achieve competitive efficiency levels.

Longitudinal Studies: Longitudinal studies provide a deeper understanding of the lasting effects of M&A. These studies assess efficiency changes over extended periods, identifying both immediate and sustained resource adjustments required to maintain optimal performance. Dynamic DEA is particularly useful in these analyses, revealing how efficiency evolves over time.

Practical Implications: The insights from these clusters have significant implications for both researchers and practitioners:

For Researchers: The dendrogram offers a roadmap for exploring how DEA and its variants are applied in M&A performance assessment. It highlights gaps in literature and suggests areas for further study, such as underexplored sectors or emerging trends in M&A and DEA research. Understanding the methodologies and sector-specific applications can help design new studies that build on existing research and explore new dimensions of M&A efficiency.

For Practitioners: Practitioners can leverage insights from these studies to improve decision making processes in M&A transactions. The application of advanced DEA models, including Inverse DEA, can provide a comprehensive understanding of efficiency gains and potential synergies. Sector-specific findings offer industry practitioners tailored insights into expected outcomes and challenges of M&A, enhancing their ability to navigate complex transactions successfully.

11.2.5.4 Conclusion

The dendrogram serves as a comprehensive map of the research landscape where DEA and its variants, including Inverse DEA, are applied to assess M&A transactions. By analyzing these clusters, we gain valuable insights into the methodologies used, sector-specific applications, and comparative studies that contribute to our understanding of M&A performance. This structured analysis highlights the depth and breadth of current research, pointing toward future directions for both academic inquiry and practical application. Through this detailed examination, we can better appreciate the interconnectedness and nuanced perspectives within the field of M&A performance assessment using DEA.

11.3 SYNTHESIS AND POTENTIAL APPLICATIONS FOR VARIANTS OF DEA MODELING IN M&A

11.3.1 Synthesis of the Literature

In the complex landscape of corporate restructuring, DEA has emerged as a pivotal tool for assessing the efficiency gains resulting from these strategic maneuvers. This literature review explored the application of DEA within the context of M&A, emphasizing its impact on the value chain, a determinant of competitive advantage and operational performance. By scrutinizing seminal works across diverse sectors, this review presented a comprehensive understanding of DEA's role in M&A and chart potential research directions that could augment our comprehension of efficiency gains through strategic consolidations. The application of DEA in analyzing the efficiency of M&A activities within the banking sector has provided valuable insights into the value creation dynamics.

Furthermore, the innovative application of inverse DEA (InvDEA) in modeling the restructuring processes of firms post-M&A presents a novel approach to understanding efficiency gains. By evaluating resource and operation redistribution within the value chain, this methodology offers deep insights into optimizing M&A activities for better performance in pre-mergers as well as in post-mergers.

The value chain analysis, integrated within the DEA framework, facilitates a holistic understanding of efficiency improvements across various operational segments of merged entities as suggested in Amin and Oukil (2019). In line with the increasing awareness about global warming and the increasing interest in green and circular economy, boosted by public authorities and international organizations, DEA and its variants can help identifying targets through the value chain in vertical as well as in horizontal integrations. The restructuring and strategic reorientation necessitated by M&A activities have significant implications for the value chain. Studies reviewed in this literature emphasize the transformational role of M&A in reshaping the operational frameworks and strategic approaches of organizations, leading to enhanced performance and efficiency across their value chains and minimizing "side effects" on the environment and DEA with undesirable outputs is strongly recommended in this domain. McKensey & Company (2024) is in fact suggesting that a new wave of integration is expected in sectors that are causing the most environmental harm.

Despite the advancements in understanding DEA's application in M&A efficiency analysis, vast avenues for future research remain untapped. Investigating the differential impact of M&A on various segments of the value chain, considering dynamic and sector-specific factors, could offer more granular insights into efficiency gains and earnings management. Additionally, integrating DEA with other analytical frameworks and data science tools like machine learning or deep learning might provide, as in Guijarro et al. (2020), a more comprehensive view of the strategic and operational synergies realized through M&A.

11.3.2 Venues for Future Research and Applications

As potential variants of DEA applications in M&A, window DEA is strongly recommended to consider delayed effects as an alternative to what was suggested in Amin et al. (2019). Alternatively, the concept of congestion in DEA can provide a useful tool to identify the cuts in resources in an M&A transaction. Dynamic and network are strongly recommended variants of DEA particularly adapted in vertical integration through the value chain and advanced AI for data can be used to handle the complications and big data nature of the characteristics of the targets. Recently, an interest has been noticed in this direction. Finally, to analyze the changes over time and assess whether the improvement and efficiency gains realized by each DMU were higher than the efficiency gains realized by the sector to which it belongs, by decomposing the Malmquist index, like Gattoufi et al. (2009). This can help assessing whether the M&A transaction is beneficial for the acquirer/acquiring by comparing the individual gain with the sectorial gain using Malmquist Index decomposition. Another interesting combination, inexistant for the time being, would be combining InvDEA with "system dynamics", explanation about which can be found in Dangerfield (2020), to analyze the steady state of an M&A transaction and analyze its long run impact.

It is worth mentioning that DEA and its variants need new theoretical developments to boost its influence and provide more operational implications and applications benefiting from the recent developments in data science theory and its applications in artificial intelligence, since there was a lack of theoretical developments recently. One suggestion is to think about going a step ahead in artificial intelligence toward "teaching machine" that not only learns, but trains humans and provides them guidance and aid in their decision making processes. One practical way is to consider "serious games Pistono et al. (2024)" as a framework and "inverse optimization, Chan et al. (2023)," including InvDEA, as a tool. These combinations of methodologies can enrich the DEA body of knowledge and boost their presence in the real world of services and goods businesses and industries. Finally, there is a potential to explore new domains for applications of DEA and its variants, since that there is was concentration on a limited number of sectors, particularly banking and financial services that can be explained by the homogeneity of the set of economic entities and by the easiness of access to the information needed by the analysis.

11.4 CONCLUSION

In conclusion, the role of DEA in elucidating the nuances of efficiency gains in M&A, particularly regarding the value chain, is invaluable. The literature reviewed herein not only highlights the positive impacts of M&A on operational efficiency but also brings to the fore the complexities and sector-specific nature of these dynamics. As organizations continue to navigate the intricacies of strategic consolidations to bolster their market positions and operational efficiencies, the insights derived from DEA applications in this domain will be crucial. Future research promises to further widen our understanding about M&A's value creation potential through vertical and horizontal integration, through the value chain, keeping in mind sustainability and circular economy, and guiding strategic decision making toward achieving optimal efficiency gains and higher competitiveness in line with strategic mission and vision.

REFERENCES

Alao, J. S., & Kuje, E. D. (2010). Determination of technical efficiency and production function for small scale furniture industry in Lafia metropolis, Nasarawa State, Nigeria. *Journal of Agriculture and Social Sciences*, 6, 64–66.

Alexandridis, G., Mavrovitis, C. F., & Travlos, N. G. (2012). How have M&As changed? Evidence from the sixth merger wave. *The European Journal of Finance, 18*(8), 663–688.

Amin, G. R., & Al-Muharrami, S. (2018). A new inverse DEA model for mergers with negative data. *IMA Journal of Management Mathematics*, 29(2), 137–149.

Amin, G. R., Al-Muharrami, S., & Toloo, M. (2019). A combined goal programming and inverse DEA method for target setting in mergers. *Expert Systems with Applications*, 115, 412–417.

Amin, G. R., Emrouznejad, A., & Gattoufi, S. (2017a). Minor and Major Consolidations in Inverse DEA: Definition and Determination. *Computers and Industrial Engineering*, 103, 193–200.

Amin, G. R., Emrouznejad, A., & Gattoufi, S. (2017b). Modelling Generalized Firms' Restructuring using Inverse DEA. *Journal of Productivity Analysis*, 48(1), 51–61.

Amin, G. R., & Ibn Boamah, M. (2020). A new inverse DEA cost efficiency model for estimating potential merger gains: A case of Canadian banks. *Annals of Operations Research*, 195(1), 21–36.

Amin, G. R., & Ibn Boamah, M. (2021). A two-stage inverse data envelopment analysis approach for estimating potential merger gains in the US banking sector. *Managerial and Decision Economics*, 42(6), 1454–1465.

Amin, G. R., & Ibn Boamah, M. (2023). Modeling Business Partnerships: A data envelopment analysis approach. *European Journal of Operational Research*, 305(1), 329–337.

Amin, G. R., & Oukil, A. (2019). Flexible target setting in mergers using inverse data envelopment analysis. *International Journal of Operational Research*, 35(3), 301–317.

Barros, C. P., Tsionas, M. G., Wanke, P., & Azad, M. A. K. (2018). Efficiency in banking of developing countries with the same cultural background: A novel distance function Frontier model. *Journal of Economic Studies, 45*(3), 638–659.

Barros, C. P., & Wanke, P. (2015). An analysis of African airlines efficiency with two-stage TOPSIS and neural networks. *Journal of Air Transport Management, 44*, 90–102.

Ben-Amar, W., & Missonier-Piera, F. (2008). Earnings management by friendly takeover targets. *International Journal of Managerial Finance, 4*(3), 232–243.

Chan, T. C. Y., Mahmood, R., & Zhu, I. Y. (2023). Inverse optimization: Theory and applications. *Operations Research Articles in Advance*, 1–29. https://doi.org/10.1287/opre.2022.0382

Charnes, A., Cooper, W. W., & Rhodes, E. (1978). Measuring the efficiency of decision making units. *European Journal of Operational Research, 2*(6), 429–444. https://doi.org/10.1016/0377-2217(78)90138-8

Cooper, W. W., Seiford, L. M., & Tone, K. (2007). *Data Envelopment Analysis: A Comprehensive Text with Models, Applications, References and DEA-Solver Software* (2nd ed.). Springer.

Dangerfield, B. (Ed.). (2020). *System Dynamics*. Springer US. https://doi.org/10.1007/978-1-4939-8790-0

Durga, A., Wollants, P., & Moelans, N. (2013). Evaluation of interfacial excess contributions in different phase-field models for elastically inhomogeneous systems. *Modelling and Simulation in Materials Science and Engineering*, *21*(5), 055018.

Emrouznejad, A., & Amin, G. R. (2023). Advances in Inverse Data Envelopment Analysis: Empowering Performance Assessment. *IMA Journal of Management Mathematics*, 34(3), 415–419.

Emrouznejad, A., Amin, G. R., Ghiyasi, M., & Michali, M. (2023). A review of inverse data envelopment analysis: Origins, development and future directions. *IMA Journal of Management Mathematics*, *34*(3), 421–440.

Emrouznejad, A., & Yang, G. L. (2018). A survey and analysis of the first 40 years of scholarly literature in DEA: 1978–2016. *Socio-Economic Planning Sciences*, *61*, 4–8.

Fakhfakh, H., & Nasfi, F. (2012). The determinants of earnings management by the acquiring firms. *Journal of Business Studies Quarterly*, *3*(4), 43–57.

Farrell, M.J. (1957). The measurement of productive efficiency. *Journal of the Royal Statistical Society. Series A (General)*, *120*(3), 253–290.

Fatima, T., & Shehzad, A. (2014). An analysis of impact of merger and acquisition on financial performance of banks: A case of Pakistan. *Journal of Poverty, Investment and Development*, *5*, 29–36.

Focarelli, D., Panetta, F., & Salleo, C. (2002). Why do banks merge? *Journal of Money, Credit and Banking*, 34(4) 1047–1066.

Førsund, F. R. (2015). *Productivity Interpretations of the Farrell Efficiency Measures and the Malmquist Index and Its Decomposition* (Memorandum No. 14/2015). University of Oslo, Department of Economics.

Gattoufi, S., Al-Muharrami, S., & Shamas, G. (2014a). Assessment of mergers and acquisitions in GCC banking. *International Journal of Accounting and Finance*, *4*(4), 358–377. http://doi.org/10.1504/IJAF.2014.066088

Gattoufi, S., Amin, G. R., & Emrouznejad, A. (2014b). A new inverse DEA method for merging banks. *IMA Journal of Management Mathematics*, *25*(1), 73–87. http://doi.org/10.1093/imaman/dps027

Gattoufi, S., Oral, M., Kumar, A., & Reisman, A. (2004a). Epistemology of data envelopment analysis and comparison with other fields of OR/MS for rational decision making. *Journal of the Operational Research Society*, *55*(9), 911–935. https://doi.org/10.1057/palgrave.jors.2601767

Gattoufi, S., Oral, M., & Reisman, A. (2004b). A Taxonomy for Data Envelopment Analysis. *Socio-Economic Planning Sciences* 38(2–3), 141–158. http://doi.org/10.1016/S0038-0121(03)00022-3

Gattoufi, S., Oral, M., Kumar, A., & Reisman, A. (2004c). Epistemology of Data Envelopment Analysis and Comparison with Other Fields of OR/MS for Relevance to Applications. *Socio Economic Planning Sciences* 38(2–3), 123–140. http://doi.org/10.1016/S0038-0121(03)00021-1

Gattoufi, S., Sakr, S. & Omran, M., (2009). Tracking the impact of change in ownership on the efficiency of commercial banks in MENA countries using a Malmquist index-based approach. *Economic Research Forum Working Paper Series*, No. 465. https://erf.org.eg/app/uploads/2014/08/465.pdf

Gattoufi, S., Oral, M., & Reisman, A. (2004b). A data envelopment analysis bibliography update (1951-2001). *Socio-Economic Planning Sciences, 38*(2–3), 159–229. https://doi.org/10.1016/S0038-0121(03)00021-1

Georgios, K., & Georgios, H. (2011). Du Pont analysis of a bank merger and acquisition between Laiki bank from Cyprus and Marfin investment group from Greece. Is there an increase of profitability of the new bank. *Kyriazopoulos-Hadjimanolis, MIBES 2011-Oral*, 5(1) 157–176.

Golubov, A., Petmezas, D., & Travlos, N. G. (2013). Empirical mergers and acquisitions research: A review of methods, evidence and managerial implications. *Handbook of Research Methods and Applications in Empirical Finance* (pp. 287–313). Edward Elgar Publishing.

Gonçalves, T., & Coelho, G. (2019). Earnings management during mergers and acquisitions—European evidence. *European Journal of Economics, Finance and Administrative Sciences, 102*, 22–43.

Goyal, K. A., & Joshi, V. (2011). A study of social and ethical issues in the banking industry. *International Journal of Economics and Research, 2*(5), 49–57.

Guijarro, F., Martínez-Gómez, M., & Visbal-Cadavid, D. (2020). A model for sector restructuring through genetics algorithm and inverse DEA. *Expert Systems with Applications, 154*(15), 113422.

Ibn Boamah, M., & Amin, G. R. (2024). Partner selection in business mergers: A data envelopment analysis approach. *Rairo Operations Research, 58*, 3107–3117.

Johnson, G., Scholes, K., & Whittington, R. (2008). *Exploring Corporate Strategy.* Financial Times Prentice Hall.

Kouser, R., & Saba, I. (2011). Effects of business combination on financial performance: Evidence from Pakistan's banking sector. *Australian Journal of Business and Management Research, 1*(8), 5–64.

Lozano, S., & Villa, G. (2010). DEA-based pre-merger planning tool. *Journal of the Operational Research Society*, 61(10), 1485–1497.

Mahla, D., Agarwal, S., Amin, G.R., & Mathur, T (2023). An inverse data envelopment analysis model to consider ratio data and preferences of decision-makers. *IMA Journal of Management Mathematics, 34*(3), 441–464.

McKinsey & Company. (2024). *Top M&A trends in 2024: Blueprint for success in the next wave of deals.* McKinsey & Company. https://www.mckinsey.com

Oukil, A. (2023). Investigating prospective gains from mergers in the agricultural sector through reverse DEA. *IMA Journal of Management Mathematics, 34*(3), 465–490.

Pistono, A., Santos, A., Baptista, R. J. V., & São Mamede, H. (2024). Framework for adaptive serious games. *Computer Applications in Engineering Education, 32*(4), 1–34. https://doi.org/10.1002/cae.22731

Porter, M. E. (1985a). *Competitive Advantage.* The Free Press.

Porter, M. E. (1985b). Technology and competitive advantage. *Journal of Business Strategy, 5*(3), 60–78.

Porter, M. E. (1990). New global strategies for competitive advantage. *Planning Review, 18*(3), 4–14.

Porter, M. E. (1996). What is strategy? *Harvard Business Review*, *74*(6), 61–78.

Proença, C., Augusto, M., & Murteira, J. (2023). The effect of earnings management on bank efficiency: Evidence from ECB-supervised banks. *Finance Research Letters*, *51*(3–4), 103450.

Puyt, R. W., Lie, F. B., & Wilderom, C. P. M. (2023). The origins of SWOT analysis. *Long Range Planning*, *56*(3), 102304.

Soltanifar, M., Ghiyasi, M., Emrouznejad, A., & Sharafi, H. (2024). A novel model for merger analysis and target setting: A CSW-inverse DEA approach. *Expert Systems With Applications-Part A*, 249, 1–9.

Soltanifar, M., Ghiyasi, M., & Sharafi, H. (2023). Inverse DEA-R models for merger analysis with negative data. *IMA Journal of Management Mathematics*, *34*(3), 491–510.

Valiya Purayil, P., & Lukose, P. J. (2022). Does cross-border acquisition reduce earnings management of emerging market acquirers? Evidence from India. *International Review of Finance*, *22*(1), 143–168.

Varian, H. R., & Melitz, M. (2024). *Intermediate Microeconomics: A Modern Approach* (10th ed.). W. W.

Vo, X. V., & Chu, T. K. H. (2019). Do foreign shareholders improve corporate earnings quality in emerging markets? Evidence from Vietnam. *Cogent Economics & Finance*, *7*(1), 1–22.

Wanke, P., Barros, C. P., & Emrouznejad, A. (2016). Assessing productive efficiency of banks using integrated Fuzzy-DEA and bootstrapping: A case of Mozambican banks. *European Journal of Operational Research*, *249*(1), 378–389.

Wanke, P., Maredza, A., & Gupta, R. (2017). Merger and acquisitions in South African banking: A network DEA model. *Research in International Business and Finance*, *41*, 362–376.

Weston, J. F., & Jawien, P. S. (1999). Perspectives on mergers and restructuring. *Business Economics: The Journal of the National Association for Business Economists*, *34*(1), 29–33.

CHAPTER 12

Two-Stage Data Envelopment Analysis With Application of an Assurance Region to the Best-Practice Frontier

Jon A. Chilingerian
Heller School, Brandeis University, USA

Mitchell P.V. Glavin
Meehan School of Business, Stonehill College, USA

12.1 INTRODUCTION

Building upon Farrell's (1957) early work on productive efficiency, Charnes, Cooper, and Rhodes in 1978 introduced the data envelopment analysis (DEA) methodology that was based on a linear programming approach. The ensuing period of over 45 years has yielded an enormous amount of scholarly work and practical application of DEA across an array of industries, government agencies, and non-profit organizations (Cook & Seiford, 2009; Panwar *et al.*, 2022).

DOI: 10.1201/9781003345206-12

As part of this evolution, DEA came to be utilized not just to form a production frontier to compare the relative efficiency of decision making units (DMUs) in converting multiple inputs into outputs but also to form a best-practice frontier for benchmarking the relative performance of similar DMUs according to a set of criteria identified by management or regulators, such as via a balanced scorecard approach (Cook *et al.*, 2014; Eilat *et al.*, 2006; Amado *et al.*, 2012). The result has been the mobilization of DEA-based approaches to assess progress toward key government and industry aims, such as improved supply chain performance and environmentally sustainable production (Wong & Wong, 2007; Dutta *et al.*, 2022; Shokri Kahi *et al.*, 2017; Zhou *et al.*, 2008; Dakpo *et al.*, 2016).

While DEA can be used to assess the relative performance of units (such as factories, retail stores, or passport offices) within a single organization, it can also be used to assess the relative performance of similar units across multiple organizations. A DEA-based comparison of the relative efficiency among competitors' sites can establish which DMUs are operating on the best-practice frontier and the potential performance gains (either reduced use of inputs for given levels of outputs or greater production of outputs for given levels of inputs) for the DMUs not on this frontier. The management of lower-performing DMUs would be expected to adopt, to the extent possible, the operating structures and practices of the higher-performing DMUs. Limitations in the availability of technology, specialized knowledge (e.g., trade secrets), or market share/scale of operations may prevent lower-performing DMUs from emulating the higher-performing DMUs. In such situations, mergers and acquisitions may be the route to improvement across the broad array of competitors, but other partnership or alliance arrangements between firms may yield an alternate route.

Amin and Ibn Boamah (2023) presented novel applications of DEA to identify performance gains that could be achieved via partnerships between DMUs. These authors utilized an illustrative example drawn from the Canadian banking industry. This chapter will provide an example of an adaptation in DEA to constrain the production frontier that can be incorporated into both the approach used by Amin and Ibn Boamah and other applications of DEA for carrying out assessments of relative performance across DMUs. The chapter begins with a brief explanation of the traditional DEA methodology, followed by a discussion of some limitations in this approach. In an effort to overcome these limitations, the authors present the use of assurance regions to refine the best-practice frontier in a manner consistent with managerial objectives in order to compare the relative performance of DMUs. An illustrative example drawn from healthcare of DEA with assurance regions that examines the delivery of primary care services by pediatricians in the United States is provided.

12.2 EVALUATING PRIMARY CARE PEDIATRICIANS USING DATA ENVELOPMENT ANALYSIS

To address concerns about increasing expenditures for healthcare services, the Health Maintenance Organization (HMO) form of managed care was implemented in the United States, whereby under a prepaid arrangement, all of the healthcare services needed by a defined population would be provided through the HMO's delivery network of physicians and hospitals (Falkson and Srinivasan, 2023). The prepaid nature of this arrangement was a key component as it established for the HMO up front the pool of funds available to pay for the clinical care services that would be mobilized to provide effective care for the defined population. This incentive for efficiency for HMOs was in contrast to the fee-for-service reimbursement of clinical care services that was the dominant mode in the United States in the 20th century with physicians and hospitals earning more revenues for providing a higher volume of services.

It is well-established that physicians treating patients through inefficient clinical practice patterns have important consequences for the cost and quality of health care. There exists a sizeable body of evidence that physicians vary widely in their use of clinical (e.g., office visits) and technical resources (e.g., tests and hospitalizations) for treating patients with similar problems, even after case mix and severity of the patient's condition have been accounted for (Wennberg & Gittelsohn, 1982; Eisenberg, 1986). Eisenberg maintains that resource utilization by physicians depends, at least in part, on practice styles and that "physicians adopt different styles of practice as a result of their different attitudes, values, tastes, and habits" (Eisenberg, 1986, p. 463). These studies point out that treatment patterns or styles of practice that result in unnecessary tests, surgical procedures, and days in the hospital contribute to higher costs even in "efficient" managed care organizations such as HMOs.

If a well-managed HMO is to have a cadre of physicians practicing the most cost-effective approach to patient care, an important issue for HMO managers is learning how to evaluate physician practice styles so HMO physicians can manage resources more efficiently. Given that variations in practice styles have marked economic consequences for HMOs, an essential task for HMO managers is to foster within the physician ranks practice styles with a strong orientation toward the basic service concept of wellness promotion through primary care and preventive medicine. To connect that service concept to an operating strategy aimed at the economical use of medical services, HMO managers must have analytic tools available to measure and evaluate the relative efficiency of physicians in the context of such practice styles. Only then can these managers properly target initiatives to improve efficiency of care and the financial viability of HMOs.

Since children generally receive primary care either from a pediatrician or a general practitioner (Fishbane & Starfield, 1981), successful HMOs use primary care pediatric services carefully and economically. By stressing efficiency as well as quality, HMO pediatricians have the potential to contain costs by reducing the quantity of unnecessary or excessive services received by their patients. Although efficient and timely pediatric care patterns are essential to an operating strategy aimed at controlling utilization and quality, only crude techniques are available to analyze the efficiency of primary care pediatricians.

Evaluating a primary care pediatrician's efficiency requires an ability to find "best practices"—i.e., the minimum set of inputs to care for a given panel of young patients. Since the mathematical relationship between primary care inputs (i.e., medical resources) and multiple outputs (i.e., the number of patients treated, by age and gender) is unclear, a method of analyzing efficiency and the nature of efficient relationships is needed. DEA, a non-parametric technique developed by Charnes *et al.* (1978), employs linear programming to measure the magnitude of departure from production efficiency frontiers for each DMU[1] based on their use of resources to produce multiple outputs.

In DEA, the unit of analysis is the DMU, which in health care applications may be defined as narrowly as an individual physician or group of physicians, or as broadly as a hospital, nursing home, or physician organization.[2] To use DEA, the comparison group of DMUs must be homogeneous—i.e., they must use "similar" inputs to produce "common" outputs. Accordingly, it is improper to use DEA when comparing the efficiency of different types of DMUs facing different task environments. For example, a comparison of surgeons and interns is not a valid use of DEA, since these physicians treat dissimilar patients with different techniques (see Chilingerian & Sherman, 1990). Health care researchers can use the DEA method to conduct in-depth studies on the utilization behavior of primary care physicians (PCPs). While it is customary to use DEA to evaluate the relative efficiency of one physician versus the others, in this chapter, the authors will illustrate how DEA can also be used to explore and compare the practice styles of physicians.

In particular, the authors have developed an empirical demonstration of how such techniques can be used to identify and analyze practice styles being used by a sample group of pediatricians. The study also presents a novel method and approach for evaluating the efficiency of physicians practicing within particular practice styles of interest to an HMO's managers. These ideas are illustrated in an evaluation of the clinical efficiency of 86 pediatricians in an independent practice association (IPA) HMO. First, a standard DEA model is used to estimate which pediatricians are on the best practices production frontier (BPPF) and which are off the frontier. Next, a second version of the DEA model is used to limit the frontier to regions that are relevant to the management of an HMO, and the pediatricians are reevaluated according to this limited frontier.

Next in this chapter, we provide a brief explanation of the traditional DEA methodology, followed by a discussion of some limitations in this approach. In an effort to overcome these limitations, the authors present a novel modification to DEA—"quality" assurance regions. This modified approach is then applied to an actual HMO's utilization data. The results of this analysis are compared to the standard DEA approach, followed by a discussion of future avenues of research.

12.3 HOW DEA WORKS[3]

DEA relies on the Pareto-Koopmans definition of efficiency, a derivation of the social welfare criterion of Pareto optimality. To construct the "best practices" production frontier, observed behavior is evaluated by using the following input-output criteria:

a. A DMU is not efficient if it is possible to decrease any input without increasing any other input and without decreasing output;

b. A DMU is not efficient if it is possible to increase any output without decreasing any other output and without increasing any input.

A DMU will be characterized as perfectly efficient only when both criteria are satisfied (Charnes *et al.*, 1978).

A brief example with a small group of pediatricians illustrates how DEA can be used to define a BPPF for physicians. Consider 14 primary care pediatricians who cared over one year for 1,000 female children 1–4 years old from a homogeneous socio-economic background. Table 12.1 summarizes one year of input and output data for each of the primary care pediatricians, and Figure 12.1 plots the input data {physician office visits and days of hospitalization for medical and surgical care}.

TABLE 12.1 Hypothetical Input and Output Values for 14 Physicians

Physician	Panel Size	Office Visits	Med-Surg Days
D_{01}	1,000	4,000	121
D_{02}	1,000	6,300	169
D_{03}	1,000	7,300	218
D_{04}	1,000	2,400	175
D_{05}	1,000	5,500	440
D_{06}	1,000	5,751	434
D_{07}	1,000	4,800	241
D_{08}	1,000	4,800	188
D_{09}	1,000	3,000	385
D_{10}	1,000	6,000	325
D_{11}	1,000	8,300	310
D_{12}	1,000	1,200	480
D_{13}	1,000	7,000	80
D_{14}	1,000	1,400	380

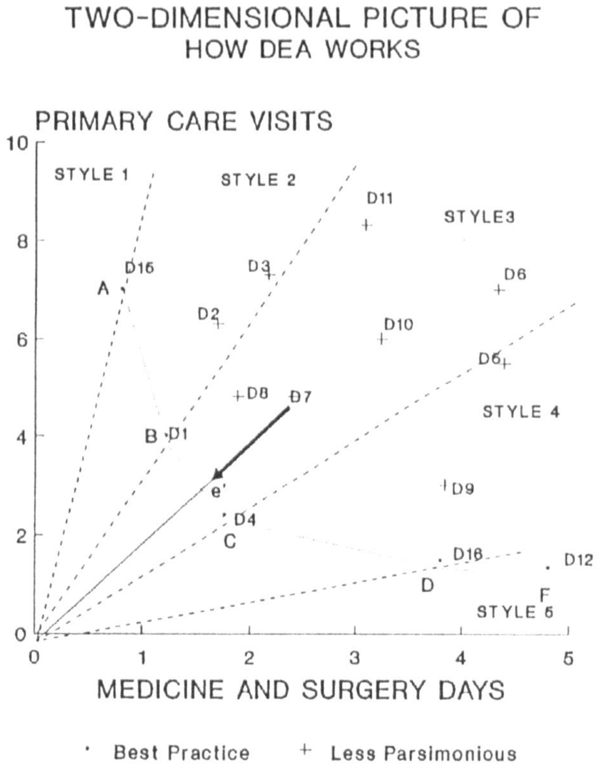

FIGURE 12.1

The Pareto–Koopmans efficiency criteria suggests that any pediatrician in Figure 12.1 who was lower and to the left of another pediatrician is more parsimonious because he or she used fewer clinical inputs to care for the "same" type of children. By floating a piecewise linear surface on top of the actual observations, Figure 12.1 also plots a BPPF for each of the 14 pediatricians. As shown in Figure 12.1, physicians D_{15}, D_1, D_4, D_{16}, and D_{12} dominate D_2, D_3, D_8, D_7, D_{11}, D_{10}, D_9, D_5, and D_6. Accordingly, D_{15}, D_1, D_4, D_{16}, and D_{12}, which are not dominated by any other DMUs, lie on a BPPF and have been designated dominant frontier points A, B, C, D, and F, respectively.

DEA defines a best practices frontier by constructing a set of piecewise linear curves, such that any point on the BPPF is a weighted sum of observed DMUs (Charnes *et al.*, 1978). DEA measures the relative efficiency of each physician by the relative distance of that physician from the frontier. Physicians on the frontier (points A, B, C, D, and F) have a score of 1 and physicians off the frontier have a value between 0.0 and 1.0. For example, in Figure 12.1, the efficiency score of D_7, who was less parsimonious than D_1 and D_4, can be measured by constructing a line segment from the origin to point D_7. This line segment crosses the BPPF at point e'. The efficiency rating of D_7 is equal to the ratio of the length of line segment Oe' to the length of OD_7. Since the numerator Oe' is less than the denominator OD_7, the efficiency rating of physician D_7 will be less than 1.0.

12.4 SOME PROBLEMS ON THE BEST PRACTICES FRONTIER

Each physician adopts a "style of practice"[4] that can be represented by the mix of clinical inputs used to treat patients. The geometry of convex cones provides a pictorial language for interpreting the practice style of physicians.[5] In mathematics, a cone is a collection of points such that if a physician P is at any point in the collection, then μP is also in the collection for all real scalars ≥ 0 (Charnes, 1953). However, cones can also be interpreted as a linear partition of physician practice styles based on a set of linear constraints such as a range of substitution ratios.

For example, in Figure 12.1, the ray that begins at the origin (0, 0) and intersects the frontier at D_1 and the ray that begins at the origin (0, 0) and intersects the frontier at D_4 define the upper and lower boundaries of a cone, labeled practice style 3. All points contained on the line that begins at the origin and intersects the frontier at D_1 have the same ratio of medicine and surgery days to primary care office visits. Therefore, a physician practicing inside the "style 3" cone will have a ratio of visits to hospital days that falls between the line that begins at the origin and intersects the frontier at D_1 and the line that begins at the origin and intersects the frontier at D_4.[6]

Cones can also be used to distinguish different practice styles of physicians. For example, practice style 2 represents physicians who use relatively more office visits than hospital days. In contrast, practice style 5 represents physicians who use relatively more medical-surgical days than primary care office visits. The standard DEA model can be used to locate cones or natural clusters of DMUs based on the optimal weights assigned by the linear program. The next section will illustrate how the new DEA concept of assurance regions can be used to identify and investigate practice styles that are acceptable to and desired by HMO managers.

12.5 THE CONCEPT OF A PRACTICE STYLE ASSURANCE REGION

Since the largest single expense item for an HMO is hospitalizations, it is easy to show that a lower input utilization rate for hospital days represents a more desirable operating strategy given good outcomes for patients. Physicians who succeed in substituting primary care visits and ambulatory surgeries for medical-surgical days in hospital will contribute to a more efficient delivery system. Understandably, most HMO managers would prefer their physicians to reduce hospital days and increase their primary care visits.

Microeconomics offers health care managers a way to quantify the clinical guideline that PCPs should trade-off medical-surgical days (input X_2) for more primary care visits (input X_1) while maintaining a constant output rate. Economists refer to the change in medical-surgical days (ΔX_2) per unit change in primary care visits (ΔX_1) as the marginal rate of technical substitution—i.e., $-dx_2/dx_1$, which equals the marginal product of primary care visits (ω_v) divided by the marginal product of medical-surgical days (ω_h).[7]

In Figure 12.1, every physician on the BPPF practices an input-minimizing style of care—i.e., their DEA efficiency score = 1. However, not all of these physicians on the BPPF are practicing minimum-cost care. In other words, the mix of inputs used by some of these physicians is not the most cost effective. Consider physician D_{12} in practice style 5 and physician D_{16} in practice style 4. Since D_{16} can care for the same patients with fewer hospital days, cost-conscious HMO managers would probably want their PCPs practicing to the left of style 5.

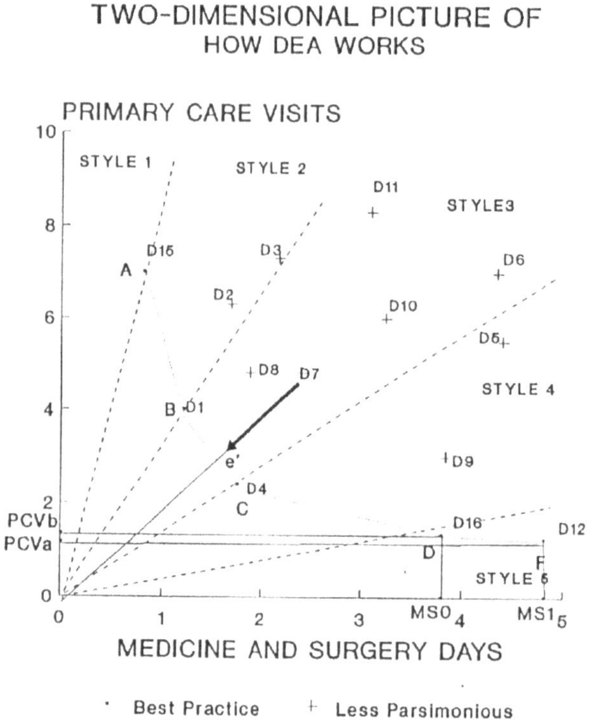

FIGURE 12.2

Now, consider the BPPF in Figure 12.2, which plots the same data from Table 12.1. If the quantity of medical-surgical days is increased from MS_0 to MS_1, the same panel of patients will be cared for with slightly fewer of the less expensive primary care visits and much more of the expensive hospital medical and surgical days. This points out a weakness in using DEA to estimate the BPPF. From an HMO's perspective, although the practice style of physician D_{12} represents an undesirable practice region, the standard DEA model would allow D_{12} to be on the BPPF.

Implementing assurance regions as an extension of the standard DEA model allows managers to specify bounds on the ratio of inputs such as the ratio of hospital days to primary care office visits (Thompson *et al.*, 1990). This two-stage approach to DEA bounds the optimal weights (or marginal productivities) and narrows the set of technically efficient behaviors. By incorporating available managerial knowledge into a standard model, an assurance region truncates the BPPF frontier—i.e., tightens the production possibility set—and opens up the potential to find more inefficiency (Thompson *et al.*, 1990).

Through the addition of a new constraint to the linear program, some of the DEA scores and the projected adjustments may change in a direction in which medical-surgical days are reduced more than primary care visits.[8] The effect of adding this constraint is to present an operating strategy that is more palatable to the HMO manager.

To state the research hypothesis more formally, the DEA score for any physician not originally located within an acceptable practice style cone would decline under an assurance region model. In addition, an assurance region model should find a higher quantity of excess hospital days, but a lower quantity of excess office visits than the standard Charnes, Cooper and Rhodes (CCR) model. An example of a clinical application of the assurance region approach is discussed in the next section.

12.6 METHODS

12.6.1 The Study Population

This study was conducted at Alpha Health Plan,[9] a large IPA with a network of 3,000 physicians located on the east coast of the United States. Unlike a pure IPA where the physician contracts with the HMO, Alpha contracted with individual provider units each composed of several physicians. As a demographically based risk-sharing unit, each provider unit was given a capitated, per member per month budget. At year's end, if a provider unit was under-budget, the surplus was shared between Alpha and that provider unit. The provider unit decided how to distribute its share of the surplus among its physicians.

The data on which this study is based included physician-level observations of every prepaid primary care service provided to children (except telephone calls and prescription drugs) for a 12-month period in 1990. The physicians comprising the study population were all PCPs who had been associated with Alpha Health Plan for at least one full year and who had average panel sizes of 50 or more. Preliminary cleaning of the data resulted in deletion of physicians who dropped out of the plan or dropped their primary care status during 1990. The study population was further limited to board-certified pediatricians. After applying these criteria, the study population included 86 physicians.

The mix of patients in a panel (sometimes referred to as the PCP's case mix) influences the utilization of clinical and technical resources. According to the literature, the utilization of primary care resources is viewed as an artifact of the various medical care experiences of the panel populations. Studies have demonstrated that risk profiles of a panel can be captured (by proxy) by the age and gender mix. Studies have found that three variables largely drive HMO costs: age, gender, and geographic location. Consequently, most HMOs set their prices based on these variables (InterStudy Edge, 1989). Given that the patients were located in a homogeneous service delivery area, only age and gender were used in this study to control for case mix.[10]

12.6.2 Defining Clinical Inputs and Outputs

Defining the inputs and outputs of each PCP is central to the development of the conceptual framework for this study. Each panel member is entitled to and guaranteed to receive the delivery of one year of comprehensive primary care. For a PCP, the outputs are the quantity of children cared for during the year. Comparing physician utilization patterns requires controlling for the case mix. The output mix (i.e., panel of patients) can be measured by the quantity of patients in each age-sex bucket. Thus, the delivery of primary care services, as it is used here, refers to utilization of primary care visits, tests, hospital days, and ambulatory surgeries to care for a given panel of children.

Table 12.2 presents the mean values for the four clinical input measures and seven patient output measures used to construct the frontier analyses of 86 pediatricians at Alpha HMO. Both the standard CCR and applied

TABLE 12.2 Summary of Input and Output Measures for 86 Physicians

Variable	Mean	Std. Dev.	Minimum	Maximum
Outputs				
Females/Males Ages < 1	14.08	7.74	0.75	35.92
Females/Males Ages 1–4	55.91	27.12	2.75	122.75
Females/Males Ages 5–15	106.76	58.93	32.50	257.92
Females Ages 16–19	8.77	8.04	0.00	52.92
Males Ages 16–19	9.32	9.74	0.00	62.92
Females Ages ≥ 20	2.25	4.41	0.00	28.83
Males Ages ≥ 20	2.55	5.34	0.00	43.00
Inputs				
Hospital Days Provided	24.45	22.91	0.00	92.70
Ambulatory Surgeries Performed	3.52	3.15	0.00	15.82
Office Visits Provided	936.29	481.73	228.98	2,429.56
Tests Performed	886.07	480.69	178.85	2,602.39

Note: Tests Performed = All Diagnostic, Radiologic, and Laboratory Tests Performed.

assurance region models used this same set of measures. On the output side, the majority of patients are found in the lower age groups which is the expected result given the fact that the physicians being evaluated are pediatricians. In fact, some physicians have no patients in the upper age categories. The separation of male and female patients in the two oldest output groups allows for potential gender-specific differences in health care utilization arising with the onset of adolescence.

Discussions with the medical director at Alpha Health Plan revealed that the four inputs listed in Table 12.2 are considered the essential set of clinical resources available to the pediatrician for managing patient care. These four inputs were thus used to construct the two models. A health maintenance organization such as Alpha has a primary care orientation where patient contacts with PCPs are promoted (office visits) in order to reduce the necessity of employing more technical clinical resources (hospitalizations, ambulatory surgeries, and tests). However, HMOs prefer their physicians to substitute ambulatory surgery for full hospitalizations when possible. The mean input measures presented in Table 12.2 are not surprising given the primary care orientation of the physicians being evaluated. Indeed, examination of the minimum values column in Table 12.2 indicates that some physicians managed to care for their patients in the study year without using any hospital days or ambulatory surgeries. Further discussions with the medical director revealed that pediatricians' practice styles should try to stay within average hospital utilization rates, without holding back on office visits (see Appendix A for a detailed explanation of how "acceptable" practice regions were specified and incorporated into DEA).

A two-stage DEA approach was employed in this analysis. First, a standard (unbounded) DEA model was run to obtain optimal weights. Based on the medical director's suggestions (see Appendix A), these weights were used to define a "quality" assurance region in the second model. For both stages, a seven-output four-input DEA model was used to locate the primary care pediatricians on the BPPF based on their observed utilization behavior in 1990. A comparison of DEA findings with and without an assurance region is reported in the next section.

12.7 RESULTS

In this analysis, the standard CCR model was used to construct a BPPF for 86 Alpha HMO pediatricians. As described above, this frontier can be divided into practice style cones. The performance efficiency of each pediatrician is then evaluated relative to the other pediatricians located within the same practice cone. In Table 12.3, the standard CCR model

TABLE 12.3 DEA Efficiency Scores Under Both Models

Physician	Standard CCR Model Theta Score	Assurance Region Model Theta Score	Theta New-Theta Old
1	0.639	0.621	−0.018
2	0.735	0.733	−0.002
3	1.000	1.000	0.000
4	1.000	1.000	0.000
5	0.984	0.654	−0.329
6	1.000	1.000	0.000
7	1.000	1.000	0.000
8	0.963	0.729	−0.234
9	1.000	1.000	0.000
10	0.665	0.665	0.000
11	0.633	0.629	−0.004
12	0.768	0.765	−0.003
13	1.000	1.000	0.000
14	0.628	0.580	−0.048
15	1.000	1.000	0.000
16	1.000	1.000	0.000
17	0.637	0.637	0.000
18	0.468	0.468	0.000
19	0.528	0.499	−0.029
20	0.808	0.801	−0.006
21	0.889	0.889	0.000
22	0.821	0.786	−0.035
23	1.000	1.000	0.000
24	1.000	1.000	0.000
25	1.000	0.989	−0.011
26	1.000	1.000	0.000
27	0.801	0.801	−0.001
28	1.000	1.000	0.000
29	0.665	0.652	−0.013
30	0.800	0.654	−0.146
31	0.860	0.726	−0.133
32	0.853	0.836	−0.017
33	0.751	0.751	0.000
34	1.000	1.000	0.000
35	0.502	0.496	−0.007
36	0.719	0.705	−0.013
37	1.000	1.000	0.000
38	0.565	0.565	0.000
39	0.481	0.463	−0.017

(Continued)

TABLE 12.3 (Continued)

Physician	Standard CCR Model Theta Score	Assurance Region Model Theta Score	Theta New-Theta Old
40	1.000	1.000	0.000
41	1.000	1.000	0.000
42	0.861	0.820	−0.041
43	0.486	0.449	−0.037
44	1.000	1.000	0.000
45	1.000	1.000	0.000
46	0.663	0.654	−0.009
47	1.000	1.000	0.000
48	0.921	0.878	−0.043
49	0.788	0.788	0.000
50	1.000	1.000	0.000
51	0.920	0.912	−0.008
52	1.000	1.000	0.000
53	0.532	0.460	−0.072
54	0.668	0.665	−0.003
55	0.501	0.444	−0.057
56	0.579	0.579	0.000
57	0.590	0.589	−0.001
58	0.846	0.844	−0.002
59	0.388	0.356	−0.033
60	1.000	1.000	0.000
61	0.613	0.603	−0.011
62	0.621	0.617	−0.004
63	0.617	0.588	−0.029
64	0.722	0.673	−0.049
65	0.935	0.911	−0.024
66	0.589	0.525	−0.064
67	0.854	0.850	−0.004
68	1.000	1.000	0.000
69	1.000	1.000	0.000
70	1.000	1.000	0.000
71	1.000	1.000	0.000
72	1.000	1.000	0.000
73	1.000	1.000	0.000
74	0.642	0.625	−0.017
75	0.640	0.494	−0.146
76	1.000	1.000	0.000
77	0.797	0.797	0.000
78	0.515	0.512	−0.002

(Continued)

TABLE 12.3 (Continued)

Physician	Standard CCR Model Theta Score	Assurance Region Model Theta Score	Theta New-Theta Old
79	1.000	1.000	0.000
80	0.587	0.587	0.000
81	0.514	0.404	−0.110
82	0.889	0.889	0.000
83	1.000	1.000	0.000
84	1.000	1.000	0.000
85	0.794	0.787	−0.006
86	0.597	0.568	−0.029

Standard CCR Model: 33 Physicians on BPPF
Assurance Region Model: 32 Physicians on BPPF

determined that 33 of the physicians were located on the BPPF for a DEA (theta) score of 1. The remaining 53 physicians were located off the BPPF and thus had DEA scores between 0 and 1.

Holding output constant, the standard CCR model is able to estimate the amounts of clinical inputs required to produce the given level of output if all 86 PCPs were operating on the BPPF. This information is presented in column B of Table 12.4. The first column in Table 12.4 is the actual levels of inputs used by the 86 PCPs during the year under study. Under the standard CCR model, the difference between projected input utilization (column B) and actual input utilization (column A) represents the potential resource savings available to the HMO (Column C). The data in Table 12.4 indicate that the HMO could potentially care for the same group of patients with reduced quantities of each input, i.e., actual utilization exceeded projected utilization for all four inputs.

As expected, the DEA scores under the assurance region model tended to show lower efficiency. There were 32 physicians classified as efficient (DEA score = l), compared to 33 for the standard CCR model. No pediatrician had a higher DEA score under this model than the standard CCR model. Indeed, the scores of 43 physicians declined.

Column D in Table 12.4 presents the input utilization predicted by the assurance region model if all 86 physicians were located on the truncated BPPF, holding output constant. The difference between this predicted input utilization and the actual input utilization (Column A) for the four clinical inputs is found in Column E. Once again, all four inputs were over-utilized by HMO pediatricians, indicating potential savings to the HMO. However, a comparison of Column E with Column C (standard CCR model) reveals

TABLE 12.4 Estimated Inefficiency of Resource Utilization

Inputs	**Aggregate Observed Resource Utilization**	**Standard CCR Model Projected Utilization if All physicians Efficient**	**Standard CCR Model Potential Resource Savings**	**Assurance Region Model Projected Utilization If All physicians Efficient**	**Assurance Region Model Potential Resource Savings**	**Assurance Region model Proj. Util.—Standard CCR model Proj. Util.**
	(A)	(B)	(C-B-A)	(D)	(E = D-A)	(F = D-B)
HOSP	2,103.83	1,347.71	−756.12	1,296.78	−807.05	−50.93
AMBSURG	302.47	198.12	−104.35	206.84	−95.63	8.72
VISITS	80,520.57	59,932.36	−20,588.21	62,261.90	−18,258.67	2,329.54
TESTS	76,201.70	56,073.69	−20,128.01	55,636.56	−20,565.14	−437.13

HOSP = Total Number of Hospital Days for All Patients
AMBSRG = Total Number of Ambulatory Surgeries Performed for All Patients
VISITS = Total Number of Office Visits Provided to All Patients
TESTS = Total Number of Diagnostic, Radiologic, and Laboratory Tests Performed for All Patients

the impact of applying an assurance region to the BPPF. Under the assurance region model, inefficiency in the use of hospital days is determined to be greater (807 excess days vs. 756 excess days), yet inefficiency in the use of office visits is reduced (18,259 excess visits vs. 20,588 excess visits). The assurance region model projected the utilization of 51 fewer hospital days, but 2,330 more office visits than the standard CCR model (see Column F of Table 12.4). The imposition of the assurance region on the standard CCR model resulted in an appropriate adjustment (per management goals) in the relative evaluations of the 86 pediatricians.

Table 12.5 provides a list of seven pediatricians whose DEA efficiency scores changed the most notably following the application of the assurance region. In particular, physician number 25 ended up slightly off the BPPF so that this pediatrician's theta score declined to 0.989 from an initial value of 1. Table 12.6 details the estimated inefficiency of these seven pediatricians under the two models. Examination of Column D reveals that for every one of these pediatricians, the assurance region model projected a greater excess utilization of hospital days than the standard CCR model. On the other hand, the standard CCR model projected a higher excess utilization of office visits for these seven physicians than the assurance region model. In fact, the assurance region model determined that three pediatricians—numbers 5, 8, and 25—actually underutilized office visits in caring for their patients (Columns C and A). In other words, these physicians did not perform as many office visits as they should have.

This last result points to the potential power of the assurance region technique. Rather than simply equating efficiency in clinical care to parsimonious use of inputs, the application of assurance regions to the standard CCR model may allow HMO managers to identify and measure

TABLE 12.5 Changes in Efficiency Rating Under Assurance Region

Physician	Standard CCR Model Theta Score	Assurance Region Model Theta Score	Theta New—Theta Old
5	0.984	0.654	−0.329
8	0.963	0.729	−0.234
25	1.000	0.989	−0.011
30	0.800	0.654	−0.146
31	0.860	0.726	−0.133
75	0.640	0.494	−0.146
81	0.514	0.404	−0.110

TABLE 12.6 Differences in Projected Efficiency Under Both Models

Physician	Inputs	Aggregate observed resource utilization	Standard CCR Model Projected Utilization if Efficient	Assurance Region Model Projected Utilization if Efficient	Assurance Region Model Proj. Util.– Standard CCR Model Proj. Util.
		(A)	(B)	(C)	(D = C-B)
5	HOSP	71.78	31.52	29.28	−2.24
	AMBSURG	7.73	7.60	5.06	−2.54
	VISITS	828.74	815.12	1,017.62	202.50
	TESTS	1,406.13	1,064.98	919.78	−145.20
8	HOSP	44.84	23.52	21.91	−1.61
	AMBSURG	2.93	2.30	2.14	−0.16
	VISITS	649.85	625.89	763.78	137.89
	TESTS	983.67	678.50	717.04	38.54
25	HOSP	11.02	11.02	4.80	−6.22
	AMBSURG	2.96	2.96	2.75	−0.21
	VISITS	271.99	271.99	433.15	161.16
	TESTS	374.78	374.78	370.76	−4.02
30	HOSP	69.02	39.36	33.37	−5.99
	AMBSURG	5.72	3.95	3.74	−0.21
	VISITS	1,167.95	934.42	1,080.92	146.50
	TESTS	1,743.53	1,076.88	1,140.55	63.67
31	HOSP	15.93	13.69	7.70	−5.99
	AMBSURG	3.83	3.26	2.04	−1.22
	VISITS	585.93	503.64	529.71	26.07
	TESTS	693.83	552.79	504.02	−48.77
75	HOSP	57.84	16.55	14.11	−2.44
	AMBSURG	3.89	0.81	1.92	1.11
	VISITS	1,075.86	688.09	745.23	57.14
	TESTS	1,045.65	568.83	516.42	−52.41
81	HOSP	36.94	13.21	12.41	−0.80
	AMBSURG	4.96	1.99	2.01	0.02
	VISITS	429.94	220.94	240.69	19.75
	TESTS	835.89	324.92	337.30	12.38

HOSP = Total Number of Hospital Days for All Patients; AMBSRG = Total Number of Ambulatory Surgeries Performed for All Patients; VISITS = Total Number of Office Visits Provided to All Patients; TESTS = Total Number of Diagnostic, Radiologic, and Laboratory Tests Performed for All Patients deviations from acceptable practice styles. These managers could then move to improve not just efficiency, but practice style quality as well.

12.8 DISCUSSION AND CONCLUSIONS

This chapter has illustrated how, with the help of a standard DEA CCR model, the best practice production frontier (BPPF) for physicians can be divided into different conic regions which can be interpreted as different physician practice styles. From the HMO manager's perspective some of these practice styles are more cost-effective and/or clinically appropriate than others. However, the standard DEA model does not incorporate such information into either the construction of the BPPF or the relative evaluations of physician performance. Therefore, the standard CCR model informs the HMO manager how efficiently such a subset of physicians is pursuing the wrong practice style. This is far from the information a manager needs to know and could be damaging to an HMO from a service delivery and competitive standpoint.

The goal of this research, then, was to address this problem by imposing assurance regions on the standard CCR model, thereby truncating the BPPF to practice styles acceptable both to clinicians and HMO management. The performance of the HMO's pediatricians would then be evaluated only relative to pediatricians practicing within a range of preferred HMO practice styles which emphasized the use of office visits with PCPs and de-emphasized hospitalizations. The assurance region for the second stage of DEA was constructed accordingly as described above. The hypothesis was that the DEA score for any pediatrician not originally located within an acceptable practice style cone would decline under the assurance region model. In addition, given the constraints imposed in the assurance region model, this model should have found a greater level of excess use of hospital days, but a lower level of excess use of office visits than the standard CCR model.

In an empirical test on 86 pediatricians, slightly more than one-third (32 out of 86) were on the best practice production frontier (BPPF). The slack associated with the pediatricians not practicing on the BPPF suggests that a substantial amount of savings could be achieved if every pediatrician practiced exactly like the 32 pediatricians on the frontier. Relying on estimates of a bounded DEA model, it was shown that more than a proportional savings could come from reducing expensive hospital days and tests rather than reducing ambulatory surgeries and office visits. During the second stage of the analysis (when the DEA model was run with an assurance region), the DEA efficiency scores of 43 pediatricians declined; moreover, six pediatricians' scores declined by greater than 10%. Hence, DEA succeeded in identifying a small group of physicians whose practice styles should be monitored more closely.

This finding not only supports the notion that there are variations in the practice style of this group of pediatricians but also that these variations have consequences for the cost and quality of delivering health care to the children who are members of the HMO. DEA provided some quantitative evidence that these pediatricians were caring for the children in their panels by using more medical-surgical days in hospital than primary care visits. Health management researchers need to pay attention to the rates of substitution implied by the so-called "efficient" or "best practicing" DMUs on the frontier.

While these results provide a preliminary exploration of how a DEA evaluation with an assurance region can sharpen an analysis of "best practices," several limitations are noted here. First, the study was conducted in only one HMO employing a sampling rule that only looked at one year of data. Second, the results of this study are very dependent on the choice of inputs and outputs, and the way that DEA measures efficiency. The researchers relied upon the expert advice of the medical director and the service concept of " gate keeping" to guide the choice of inputs and outputs. Third, although a standard CCR DEA estimate presumes that inefficiency or excess utilization has been found, in fact only parsimony has been found. A highly focused follow-up study by clinician managers would be needed before the departures from the BPPF could be called utilization of unnecessary tests, procedures, or hospital days. In other words, only further research will determine whether or not DEA is producing clinically meaningful information at the physician level. Lastly, when better measures of quality of care and health status of panels are developed, HMO managers will be able to argue that they are locating best practices because output is being measured in constant quality terms.

Despite these limitations, a DEA model with an assurance region (or regions) appears to provide a useful way to frame the issue of primary care practice style. DEA can be used as an attention-directing managerial device to identify styles of practice based on substitution of inputs. However, as was pointed out, more research is needed on what are reasonable upper and lower bounds—i.e., what are the correct marginal rates of substitution? In any case, using DEA with an assurance region appears a promising approach for future research in healthcare.

Healthcare expenditures and activities typically represent a sizable portion of a country's economic activity. The production of clinical care services and the essential products of the medical supply chain depend on the mobilization of a wide array of personnel and technologies and entail interactions and ongoing relationships across multiple organizations. Both horizontal integration (groups of hospitals, nursing homes, vision care centers, pharmaceutical firm mergers, and so forth) and vertical integration (integrated delivery networks, health systems with community-based care and institution-based care sites, mergers of manufacturing and distribution entities in the medical supply chain, and more) can be readily observed in healthcare.

Improving the information used to make decisions about mergers, joint ventures, and partnerships can generate better paths for generating more efficient, high-quality healthcare while also devoting more attention to sustainability (The Joint Commission, n.d.). The DEA models with assurance regions can be a valuable tool for such efforts.

12.8.1 Applications Beyond Healthcare

Two-stage DEA incorporating an assurance region certainly can be a useful approach beyond healthcare, as evidenced by the initial work in agriculture by Thompson *et al.* (1990). Pitchipoo *et al.* (2013) used DEA as a tool for suppler selection in the chemical processing industry where suppliers were evaluated in terms of their efficiency in converting three inputs into two outputs. The inputs were days for delivery, quantity of goods delivered, and days of warranty protection provided. The outputs were cost (total cost of goods delivered) and quality (percentage of good delivered with acceptable quality). In an analogous manner to the illustrative example of PCPs described above, a second-stage DEA model incorporating assurance regions could have been applied. For example, desirable characteristics of potential suppliers, such as ability to deliver greater than average quantities of good in a shorter than average delivery time, could be used to establish an assurance region with two-stage DEA.

There is potential to look beyond manufacturing to production of services as well in terms of employing two-stage DEA that incorporates an assurance region. For example, Athanassopoulos (2004) examined the efficiency in providing public services of branch offices of the Hellenic Tobacco Organization in Greece. Each branch is located in a tobacco-producing region of Greece and is responsible for a range of activities that center on supporting the production of tobacco to meet production quotas and distributing European Union subsidies to local farmers. A DEA model with four inputs and six outputs was used to assess the efficiency of the branch offices as the DMUs, with the specific inputs as follows: (1) supervisory staff with higher education degree, (2) technical staff, (3) administrative staff, and (4) technology infrastructure (Athanassopoulos, 2004). A plausible goal for management might be reducing the level of (expensive) supervisory staff by making expanded investments in technology infrastructure (e.g., information technology and artificial intelligence systems). DEA with an assurance region for these two inputs consistent with this management goal will identify the branch offices most efficient which are employing the desired tradeoff between these two inputs, similar to the practice style of physicians desired by the HMO discussed previously in this chapter.

For the production of both goods and services, management increasingly can strive to operate in a more environmentally friendly manner through approaches such as using energy drawn from renewable sources. Accordingly, management may embrace the tradeoff of reducing the use of non-renewable energy by locating operations in areas with higher property costs (rent and real estate taxes) where renewable energy sources are more available. Once again, DEA with an assurance region consistent with the tradeoff between these two inputs could be a valuable approach.

The incorporation of assurance regions in DEA can provide firms with valuable insight to identify from the set of DMUs the potential partners for meaningful resource sharing and collaboration that are achieving efficiency in production at a high level in a manner consistent with management objectives. These objectives can certainly include cost minimization, but others could be minimization of disruptions (reliability) or maximization of sustainable practices. The potential for such applications is notable.

APPENDIX A: SPECIFYING AN ACCEPTABLE PRIMARY CARE ASSURANCE REGION

The assurance region was estimated based on some conversations with the medical director of the Alpha HMO. Discussions with the medical director revealed that (under their somewhat loose clinical guidelines), pediatricians should try to stay within average hospital utilization rates, but not hold back on office visits. An analysis of one year of hospital utilization data and office visit utilization data for the 86 physicians revealed that the mean medical-surgical days per thousand children was 121 days and the mean primary care visits per thousand children was 4,868. The researchers decided that bounds on the marginal rates of substitution (MRSs) for the 86 pediatricians could be developed based on the original DEA estimates of the marginal productivities of the medical-surgical days and office visits.[11] The upper and lower bounds would be based on the MRS of physicians on the BPPF who were: (1) at or below mean medical surgical days per thousand children; and (2) at or above mean primary care visits per thousand children.

Step one was to run the standard DEA model (CCR). Based on the result of the first DEA run, an estimate of the MRS of hospital days (ω_2) to primary care office visits (ω_1) for each physician was obtained. After the model was run, 33 physicians were on the BPPF. Of those 33 physicians, nine physicians had utilized less than 121 medical-surgical days per thousand children and more than 4,868 primary care office visits per thousand children. An assurance region was defined by taking the minimum and maximum MRS for these nine physicians. The MRS values (based on the ratio of the marginal productivity of medical-surgical days ω_h to the marginal productivity of primary care office visits ω_v) for the nine physicians ranged from 27 to 422. This inequality constraint—$27 \leq \omega_h/\omega_v \leq 422$—was added to the standard CCR model as follows:

a. $v_h - 422v_v \leq 0$

b. $v_h - 27v_v \geq 0$

where $\omega_h = v_h$ and $\omega_v = v_v$

The objective of testing this range was to see if DEA could force more than a proportional reduction in hospital days in order to define a practice style assurance region acceptable to the medical director of Alpha HMO.

NOTES

1 This term was introduced by Charnes *et al.* (1978) to refer to entities (individuals, groups, organizations, and so on) that utilize resource inputs to produce outputs.

2 For reviews of DEA applications in health care, see Rosko (1990) and Cantor and Poh (2017).

3 The next section draws heavily on Rosko (1990) and Charnes *et al.* (1978).

4 Practice style, as it used here, refers to the treatment patterns defined by the specific mix of resource inputs used by a physician (or group of physicians) to care for a given mix of patients.

5 Most of this section is taken from published lectures on mathematical theory and linear programming by the late Abraham Charnes (1953).
6 Although the two-input single output problem can be solved graphically, multi-input multi-output problems require a mathematical formulation that can only be solved by using a linear programming model.
7 In economics, holding constant all other inputs, the marginal product of an input is the addition to total output resulting from using the last unit of input. The ratio of the marginal products is the marginal rate of technical substitution, defined as $-dx_2/dx_1$.
8 In the standard DEA model (CCR), the solution estimates the amount of inefficiency by proportionate reduction of inputs. A DEA model with an assurance region no longer estimates the amount of excess input utilization in matching proportions, but in accordance with the range imposed for the marginal rates of substitution.
9 Disguised name for the Health Maintenance Organization.
10 Panel mix can be classified in terms of patient demographics, diagnosis, organ system, severity of illness, and so on. While disease-specific clinical severity measures for office practice are preferable to the crude age and gender measures, individual PCPs rarely have enough cases to use diagnostic groupings or severity classification systems.
11 Estimates of the marginal rates of input substitution (MRS) for these physicians were obtained by dividing the optimal weights for medical surgical days (marginal product of medical surgical days ω_h) by the optimal weights for office visits (marginal product of primary care visits ω_v). The optimal weights were obtained from the solution to the standard (unbounded) DEA model.

REFERENCES

Amado, C.A.F., Santos, S.P., & Marques, P.M. (2012). Integrating the Data Envelopment Analysis and the Balanced Scorecard approaches for enhanced performance assessment, *Omega*, *40*(3):390–403. https://doi.org/10.1016/j.omega.2011.06.006

Amin, G.R. & Ibn Boamah, M. (2023). Modeling business partnerships: A data envelopment analysis approach, *European Journal of Operational Research*, *305*(1):329–337. https://doi.org/10.1016/j.ejor.2022.05.036

Athanassopoulos, A.D. (2004). Using frontier efficiency models as a tool to re-engineer networks of public sector branches: An application to the Hellenic Tobacco Organization, *European Journal of Operational Research*, *154*(2): 533–547. https://doi.org/10.1016/S0377-2217(03)00187-5

Cantor, V. J. M., & Poh, K. L. (2017). Integrated Analysis of Healthcare Efficiency: A Systematic Review. *Journal of medical systems*, *42*(1), 8. https://doi.org/10.1007/s10916-017-0848-7

Charnes, A. (1953). Lectures on the Mathematical Theory of Linear Programming, in Charnes, A., Cooper, W.W. & Henderson, A. (Eds.), *An Introduction to Linear Programming*. New York: John Wiley & Sons.

Charnes, A., Cooper, W.W., & Rhodes, E. (1978). Measuring the efficiency of decision making units, *European Journal of Operational Research*, *2*(6), 429–444. https://doi.org/10.1016/0377-2217(78)90138-8

Chilingerian, J. A., & Sherman, H. D. (1990). Managing physician efficiency and effectiveness in providing hospital services. *Health services management research*, *3*(1), 3–15. https://doi.org/10.1177/095148489000300101

Cook, W.D., & Seiford, L.M. (2009). Data envelopment analysis (DEA) – Thirty years on, *European Journal of Operational Research*, *192*(1), 1–17. https://doi.org/10.1016/j.ejor.2008.01.032

Cook, W.D., Tone, K. & Zhu, J. (2014). Data envelopment analysis: Prior to choosing a model, *Omega*, *44*, 1–4. https://doi.org/10.1016/j.omega.2013.09.004

Dakpo, K.H., Jeanneaux, P. & Latruffe, L. (2016). Modelling pollution-generating technologies in performance benchmarking: Recent developments, limits and future prospects in the nonparametric framework, *European Journal of Operational Research*, *250*(2), 347–359. https://doi.org/10.1016/j.ejor.2015.07.024

Dutta, P., Jaikumar, B. & Arora, M.S. (2022). Applications of data envelopment analysis in supplier selection between 2000 and 2020: a literature review. *Ann Oper Res 315*, 1399–1454. https://doi.org/10.1007/s10479-021-03931-6

Eilat, H., Golany, B., & Shtub, A. (2006). Constructing and evaluating balanced portfolios of R&D projects with interactions: A DEA based methodology, *European Journal of Operational Research*, *172*(3), 1018–1039. https://doi.org/10.1016/j.ejor.2004.12.001

Eisenberg, J. (1986). *Doctors Decisions and the Cost of Medical Care*. Ann Arbor: Health Administration Press.

Falkson, S.R. & Srinivasan, V.N. (2023, March 6). Health Maintenance Organization in *StatPearls* [Internet]. Treasure Island (FL): StatPearls Publishing. https://www.ncbi.nlm.nih.gov/books/NBK554454/

Farrell, M. J. (1957). The Measurement of Productive Efficiency. *Journal of the Royal Statistical Society. Series A (General)*, *120*(3), 253–290. https://doi.org/10.2307/2343100

Fishbane, M., & Starfield, B. (1981). Child health care in the United States: A comparison of pediatricians and general practitioners. *The New England journal of medicine*, *305*(10), 552–556. https://doi.org/10.1056/NEJM198109033051004

Interstudy. (1989) *The InterStudy Edge*. Excelsior, MN: Interstudy.

Panwar A, Olfati M, Pant M, & Snasel V. (2022). A review on the 40 years of existence of data envelopment analysis models: Historic development and current trends. *Arch Comput Methods Eng*, *29*(7), 5397–5426. https://doi.org/10.1007/s11831-022-09770-3

Pitchipoo P., Ragavan R., Venkumar P. and Rajakarunakaran S. (2013). Development of DEA Decision Model for Supplier Selection. *Conference paper from IEEE - International Conference on Research and Development Prospects on Engineering and Technology (ICRDPET 2013)*, EGS Pillay Engineering College, Nagapattinam, India.

Rosko M. D. (1990). Measuring technical efficiency in health care organizations. *Journal of medical systems*, *14*(5), 307–322. https://doi.org/10.1007/BF00993937

Shokri Kahi, V., Yousefi, S., Shabanpour, H. & Farzipoor Saen, R. (2017), How to evaluate sustainability of supply chains? A dynamic network DEA approach. *Industrial Management & Data Systems, 117*(9), 1866–1889. https://doi.org/10.1108/IMDS-09-2016-0389

The Joint Commission. (n.d.). Sustainable Healthcare. https://www.jointcommission.org/our-priorities/sustainable-healthcare/

Thompson, R.G., Langemeier, L.N., Lee, C., Lee, E. & Thrall, R.M. (1990). The role of multiplier bounds in efficiency analysis with application to Kansas farming. *Journal of Econometrics*, *46*(1–2):93–108. https://doi.org/10.1016/0304-4076(90)90049-Y

Wennberg, J., & Gittelsohn, A. (1982). Variations in medical care among small areas. *Scientific American*, *246*(4):120–134. https://doi.org/10.1038/scientificamerican0482-120

Wong, W.P. & Wong, K.W. (2007). Supply chain performance measurement system using DEA modeling, *Industrial Management & Data Systems*, *107*(3): 361–381. https://doi.org/10.1108/02635570710734271

Zhou, P., Ang, B.W., & Poh, K.L. (2008). A survey of data envelopment analysis in energy and environmental studies, European Journal of Operational Research, *189*(1):1–18. https://doi.org/10.1016/j.ejor.2007.04.042

CHAPTER 13

Inverse Data Envelopment Analysis in Mergers

Origin and Recent Development

Gholam R. Amin and Mustapha Ibn Boamah
Faculty of Business, University of New Brunswick, Saint John, Canada

13.1 INTRODUCTION

Data envelopment analysis (DEA), introduced in Charnes, Cooper, and Rhodes (1978), is a linear programming optimization approach for measuring relative efficiency of decision making units (DMUs). It is widely used for estimating potential gains of mergers and acquisitions (MAs) where a set of at least two DMUs combine their resources to create an entity that performs better than the individual DMUs. The conventional DEA has been applied to assess the potential gains of mergers, see Bogetoft and Wang (2005), Lozano and Villa (2010), Xiao et al. (2017), and Ray and Sethia (2022).

These DEA models estimate potential gains for merging DMUs individually and collectively. A common aspect of all DEA models is that they do not reveal the input–output data that is required for a merged entity to realize efficiency targets. Addressing such inverse problems falls outside the capabilities of DEA. Inverse DEA is an appropriate approach for

 DOI: 10.1201/9781003345206-13

determining optimal quantity of input–output data needed for desired efficiency targets. Gattoufi, Amin and Emrouznejad (2014) are the first study that formulated the estimation of merger gains through defining novel inverse problems. They developed inverse DEA models for solving their inverse problems. Since the first publication of inverse DEA on mergers in Gattoufi et al. (2014), the literature has been enhanced both in theoretical foundations and practical applications (Emrouznejad et al. 2023). Amin, Emrouznejad and Gattoufi (2017) extended the concept of merging DMUs to a new concept called generalized restructuring where a set of pre-restructuring DMUs create a new set of post-restructuring DMUs. They developed inverse DEA models to help post-restructuring DMUs determine their input–output data inherited from pre-restructuring DMUs for given efficiency targets. Amin and Al-Muharrami (2018) developed inverse DEA models for estimating mergers gains in the presence of negative data. Amin and Ibn Boamah (2020) developed cost efficiency-based inverse DEA models for estimating potential merger gains. Amin and Ibn Boamah (2023) introduced more general structures than mergers and introduced DEA and inverse DEA mathematical models for various business partnerships. Ibn Boamah and Amin (2024) developed inverse DEA models to determine the best merger partners when there is a list of potential candidates for a merger.

This chapter discusses mathematical basis of existing inverse DEA models for estimating mergers' gains.

13.2 INVERSE DEA

Akin to the relationship between optimization and inverse optimization, the standard DEA determines efficiency score of DMUs given their input–output data, whereas the inverse DEA determines input–output data for given efficiency targets. In inverse DEA, the objective is typically to determine the quantity of inputs and/or outputs that is required to achieve desired efficiency targets. The concept of inverse optimization is very helpful in DEA because efficiency score is an important component of the optimal solution in DEA models and determining the quantity of inputs and/or outputs is of interest to DMUs. Inverse DEA was introduced in Wei et al. (2000) after the development of inverse linear programming in Zhang and Liu (1996). It was further developed in Yan et al. (2002) by applying inverse DEA to resource allocation. The number of publications on inverse DEA was limited until its novel application in estimating merger gains in Gattoufi et al. (2014). Recent years have seen the potential of inverse

DEA in solving challenging post-DEA decision making problems in areas such as resource allocation, supply chain management, greenhouse gas emissions reduction, mergers and acquisitions, and business partnerships (Emrouznejad et al. 2023; and Ibn Boamah and Amin, 2024).

In this section, we explain the concept of inverse DEA using some simple inverse problems. It is worth noting that inverse DEA is not limited to the inverse problems introduced in the existing publications on inverse DEA. In fact, the contribution of a research article on inverse DEA mainly depends on the novelty of its inverse problem(s).

The standard DEA measures relative efficiency of DMUs based on their inputs and outputs. Let us assume there are n DMUs each observed with m inputs and s outputs. Suppose x_{ij} and y_{rj}, respectively, denote the ith input and the rth output of the jth DMU, for each $i = 1,\ldots,m, r = 1,\ldots,s$, and $j = 1,\ldots,n$. The conventional variable return to scale DEA model (Banker et al., 1984) is as follows:

$$\begin{aligned}
&\min\ \theta \\
&s.t. \\
&\sum_{j=1}^{n} \lambda_j x_{ij} \leq \theta x_{ik} \quad i = 1,..,m \\
&\sum_{j=1}^{n} \lambda_j y_{rj} \geq y_{rk} \quad r = 1,..,s \\
&\sum_{j=1}^{n} \lambda_j = 1 \\
&\lambda_j \geq 0 \quad j = 1,..,n
\end{aligned} \tag{13.1}$$

The standard DEA model shown above is our base optimization model. It is worth noting that in writing an inverse DEA, the base optimization model can be any type of DEA model. The DMU under assessment in Model (13.1) is DMU-k $(k = 1,..,n)$. Suppose we apply the above model and obtain the efficiency score of all DMUs. Let us assume F denotes all efficient DMUs on the efficiency frontier. Without loss of generality, assume the existing efficiency frontier remains unchanged in solving the inverse DEA models. Let us assume for a given DMU, let say DMU-p, there is some extra inputs denoted by $\bar{x}_{ip}$ $(i = 1,..,m)$ to be used by this unit to produce as much additional outputs as possible given an efficiency target $\bar{\theta}_p$.

Many research studies assumed that the DMU under evaluation preserves its existing efficiency score, but we extend this idea by assuming $\bar{\theta}_p \geq \theta_p^*$, where θ_p^* is the efficiency score of DMU-p. The inverse problem is to determine the level of outputs for DMU-p for a given efficiency target $\bar{\theta}_p$. Therefore, the corresponding inverse DEA model becomes as follows:

$$
\begin{aligned}
& \max\ \left(\delta_1,\ldots,\delta_s\right) \\
& s.t. \\
& \sum_{j\in F}\lambda_j x_{ij} + \lambda_p\left(x_{ip+}\bar{x}_{ip}\right) \leq \bar{\theta}_p\left(x_{ip+}\bar{x}_{ip}\right) \quad i=1,..,m \\
& \sum_{j\in F}\lambda_j x_{ij} + \lambda_p\left(y_{rp+}\delta_r\right) \geq \left(y_{rp+}\delta_r\right) \quad r=1,..,s \\
& \sum_{j\in F}\lambda_j = 1 \\
& \lambda_j \geq 0 \quad j=1,..,n,\ \delta_r \geq 0\, r=1,..,s
\end{aligned}
\tag{13.2}
$$

The above inverse DEA model (Model (13.2)) is a vector optimization problem. It can be converted to a single optimization model by assigning weight to different outputs or other optimization approaches.

Many more inverse problems can be defined for the base DEA model (Model (13.1)). As another inverse problem consider a situation when a DMU, say DMU-p, need to make some additional outputs $\bar{y}_{rp}\ \left(r=1,..,s\right)$ with the lowest additional inputs for a given efficiency target $\bar{\theta}_p$. The corresponding inverse DEA model can be formulated as follows:

$$
\begin{aligned}
& \min\ \left(\alpha_1,\ldots,\alpha_m\right) \\
& s.t. \\
& \sum_{j\in F}\lambda_j x_{ij} + \lambda_p\left(x_{ip+}\alpha_i\right) \leq \bar{\theta}_p\left(x_{ip+}\alpha_i\right) \quad i=1,..,m \\
& \sum_{j\in F}\lambda_j x_{ij} + \lambda_p\left(y_{rp+}\bar{y}_{rp}\right) \geq \left(y_{rp+}\bar{y}_{rp}\right) \quad r=1,..,s \\
& \sum_{j\in F}\lambda_j = 1 \\
& \lambda_j \geq 0 \quad j=1,..,n,\ \alpha_i \geq 0\, i=1,..,m
\end{aligned}
\tag{13.3}
$$

Akin to the inverse DEA model (Model 13.2), Model (13.3) can also be converted to a single optimization problem. In the next section, we describe the inverse DEA models for estimating merger gains, a type of business partnership, originally developed in Gattoufi et al. (2014).

13.3 ESTIMATING MERGER GAINS

As stated, Gattoufi, Amin, and Emrouznejad (2014) defined novel inverse problems for estimating potential gains of mergers and introduced inverse DEA models for solving them. Their input-oriented inverse problem can be stated as follows: for a set of pre-merger DMUs, what is the minimum quantity of inputs needed to produce the combined outputs of all merging DMUs? More precisely, how much of each input inherited from the merging DMUs can be saved to make the combined outputs of all merging DMUs if the merged entity wanted a desired efficiency target? For simplicity, they assumed there are only two merging DMUs k and l. It is also assumed that the merged entity produces the same level of outputs as the combined outputs of the two merging DMUs using the minimum quantity of inherited inputs to reach a desired efficiency target $\bar{\theta}$. Suppose α_{iq} denotes the level of the ith input from the merging DMU_q ($q = k, l$) that will be used by the merged entity M. The input-oriented inverse DEA model proposed in Gattoufi et al. (2014) can be written as below:

$$
\begin{aligned}
&\min \sum_{i=1}^{m} \left(\alpha_{ik} + \alpha_{il}\right) \\
&s.t. \\
&\qquad \sum_{j\in F} \lambda_j x_{ij} + \left(\alpha_{ik} + \alpha_{il}\right)\lambda_{\mathrm{M}} - \left(\alpha_{ik} + \alpha_{il}\right)\bar{\theta} \le 0 \quad i = 1,\cdots,m \\
&\qquad \sum_{j\in F} \lambda_j y_{rj} + \left(y_{rk} + y_{rl}\right)\lambda_{\mathrm{M}} \ge y_{rk} + y_{rl} \quad r = 1,\cdots,s \\
&\qquad \sum_{j\in F} \lambda_j + \lambda_M = 1 \\
&\qquad 0 \le \alpha_{iq} \le x_{iq} \quad q = k,l \,\&\, i = 1,\cdots,m \\
&\qquad\quad \lambda_j \ge 0, \ j \in F, \lambda_M \ge 0
\end{aligned}
\tag{13.4}
$$

where λ_j is the intensity variable $\left(j \in F\right)$ and F presents the set of available peers in the post-merger evaluation process. As discussed in Amin and Ibn Boamah (2020), the coefficient of λ_{M} in the original inverse DEA model in Gattoufi et al. (2014), Model (13.4) shown above, is not correct because the optimal value of their model is always zero. This can easily be verified by taking $\lambda_{\mathrm{M}}=0$ and all $\alpha_{iq}=0$ for all $q=k,l\,\&\,i=1,\cdots,m$. Amin and Ibn Boamah (2020) made the correction as shown in the following inverse DEA model (Model 13.5).

$$
\begin{aligned}
&\min \sum_{i=1}^{m}\left(\alpha_{ik}+\alpha_{il}\right)\\
&s.t.\\
&\qquad \sum_{j\in F}\lambda_j x_{ij}+\left(x_{ik}+x_{il}\right)\lambda_{\mathrm{M}}-\left(\alpha_{ik}+\alpha_{il}\right)\bar{\theta}\le 0 \quad i=1,\cdots,m\\
&\qquad \sum_{j\in F}\lambda_j y_{rj}+\left(y_{rk}+y_{rl}\right)\lambda_{\mathrm{M}}\ge y_{rk}+y_{rl} \quad r=1,\cdots,s\\
&\qquad \sum_{j\in F}\lambda_j+\lambda_M=1\\
&\qquad 0\le\alpha_{iq}\le x_{iq} \quad q=k,l\,\&\,i=1,\cdots,m\\
&\qquad\quad \lambda_j\ge 0,\ j\in F,\lambda_M\ge 0
\end{aligned}
\tag{13.5}
$$

One of the challenging issues that exists in the inverse DEA models, especially in developed models for estimating merger gains, is the case of infeasibility. That is when the inverse DEA models become infeasible. Unlike a general optimization problem where the attempt is to make an infeasible optimization problem feasible with minimum perturbation, the issue of infeasibility in inverse DEA is not always a bad outcome. For instance, when two merging DMUs undergo a merger, there is a chance that the merged entity may fall on the efficiency frontier. In this case, the inverse DEA model (Model 13.5) will be infeasible for all efficiency targets of $\bar{\theta}<1$, and it is feasible if and only if $\bar{\theta}=1$. In fact, it would be misleading in this case if one assumes an efficiency target of less than one or equivalently assuming the merged entity is inefficient. Another case is the situation when the merged entity falls outside the production possibility set (PPS). In this case, the only target setting that makes sense is $\bar{\theta}=1$ and hence in the optimal solution of Model (13.5) $\lambda_{\mathrm{M}}=1$, and all other Lambdas are zero. Considering the above important properties, it is easy to show that the inverse DEA model (Model 13.5) for estimating merger gains is always feasible. This statement does not mean that the inverse DEA model (Model 13.5) is always feasible for all $\bar{\theta}$ because it does not make any sense for a merger where a minimum efficiency target is, for instance 75%, and a decision maker considers an efficiency target of less than that threshold. That would obviously be infeasible, and this infeasibility is not a drawback of the model. It is instead an advantage of the inverse DEA as it indicates a merger which is not practically possible and beneficial.

Now, let us discuss the inverse DEA estimation of merger gains from the output orientation. This is the situation when the merged entity consumes the entire inputs inherited from merging DMUs and attempt to produce as much additional outputs as possible for a given desired efficiency target. The output-oriented inverse DEA model presented in Gattoufi et al. (2014) for estimating potential merger gains is as follows:

$$\begin{aligned}
&\max \sum_{r=1}^{s} \beta_r \\
&s.t. \\
&\qquad \sum_{j \in F} \lambda_j x_{ij} + \left(x_{ik} + x_{il}\right)\lambda_M \le x_{ik} + x_{il} \quad i = 1,\cdots,m \\
&\qquad \sum_{j \in F} \lambda_j y_{rj} + \left(y_{rk} + y_{rl} + \beta_r\right)\lambda_M \ge \left(y_{rk} + y_{rl} + \beta_r\right)\bar{h} \quad r = 1,\cdots,s \\
&\qquad \sum_{j \in F} \lambda_j + \lambda_M = 1 \\
&\qquad \lambda_j \ge 0,\ j \in F, \lambda_M \ge 0,\ \beta_r \ge 0, r = 1,\ldots,s
\end{aligned} \tag{13.6}$$

It is assumed that the merged entity in the above model consumes all inputs inherited from merging DMUs and produces some additional outputs in addition to making the combined outputs of merging DMUs. The objective of the above inverse DEA model (Model 13.6) is to estimate the maximum quantity of outputs that can be produced in addition to the combined outputs of merging units for a given efficiency target ($1/\bar{h}$). It is worth noting that both the inverse DEA models (13.5) and (13.6) should be written as a vector optimization problem, but they can be simplified to the current single objective forms if all inputs in Model (13.5) and outputs in Model (13.6) have identical units of measurement. Theorem 1 shows that if the merged entity falls outside the PPS the inverse DEA model (Model 13.6) is unbounded. In this case, the decision maker may set desired additional outputs for the merged entity.

Theorem 1

The output-oriented inverse DEA Model 13.6 is unbounded when the merged entity falls outside the PPS.

Proof: The merged entity falls outside the PPS if and only if the following model is infeasible.

$$\begin{aligned}
&\max \sum_{r=1}^{s} \beta_r \\
&s.t. \\
&\qquad \sum_{j \in F} \lambda_j x_{ij} \le x_{ik} + x_{il} \quad i = 1, \cdots, m \\
&\qquad \sum_{j \in F} \lambda_j y_{rj} \ge \left(y_{rk} + y_{rl} + \beta_r \right) \bar{h} \quad r = 1, \cdots, s \\
&\qquad \sum_{j \in F} \lambda_j = 1 \\
&\qquad \quad \lambda_j \ge 0, \ j \in F, \beta_r \ge 0, r = 1, \ldots, s
\end{aligned} \tag{13.7}$$

For any value of $\bar{h}$, the inverse DEA Model 13.7 is obtained from Model (13.6) after removing λ_M from the model $(\lambda_M = 0)$. If the merged entity falls outside the PPS, then Model (13.7) is infeasible for $\bar{h} = 1$. In this case, $\lambda_M = 1$ and $\lambda_j = 0$ for all $j \ne M$ is a feasible solution of Model (13.6) and β_r can be increased arbitrary, indicating that Model (13.6) is unbounded. This completes the proof. ■

The following table presents a numerical example consisting of ten DMUs each with two inputs and two outputs.

Let us consider a merger between two DMUs k and l in Table 13.1 where $k, l \in \{1,,,10\}$ and $k \ne l$. The associated input-oriented inverse DEA Model (13.5) can be written as follows:

$$\begin{aligned}
&\min \alpha_{1k} + \alpha_{1l} + \alpha_{2k} + \alpha_{2l} \\
&s.t. \\
&\qquad \sum_{j \in F} \lambda_j x_{ij} + \left(x_{ik} + x_{il} \right) \lambda_M - \left(\alpha_{ik} + \alpha_{il} \right) \bar{\theta} \le 0 \quad i = 1, 2 \\
&\qquad \sum_{j \in F} \lambda_j y_{rj} + \left(y_{rk} + y_{rl} \right) \lambda_M \ge y_{rk} + y_{rl} \quad r = 1, 2 \\
&\qquad \sum_{j \in F} \lambda_j + \lambda_M = 1 \\
&\qquad 0 \le \alpha_{iq} \le x_{iq} \quad q = k, l \,\&\, i = 1, 2 \\
&\qquad \quad \lambda_j \ge 0, \ j \in F, \lambda_M \ge 0
\end{aligned}$$

Table 13.2 presents the result of the above model for some merger scenarios.

It is obvious that for any $\bar{\theta} < 1$, the merger between DMUs 1 and 3 is not practically possible. In this case, the infeasibility of the corresponding

TABLE 13.1 Data for Ten DMUs

DMUs	Input 1	Input 2	Output 1	Output 2	Efficiency
DMU-1	4	5	6	3	1.000
DMU-2	7	6	4	3	0.833
DMU-3	6	9	5	7	1.000
DMU-4	8	10	8	2	1.000
DMU-5	11	17	2	1	0.364
DMU-6	18	10	6	6	1.000
DMU-7	9	15	1	4	0.500
DMU-8	18	14	1	2	0.357
DMU-9	8	9	3	2	0.556
DMU-10	10	8	1	1	0.625

TABLE 13.2 Merger Scenarios (Input Orientation)

Merging DMUs	Efficiency Target	Optimal Solution
$k=1, l=3$	$\overline{\theta}=1$	$\lambda_M^*=1, \lambda_j^*=0 \forall j \neq M$
$k=5, l=7$	$\overline{\theta}=0.8$	$\alpha_{15}^*=6.25, \alpha_{25}^*=8.75$
$k=5, l=7$	$\overline{\theta}=0.9$	$\alpha_{15}^*=5.56, \alpha_{25}^*=7.78$
$k=5, l=7$	$\overline{\theta}=1$	$\alpha_{15}^*=5, \alpha_{25}^*=7$
$k=8, l=10$	$\overline{\theta}=1$	$\alpha_{18}^*=4, \alpha_{2,10}^*=5$

input-oriented inverse DMU Model 13.5 is realistic. The last row in Table 13.2 shows a merger between DMUs 8 and 10 where the merged entity needs 4 units of the first input of DMU 8 and 5 units of the second input of DMU 10, assuming the merged entity being fully efficient. Akin to the input orientation, we can apply the output-oriented inverse DEA Model 13.6 to estimate the potential merger gains. The corresponding inverse DEA model is as follows:

$$\max \beta_1 + \beta_2$$

$$s.t.$$

$$\sum_{j \in F} \lambda_j x_{ij} + \left(x_{ik} + x_{il}\right)\lambda_M \leq x_{ik} + x_{il} \quad i = 1,2$$

$$\sum_{j \in F} \lambda_j y_{rj} + \left(y_{rk} + y_{rl} + \beta_r\right)\lambda_M \geq \left(y_{rk} + y_{rl} + \beta_r\right)\overline{h} \quad r = 1,2$$

$$\sum_{j \in F} \lambda_j + \lambda_M = 1$$

$$\lambda_j \geq 0, \; j \in F, \lambda_M \geq 0, \; \beta_r \geq 0, \; r = 1,2$$

TABLE 13.3 Merger Scenarios (Output Orientation)

Merging DMUs	Efficiency Target	Optimal Solution
$k=1, l=3$	$\bar{h}=1$	The problem is unbounded
$k=7, l=10$	$\bar{h}=1.25$	$\beta_1^*=2, \beta_2^*=0.5$
$k=7, l=10$	$\bar{h}=1.15$	$\beta_1^*=3.22, \beta_2^*=0.22$
$k=7, l=10$	$\bar{h}=1$	$\beta_1^*=3, \beta_2^*=2$

Table 13.3 presents some merger examples using the above output-oriented inverse DEA model.

The first row in Table 13.3 shows a merger between DMUs 1 and 3 where the efficiency target is one. Akin to the input orientation, the inverse DEA model for estimating merger gains under the output orientation for these DMUs is feasible if and only if $\bar{h}=1$. As stated in Theorem 1, the merged entity falls outside the PPS for the merger between DMUs 1 and 3, and hence, the inverse DEA model is unbounded. The decision maker may set some desired additional output values depending on the competition in the market.

The last three rows in Table 13.3 show three merger scenarios between two merging DMUs 7 and 10 with three different efficiency targets. The merged entity in the first scenario is assumed to be 80% efficient $\left(\bar{h}=1.25\right)$, requiring 2 and 0.5 additional units of outputs 1 and 2, respectively. The last merger scenario between DMUs 7 and 10 assumes the merged entity is fully efficient. In this case, the merged entity needs to produce 3 and 2 additional units of outputs 1 and 2, respectively.

Nonetheless, the number of merging DMUs in the above inverse DEA models can easily be increased to a merger with more than two DMUs.

13.4 RECENT DEVELOPMENT

The number of articles on inverse DEA and mergers has been increasing since the work of Gattoufi et al. (2014). Oukil et al. (2024a) applied a two-stage DEA and inverse DEA for unveiling the potential of hotel mergers. In the first stage they applied the inverse DEA for determining the maximum potential merger gains that could be obtained from pairwise consolidations among 58 hotels in Oman. They applied a DEA model for choosing the best pairs of hotel mergers in the second stage. Oukil et al. (2024b) applied inverse DEA and ordered weighted averaging operator to select merger plans that maximizes agricultural sustainability for a dataset consisting of 43 tomato greenhouse farms in Algeria. Their findings indicate that mergers can enhance sustainability in agricultural sector and can help energy saving significantly.

Ghiyasi et al. (2022) developed inverse DEA models for optimizing input–output data perturbations when there is ratio data. They used an application consisting of 130 public hospitals in Iran to demonstrate their developed inverse DEA models.

Boubaker et al. (2023) applied conventional inverse DEA to assess banks performance and to investigate how they could preserve their efficiency and remain resilient in the post-Covid-19 market.

Lozano and Contreras (2024) developed non-radial inverse DEA models for target setting. They applied goal programming to guarantee the feasibility and included a post-inverse DEA approach to ensure that the obtained targets from multi-objective optimization are efficient. An application from Spanish universities was used to demonstrate their novel inverse DEA approach. Emrouznejad et al. (2023) and Emrouznejad and Amin (2023) reviewed the latest development and future directions of inverse DEA. Amin and Ibn Boamah (2023) introduced DEA and inverse DEA approach for modeling business partnerships. Their study provides guidelines and helps business companies create strategic alliances and partnerships under the umbrella of DEA.

As shown in recent years, and to name but a few examples above, inverse DEA has potential to solve challenging decision making problems by defining novel inverse problems in areas such as mergers and acquisitions, greenhouse emissions reduction and addressing climate change issues, and environmental efficiency.

13.5 CONCLUDING REMARKS

Inverse DEA is a post-DEA optimization approach for solving inverse problems. It empowers the productivity and efficiency analysis of decision making units (DMUs) by means of DEA. This chapter discusses the origin and mathematical basis of inverse DEA for estimating potential gains of mergers. We show that applying inverse DEA reveals insightful information for mergers and other business partnerships that is beyond the scope of DEA. The chapter also discusses the case of infeasibility and some of recent developments in inverse DEA for business mergers.

REFERENCES

Amin, G.R., and Al-Muharrami, S. (2018). A new inverse DEA model for merger with negative data. *IMA Journal of Management Mathematics*, 29(2), 137–149.

Amin, G.R., Emrouznejad, A., and Gattoufi, S. (2017). Modelling generalized firms' restructuring using inverse DEA. *Journal of Productivity Analysis*, 48(1), 51–61.

Amin, G. R., & Ibn Boamah, M. (2020). A new inverse DEA cost efficiency model for estimating potential merger gains: A case of Canadian banks. *Annals of Operations Research*, 295(1), 21–36.

Amin, G.R., & Ibn Boamah, M. (2023). Modeling business partnerships: A data envelopment analysis approach. *European Journal of Operational Research*, 305(1), 329–337.

Bogetoft, P., and Wang, D. (2005). Estimating the potential gains from mergers. *Journal of Productivity Analysis*, 23, 145–171.

Boubaker, S., Le, T.D.Q., and Ngo, T. (2023). Managing bank performance under COVID-19: A novel inverse DEA efficiency approach. *International Transaction in Operational Research*, 30(5), 2436–2452.

Charnes, A., Cooper, W.W., & Rhodes, E. (1978). Measuring the efficiency of decision making units. *European Journal of Operational Research*, 2, 429–444.

Emrouznejad, A., and Amin, G.R. (2023). Advances in inverse data envelopment analysis: Empowering performance assessment. *IMA Journal of Management Mathematics*, 34(3), 415–419.

Emrouznejad, A., Amin, G.R., Ghiyasi, M., & Michali, M. (2023). A review of inverse data envelopment analysis: Origins, development, and future directions. *IMA Journal of Management Mathematics*, 34(3), 421–440.

Gattoufi, S., Amin, G.R., & Emrouznejad, A. (2014). A new inverse DEA method for merging banks, *IMA Journal of Management Mathematics*, 25, 73–87.

Ghiyasi, M., Soltanifar, M., and Sharafi, H. (2022). A novel inverse DEA-R model with application in hospital efficiency. *Socio-Economic Planning Sciences*, 84, 101427.

Ibn Boamah, M., & Amin, G.R. (2024). Partner selection in business mergers: A data envelopment analysis approach. *Rairo Operations Research*, 58, 3107–3117.

Lozano, S., and Contreras, I. (2024). Target setting using an enhanced inverse DEA approach: application to Spanish universities. *Journal of the Operational Research Society*, https://doi.org/10.1080/01605682.2024.2306176

Lozano, S., and Villa, G. (2010). DEA-based pre-merger planning tool. *Journal of the Operational Research Society*, 61, 1485–1497.

Oukil, A., Kennedy, R.E., Al-Hajri, A., and Soltani, A.A. (2024a). Unveiling the potential of hotel mergers: A hybrid DEA approach for optimizing sector-wide performance in the hospitality industry. *International Journal of Hospitality Management*, 116, 103620.

Oukil, A., Nourani, A., Soltani, A.A., Boulassel, M.R., and Bencheikh, A. (2024b). Improving agricultural sustainability through farm mergers: an energy efficiency perspective. *International Journal of Agricultural Sustainability*, 22(1), 1–24.

Ray, S., and Sethia, S. (2022). Nonparametric measurement of potential gains from mergers: an additive decomposition and application to Indian bank mergers. *Journal of Productivity Analysis*, 57(2), 115–130.

Wei, Q., Zhang, J., and Zhang, X. (2000). An inverse DEA model for inputs/outputs estimate. *European Journal of Operational Research*, 121(1), 151–163.

Xiao, S., Li, Y., Emrouznejad, A., Xie, J., and Liang, L. (2017). Estimation of potential gains from bank mergers: A novel two-stage cost efficiency DEA model. *Journal of the Operational Research Society*, 68(9), 1045–1055.

Yan, H., Wei, Q., and Hao, G. (2002). DEA models for resource reallocation and production input/output estimation. *European Journal of Operational Research*, 136(1), 19–31.

Zhang, J., and Liu, Z. (1996). Calculating some inverse linear programming problems. *Journal of Computational and Applied Mathematics*, 72, 261–273.